CONTINUUM MODELS OF DISCRETE SYSTEMS

PROCEEDINGS OF THE FIFTH INTERNATIONAL SYMPOSIUM ON CONTINUUM MODELS OF DISCRETE SYSTEMS / NOTTINGHAM / 14-20 JULY 1985

Continuum Models of Discrete Systems

Edited by
A.J.M.SPENCER
Department of Theoretical Mechanics, University of Nottingham

A.A.BALKEMA / ROTTERDAM / BOSTON / 1987

CIP-DATA KONINKLIJKE BIBLIOTHEEK, DEN HAAG

Continuum

Continuum models of discrete systems: proceedings of the fifth international symposium, Nottingham, 14-19 July 1985 / ed. by A.J.M. Spencer. – Rotterdam [etc.]: Balkema. – Ill.
ISBN 90-6191-682-8 bound
SISO 533 UDC 531(063)
Subject heading: mechanics

The texts of the various papers in this volume were set individually by typists under the supervision of each of the authors concerned.

ISBN 90 6191 682 8

Distributed in USA & Canada by: A.A.Balkema Publishers, P.O.Box 230, Accord, MA 02018

Printed in the Netherlands

Contents

Preface

This volume contains the texts of most of the lectures presented at the Fifth International Symposium on Continuum Models of Discrete Systems (CMDS 5) which was held in Derby Hall, University of Nottingham, England from 14th-19th July 1985.

As with earlier symposia in this series, which began in Poland in 1975, the purpose of the meeting was to bring together researchers concerned with various aspects of continuum modelling of discrete systems, to discuss topics of common interest and to disseminate information about recent developments. The lively discussions which took place both inside and outside the lecture room suggest that this objective was achieved. The scientific programme comprised six general lectures and twenty-eight sessional lectures. Twenty-nine of these lectures, and abstracts of another three, are recorded in this volume.

The Scientific Committee for the Symposium consisted of

O. Brulin (past-Chairman)	Sweden
G. H. A. Cole	U.K.
G. Duvaut	France
G. Herrmann	U.S.A.
E. Kröner (past-Chairman)	FR Germany
I. Kunin	U.S.A.
A. J. M. Spencer (Chairman)	U.K.
H. Zorski (past-Chairman)	Poland

It is a pleasure to record my thanks to the members of the Scientific Committee for their advice and encouragement; to all who participated, and especially to Henryk Zorski, David Broomhead, Kerry Havner, Marcel Crochet, Gérard Maugin and Masao Satake for an excellently prepared and presented series of general lectures; and to the participants who acted as chairmen of the various sessions. I am most grateful to Dr. Carol McTaggart and Mrs. Judith Hind for invaluable help with the local organisation, to my wife, Margaret, for arranging the social programme, and to the Nottingham City Council for granting us the use of the George Green Windmill and Science Centre as a most appropriate venue for a reception. Derby Hall and its staff provided excellent facilities for the Symposium and my colleagues and students in the Department of Theoretical Mechanics at Nottingham helped in many ways. Thanks are also due to Mrs. Hind for most of the work involved in preparing this volume for publication. Finally, I acknowledge with gratitude financial support from The Royal Society.

Preparations are under way for the next symposium in the series to be held in France under the chairmanship of Professor G. A. Maugin. I look forward to CMDS 6 and wish it every success.

A. J. M. SPENCER

5th International Symposium on Continuum Models of Discrete Systems / Nottingham / 14-20 July 1985

Elastic behaviour of textured aggregates of cubic crystals

L.J.Walpole
University of East Anglia, Norwich, UK

ABSTRACT: In evaluating the overall elastic properties of a polycrystal, for instance by ultrasonic non-destructive testing, the anisotropy induced by preferred alignments of the grains must be allowed for. Cubic grains present a simplification that reduces the number of independent elastic moduli in the stress-strain relation. For a given description of the overall elastic symmetry and of the underlying texture, the moduli can be bounded (sometimes to a coincidence) in terms of the elastic moduli of the grains.

1 INTRODUCTION

A polycrystalline material is a compact aggregate of irregularly shaped crystals (the "grains"), each of which has its own particular orientation of its crystallographic axes, differing abruptly from those of its neighbours, not necessarily in a purely random fashion. A particular preparation or treatment ensures in practice that each set of axes tends rather to be aligned in a preferred manner of "texture", as described in a variety of examples by Barrett & Massalski (1980). An anisotropic constitutive relation between stress and infinitesimal strain is called for firstly, if the elastic response of the polycrystal as a whole is to be evaluated adequately, for instance, when ultrasonic body or surface waves are propagated (Atthey 1985, Bonilla & Keller 1985, Sayers 1985).

2 STRESS-STRAIN RELATIONS

When it is subjected to an infinitesimal elastic straining, the aggregate and each of its representative volumes is understood to respond as if it were an elastically homogeneous and anisotropic whole, obedient to the "macroscopic" linear relation

$$\sigma_{ij} = C_{ijk\ell}\, \varepsilon_{k\ell} \qquad (1)$$

between the fields of stress σ_{ij} and strain ε_{ij} (relative to fixed cartesian axes). The "overall" (or "effective") tensor $C_{ijk\ell}$ of elastic moduli is characteristic of the particular aggregate and (Hill 1963) takes on the symmetries

$$C_{ijk\ell} = C_{jik\ell} = C_{ij\ell k} = C_{k\ell ij}$$

similar to those found in the corresponding tensor of an individual crystal. The definition

$$Q = \sigma_{ij}\, \varepsilon_{ij} = \varepsilon_{ij}\, C_{ijk\ell}\, \varepsilon_{k\ell} \qquad (2)$$

introduces a scalar Q as a quadratic function of the strain components; it can be interpreted (Hill 1963, Hashin 1964) as twice the strain energy per representative volume. The tensor $C_{ijk\ell}$ is required to be "positive definite", to ensure that Q is always positive. The relation (1) is then certainly invertible as

$$\varepsilon_{ij} = S_{ijk\ell}\, \sigma_{k\ell}\ , \qquad (3)$$

and the compliance tensor $S_{ijk\ell}$ has the symmetry and positive definite properties of $C_{ijk\ell}$. By the introduction

$$\sigma'_{ij} = \sigma_{ij} - \sigma_{kk}\, \delta_{ij}/3\ ,$$

$$\varepsilon'_{ij} = \varepsilon_{ij} - \varepsilon_{kk}\, \delta_{ij}/3\ ,$$

of the deviatoric ("traceless") parts of the stress and strain tensors, (2) may be recast as

$$Q = \sigma_{jj}\, \varepsilon_{kk}/3 + Q'\ ,$$

where

$$Q' = \sigma'_{ij}\, \varepsilon'_{ij}$$

$$= (\sigma_{11} - \sigma_{22})(\varepsilon_{11} - \varepsilon_{22})/2 + 3\sigma'_{33}\,\varepsilon'_{33}/2 + 2(\sigma_{12}\varepsilon_{12} + \sigma_{23}\varepsilon_{23} + \sigma_{31}\varepsilon_{31}) , \quad (4)$$

in preparation especially (but not only) for circumstances that give a distinction to the x_3-axis.

It is supposed henceforth that the grains are all made of the same cubic substance whose bulk modulus is κ_c say. Each such crystal is compressed isotropically on its own by a hydrostatic stress, to the extent allowed by the bulk modulus. Uniform fields of hydrostatic stress and purely dilatational strain can be established therefore, self-compatibly, throughout the aggregate (Hill 1952), and κ_c can be identified as an overall as well as a local bulk modulus. Therefore a second-rank isotropic tensor (the appropriate multiple of the Kronecker delta) is produced when the third and fourth (or the first and second) suffixes of $C_{ijk\ell}$ and $S_{ijk\ell}$ are contracted (Walpole 1985). These properties

$$C_{ijkk} = C_{kk\ell\ell}\,\delta_{ij}/3 , \qquad (5)$$

$$S_{ijkk} = S_{kk\ell\ell}\,\delta_{ij}/3 ,$$

are equivalent algebraically to each other (since $C_{ijk\ell}$ and $S_{ijk\ell}$ are mutually inverse). The decompositions

$$C_{ijk\ell} = \kappa_c\,\delta_{ij}\,\delta_{k\ell} + C'_{ijk\ell} ,$$

$$S_{ijk\ell} = \delta_{ij}\,\delta_{k\ell}/9\kappa_c + S'_{ijk\ell} ,$$

are made available, where there is the connection

$$C_{kk\ell\ell}/9 = 1/S_{kk\ell\ell} = \kappa_c ,$$

and where the prime denotes "traceless" tensors that are reduced to the zero second-rank tensor when their first and second or when their third and fourth suffixes are contracted. There is a corresponding decomposition of the overall relation (1) as

$$\sigma_{kk} = 3\kappa_c\,\varepsilon_{kk} , \qquad (6)$$

$$\sigma'_{ij} = C'_{ijk\ell}\,\varepsilon'_{k\ell} , \qquad (7)$$

and of the inverse relation (3) similarly, in which the traceless parts of the stress and strain tensors are not coupled to the remaining isotropic (hydrostatic and dilatational) parts. There are generally five independent relations (7) between the traceless parts, whereby (for instance) each of σ_{12}, σ_{23}, σ_{31}, $\sigma_{11} - \sigma_{22}$ and σ'_{33} can be expressed as a linear combination of the corresponding strain components. By substitution in (4), a quadratic form

$$Q' = h(\varepsilon_{11} - \varepsilon_{22})^2 + 3g\varepsilon'^2_{33} + 4(m\varepsilon^2_{12} + p\varepsilon^2_{23} + q\varepsilon^2_{31}) + 4(\varepsilon_{11} - \varepsilon_{22})(s\varepsilon_{12} + u\varepsilon_{23} + v\varepsilon_{31} + 3\ell\varepsilon'_{33}/4) + 8(n\varepsilon_{12}\varepsilon_{23} + r\varepsilon_{23}\varepsilon_{31} + f\varepsilon_{12}\varepsilon_{31}) + 12\varepsilon'_{33}(t\varepsilon_{12} + w\varepsilon_{23} + y\varepsilon_{31}) \qquad (8)$$

is obtained in terms of a total of fifteen elastic moduli (h, g, ... , y). The five stress-strain relations (7) can be recovered by means of the partial differentiations

$$\sigma_{11} - \sigma_{22} = \partial Q'/\partial(\varepsilon_{11} - \varepsilon_{22}) ,$$

$$3\sigma'_{33} = \partial Q'/\partial\varepsilon'_{33} , \qquad 4\sigma_{12} = \partial Q'/\partial\varepsilon_{12} ,$$

$$4\sigma_{23} = \partial Q'/\partial\varepsilon_{23} , \qquad 4\sigma_{31} = \partial Q'/\partial\varepsilon_{31} , \qquad (9)$$

and they combine with (6) to make up the full complement of six relations. However, because of the five connections (5), they contain a maximum of sixteen independent elastic moduli (in place of the usual twenty one), one of which is identified as a bulk modulus coincident with that of the grains. By returning to the original quadratic form (2), we may retrieve the twenty one equations

$$c_{11} + c_{12} + c_{13} = c_{12} + c_{22} + c_{23} = c_{13} + c_{23} + c_{33} = 3\kappa_c ,$$

$$c_{14} + c_{24} = -c_{34} = -2w ,$$

$$c_{15} + c_{25} = -c_{35} = -2y ,$$

$$c_{16} + c_{26} = -c_{36} = -2t ,$$

$$c_{13} - c_{23} = 2\ell , \qquad c_{14} - c_{24} = 2u ,$$

$$c_{15} - c_{25} = 2v , \qquad c_{16} - c_{26} = 2s ,$$

$$c_{44} = p , \quad c_{55} = q , \quad c_{66} = m ,$$

$$c_{45} = r , \quad c_{46} = n , \quad c_{56} = f ,$$

$$c_{33} = \kappa_c + 4g/3 , \qquad c_{11} + c_{22} - 2c_{12} = 4h ,$$

which incorporate the connections (5) and which enable the twenty one conventional double-suffixed constants to be each expressed in terms of the present sixteen. As Q' is expressed in (8) as a positive definite quadratic form in the five strain variables, necessary restrictions on the fifteen elastic coefficients can be written down in the standard way. It may be noted in particular that h, g, m, p and q are all necessarily positive and that $3\ell^2$, s^2, u^2, v^2, n^2, r^2, f^2, $3t^2$, $3w^2$ and $3y^2$ are less than 4gh, mh, ph, qh, pm, pq, mq, gm, gp and gq respectively. The inversion of the five relations (9), to expressions for strain in terms of stress, will require the inversion of a 5×5 matrix.

The number of independent elastic moduli can be reduced further when there is an overall elastic symmetry, and the stress-strain relations can be simplified if the cartesian axes are placed appropriately. First suppose that there is one plane of symmetry (taken parallel to the x_1x_2-plane) at every point, brought about by a textural preference for two (non-perpendicular) directions in that plane. For such a monoclinic symmetry, the moduli u, v, n, f, w and y all vanish, to leave a total of ten overall elastic moduli (instead of the usual monoclinic thirteen). The five stress-strain relations (9) simplify to

$$\sigma_{11}-\sigma_{22} = 2h(\varepsilon_{11}-\varepsilon_{22}) + 3\ell\varepsilon'_{33} + 4s\varepsilon_{12} ,$$

$$\sigma'_{33} = 2g\varepsilon'_{33} + \ell(\varepsilon_{11}-\varepsilon_{22}) + 4t\varepsilon_{12} ,$$

$$\sigma_{12} = 2m\varepsilon_{12} + s(\varepsilon_{11}-\varepsilon_{22}) + 3t\varepsilon'_{33} ,$$

$$\sigma_{23} = 2p\varepsilon_{23} + 2r\varepsilon_{31} ,$$

$$\sigma_{31} = 2q\varepsilon_{31} + 2r\varepsilon_{23} , \tag{10}$$

and the inversion of the first three and of the remaining two requires only the inversion of a 3×3 and of a 2×2 matrix respectively. Secondly, if the symmetry is the orthotropic one of three orthogonal planes (the coordinate planes), due to a textural preference for two perpendicular directions, then r, s and t also vanish and so leave a total of seven elastic moduli (instead of the usual nine). If thirdly there is the tetragonal symmetry in which the two directions (of the x_1 and x_2 axes) are texturally indistinguishable then ℓ vanishes as well and p is equated to q (and also to m if the symmetry is cubic). Fourthly, for transverse isotropy, for which there is a single preferred textural direction (the x_3-axis), m and h are to be equated as well, to leave four (instead of the usual five) elastic moduli, and to simplify the five stress-strain relations (10) as

$$\sigma_{11} - \sigma_{22} = 2m(\varepsilon_{11} - \varepsilon_{22}) ,$$

$$\sigma'_{33} = 2g\varepsilon'_{33} , \qquad \sigma_{12} = 2m\varepsilon_{12} ,$$

$$\sigma_{23} = 2p\varepsilon_{23} , \qquad \sigma_{31} = 2p\varepsilon_{31} .$$

Lastly, when the aggregate is completely isotropic in the overall elastic sense, g, m and p all coincide as the unique overall shear modulus.

The various elastic moduli can be related to others that have direct physical interpretations (as shear or Young's moduli or Poisson's ratios). For example, when σ_{33} is taken as the only non-vanishing stress component, the transversely isotropic stress-strain relations invert to

$$\varepsilon_{33} = \sigma_{33}/E ,$$

$$\varepsilon_{11} = \varepsilon_{22} = -\nu\sigma_{33}/E ,$$

where

$$3/E = 1/3\kappa_c + 1/g ,$$

$$\nu = (3\kappa_c - 2g)/2(3\kappa_c + g) , \tag{11}$$

in order to identify the axial Young's modulus and accompanying Poisson's ratio. On the other hand if only σ_{11} is non-zero, the inversion

$$\varepsilon_{11} = \sigma_{11}/E' ,$$

$$\varepsilon_{22} = -\nu'\sigma_{11}/E' ,$$

$$\varepsilon_{33} = -\nu''\sigma_{11}/E' ,$$

identifies the transverse Young's modulus and the two accompanying Poisson's ratios by means of the definitions

$$1/E' = 1/9\kappa_c + 1/12g + 1/4m ,$$

$$\nu' = (1/m - 1/3g - 4/9\kappa_c) / (1/m + 1/3g + 4/9\kappa_c) , \tag{12}$$

$$\nu'' = (1/6g - 1/9\kappa_c)/(1/9\kappa_c + 1/12g + 1/4m) , \tag{13}$$

which make the connection

$$\nu'' = \nu(1 - \nu')/(1 - \nu) \tag{14}$$

between the Poisson's ratios. As g, m, p and κ_c are necessarily all positive, E and E' are naturally kept positive also, and moreover the Poisson's ratios are subjected to the restrictions

$$-1 < \nu < \tfrac{1}{2} ,$$

$$-1 < \nu' < 1 , \tag{15}$$

$$-1 < \nu'' < 2 ,$$

which specify their full numerical ranges. [The corresponding three Poisson's ratios of the usual transversely isotropic stress-strain relations (with five independent moduli) can be specified independently, and, although ν' is anyway obedient to (15), there are no positive or negative limits placed upon ν or ν''.] From (12), (13) and (14), we may extract the expression

$$m = 3\kappa_c(1-\nu')(1-2\nu)/2(1+\nu')(1-\nu) ,$$

which puts m, like g in (11), in terms of the bulk modulus of the grains and of Poisson's ratios that have restricted numerical ranges.

3 PREDICTION OF THE OVERALL ELASTIC MODULI

While the elastic moduli of a given aggregate are susceptible to experimental measurement, for instance by the passage of ultrasonic waves, it remains of theoretical and practical interest to enquire to what extent their values can be predicted. Measurements on macroscopically identical aggregates can be expected to show a certain amount of scatter that will reflect the amount of anisotropy in the constituent crystals and (to a lesser extent perhaps) the variability of the geometrical structure. Given the degree of overall elastic symmetry, with or without a description of the specific texture, we may seek bounds on the overall moduli in terms of the grain moduli.

In the first place, as in other composite materials, it is to be expected (and it can be proved) that the elastic properties of the aggregate are intermediate to those of the constituents, to imply, for instance, that every overall shear modulus falls in magnitude between the greatest and least shear moduli of the single crystal. In terms of the three conventional independent elastic moduli of a cubic crystal, c_{11}, c_{12} and c_{44}, Zener (1948) defined the bulk modulus κ_c and two particular shear moduli μ_c and μ_c' as

$$\kappa_c = (c_{11} + 2c_{12})/3 ,$$

$$\mu_c = (c_{11} - c_{12})/2 , \qquad \mu_c' = c_{44} .$$

We may further identify μ_c and μ_c' as the extreme shear moduli, in that every other shear modulus lies below whichever is the larger and above the smaller of them. Numerical measurements of μ_c and μ_c' are listed for several crystals by Zener (1948), Cottrell (1964) and Shukla (1982), and in most cases it is found that μ_c' is larger than μ_c. We can prove accordingly that each of the overall moduli h, g, m, p and q lies between μ_c and μ_c'. An appeal to the "strengthening" argument of Hill (1963) shows first that each of them is an increasing function of μ_c and of μ_c', for a fixed configuration. It is then only necessary to note that if μ_c and μ_c' are imagined to be brought to a coincidence as the shear modulus of isotropic grains, then in the consequently isotropic aggregate h, g, m, p and q would all coincide with that modulus. It has been shown already that the overall bulk modulus coincides with κ_c. The remaining overall moduli all vanish (through positive or negative values) when μ_c and μ_c' reach a coincidence. More precisely, it follows that the magnitude of ℓ is less than $|\mu_c' - \mu_c|/\sqrt{3}$, that each of s, u, v, n, r and f has a magnitude less than $|\mu_c' - \mu_c|/2$, and that each of t, w and y has a magnitude less than $|\mu_c' - \mu_c|/2\sqrt{3}$.

These elementary predictions can be improved upon firstly by taking account of the distribution of grain orientations, as will be exemplified here for the transversely isotropic stress-strain relations. Unit "orientation" vectors $\underline{a}^r$, $\underline{b}^r$ and $\underline{c}^r$ can be placed along the three mutually perpendicular crystallographic axes, the edges of the cube, in the rth grain. It suffices to introduce one local orientation factor in each grain and one overall factor, in relation to the x_3-axis of overall symmetry, by means of the definitions

$$\phi_r = (a_3^r\, b_3^r)^2 + (b_3^r\, c_3^r)^2 + (c_3^r\, a_3^r)^2 ,$$

$$\phi = \sum v_r\, \phi_r , \tag{16}$$

where v_r is the relative volume of the rth grain and where the summation extends in any order over all the grains (so that $\sum v_r = 1$). Then the inequalities

$$g \leqslant \mu_c + 3\phi(\mu_c' - \mu_c) ,$$

$$1/g \leqslant 1/\mu_c + 3\phi(1/\mu_c' - 1/\mu_c) ,$$

$$2m \leqslant \mu_c + \mu_c' + \phi(\mu_c' - \mu_c) ,$$

$$2/m \leqslant 1/\mu_c + 1/\mu_c' + \phi(1/\mu_c' - 1/\mu_c) ,$$

$$p \leqslant \mu_c' + 2\phi(\mu_c - \mu_c') ,$$

$$1/p \leqslant 1/\mu_c' + 2\phi(1/\mu_c - 1/\mu_c') ,$$

of Walpole (1986), place upper and lower bounds on the uncertainties left in the values of the overall moduli g, m and p. When ϕ is equal to one third (its maximum value), the bounds on g coincide with one another and with μ_c'. This value is achieved by ϕ (and by every ϕ_r) when every grain has a cube diagonal (a <111> direction) parallel to the x_3-axis, while being otherwise orientated at random. On the other hand, when ϕ takes its minimum value of zero, that is, when every grain has a cube edge (a <100> direction) parallel to the x_3-axis, the coincidences $g = \mu_c$, $p = \mu_c'$ are reached and only the modulus m is left to the uncertainty allowed by the inequalities. When ϕ takes the intermediate value of one fifth, the inequalities give the same upper bound to each modulus g, m and p, and the same lower bound as well, identical to the known upper and lower bounds (Hill 1952) on the shear modulus of an isotropic aggregate. It is suggested therefore that ϕ is equal to one fifth if and only if the aggregate is elastically isotropic. An anisotropy factor γ can be defined then by

$$\phi = (1 - \gamma)/5 .$$

The numerical range for ϕ carries over to a corresponding one for γ, namely

$$-\frac{2}{3} \leqslant \gamma \leqslant 1 .$$

It is usually assumed that a summation such as (16) can be replaced by a triple integration over all possible grain orientations (the Euler angles), weighted by means of the so-called "crystallite orientation distribution function" of Roe (1965). When this function is expanded in a series of generalised spherical harmonics, it is found (Pursey & Cox 1954, Bunge 1968) that only one of its coefficients, namely w_{400}, remains in the integration to leave the connection

$$\gamma = 16\sqrt{2}\pi^2 w_{400}/7 .$$

The coefficient w_{400} can be measured directly by x-ray or neutron diffraction (Roe 1965, Bunge 1965, Allen et al 1983).

REFERENCES

Allen, A.J., M.T.Hutchings, C.M.Sayers, D.R.Allen & R.L.Smith 1983. Use of neutron diffraction texture measurements to establish a model for calculation of ultrasonic velocities in highly oriented austenitic weld material. J.Appl.Phys. 54:555-560.

Atthey, D.R. 1985. The propagation of elastic waves through a textured granular material. Int.J.Engng.Sci. 23:937-951.

Barrett, C.S. & T.B.Massalski 1980. Structure of metals. Oxford: Pergamon Press.

Bonilla, L.L. & J.B.Keller 1985. Acoustoelastic effect and wave propagation in heterogeneous weakly anisotropic materials. J.Mech.Phys.Solids 33:241-261.

Bunge, H.J. 1965. Zur Darstellung allgemeiner Texturen. Z.Metallk. 56:872-874.

Bunge, H.J. 1968. Über die elastischen Konstanten kubischer Materialien mit beliebiger Textur. Krist.Tech. 3:431-438.

Cottrell, A.H. 1964. The mechanical properties of matter. New York: Wiley.

Hashin, Z. 1964. Theory of mechanical behaviour of heterogeneous media. Appl.Mech.Rev. 17:1-9.

Hill, R. 1952. The elastic behaviour of a crystalline aggregate. Proc.Phys.Soc. A 65:349-354.

Hill, R. 1963. Elastic properties of reinforced solids: Some theoretical principles. J.Mech.Phys.Solids 11:357-372.

Pursey, H. & H.L.Cox 1954. The correction of elasticity measurements on slightly anisotropic materials. Phil.Mag. 45: 295-302.

Roe, R.J. 1965. Description of crystallite orientation in polycrystalline materials. III. General pole figure inversion. J.Appl.Phys. 36:2024-2031.

Sayers, C.M. 1985. Angular dependence of the Rayleigh wave velocity in polycrystalline metals with small anisotropy. Proc.R.Soc.Lond. A 400:175-182.

Shukla, M.M. 1982. The shear bounds of the cubic polycrystal and its experimental shear modulus. J.Phys.D 15:L177-L180.

Walpole, L.J. 1985. The stress-strain law of a textured aggregate of cubic crystals. J.Mech.Phys.Solids 33:363-370.

Walpole, L.J. 1986. Evaluation of the elastic moduli of a transversely isotropic aggregate of cubic crystals. J.Mech.Phys. Solids, in press.

Zener, C. 1948. Elasticity and anelasticity of metals. Chicago: University Press.

5th International Symposium on Continuum Models of Discrete Systems / Nottingham / 14-20 July 1985

Simulation of crystal defects by finite element techniques

C.Teodosiu
Max-Planck-Institut für Metallforschung, Institut für Physik, Stuttgart, FR Germany

ABSTRACT: A new semidiscrete method for studying the atomic configurations around lattice imperfections is proposed, which makes use of finite element techniques for the continuum part of the model. This approach does not require any analytical solution and permits, therefore, the investigation of relatively complex lattice imperfections. General non-linear anisotropic material properties, as well as free boundaries and interfaces are easily taken into account.

1 INTRODUCTION

The understanding of the strength of crystalline materials rests ultimately on knowledge of the atomic arrangement around dislocations, point defects, grain and phase boundaries, since it is these lattice imperfections that make possible the plastic flow and the nucleation and propagation of cracks at applied forces that are several orders of magnitude lower than those necessary to fracture a perfect crystal. Such atomic configurations are most effectively studied by means of semidiscrete methods, which employ the lattice theory for the heavily distorted region at close proximity of crystal defects (region I, fig. 1) and the elasticity theory for the remaining of the crystal (region II).

Semidiscrete simulations have been based so far on knowledge of analytic, linear or non-linear elastic solutions, for arbitrarily prescribed boundary conditions on the separation surface Σ_0 between regions I and II. Consequently, the applicability of these methods has been generally limited to infinite media and to relatively uninvolved configurations of the imperfections (see e.g. Teodosiu 1982: 244-263, where further references on this subject may be found).

In this paper, an alternative approach is proposed, which makes use of finite element techniques for the elastic region. Accordingly, this region is divided into finite elements and the elastic displacement field is approximated by means of piecewise polynomial functions in terms of the nodal displacements. Then, the configuration of the imperfect crystal is determined by using the principle of virtual work for region II and simultaneously searching the minimum of the potential energy of the crystal in region I, considered as a function of the individual atomic positions in this region. The physical and numerical coupling between these two regions is provided by the atoms of region II that interact with region-I atoms and whose current positions must be also stored throughout the calculation. Clearly, this approach does not require any analytical solution and hence permits the study of relatively complex configurations of lattice defects and boundary conditions (applied stress, free boundaries and interfaces). Moreover, the incorporation of anisotropic and non-linear elastic effects - which is a standard facility of evolved finite element codes - allows a substantial reduction of the number of atoms in region I, and hence of the total computing time.

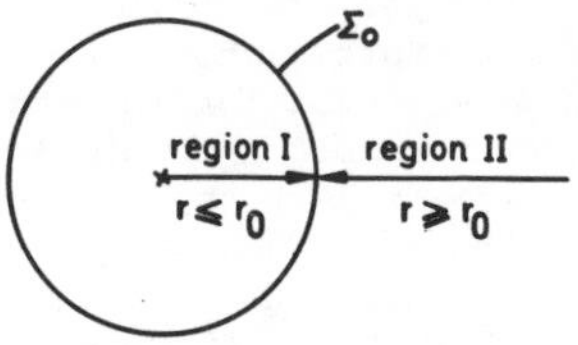

Fig. 1 Concentric arrangement of the atomistic and continuum regions for the semidiscrete simulation of crystal defects

2 THE ATOMISTIC REGION

Denote by $\underset{\sim}{x}_1,\ldots,\underset{\sim}{x}_n$ the position vectors of the atoms in region I and by $\underset{\sim}{y}_1,\ldots,\underset{\sim}{y}_m$ the position vectors of those atoms in region II that interact with at least one region-I atom. The number of these 'transition' atoms is obviously determined by the (finite) range of the assumed interatomic potential. We shall also make use of the concise notation $\underset{\sim}{X} = \{\underset{\sim}{x}_1,\ldots,\underset{\sim}{x}_n\}$, $\underset{\sim}{Y} = \{\underset{\sim}{y}_1,\ldots,\underset{\sim}{y}_n\}$. Assume that the total potential energy Φ of region-I atoms for fixed $\underset{\sim}{Y}$ is known as a function of $\underset{\sim}{X}$ and $\underset{\sim}{Y}$, i.e. $\Phi = \Phi(\underset{\sim}{X},\underset{\sim}{Y})$. The force acting on atom i is then defined by

$$\underset{\sim}{F}_i = -\frac{\partial\Phi(\underset{\sim}{X},\underset{\sim}{Y})}{\partial\underset{\sim}{x}_i} \equiv \underset{\sim}{F}_i(\underset{\sim}{X},\underset{\sim}{Y}). \qquad (1)$$

If $\underset{\sim}{X} = \underset{\sim}{X}^{(e)}$ is an equilibrium configuration which corresponds to a minimum of the potential energy Φ for fixed $\underset{\sim}{Y}$, then obviously,

$$\underset{\sim}{F}_i(\underset{\sim}{X}^{(e)},\underset{\sim}{Y}) = \underset{\sim}{0}, \quad i = 1,\ldots,n. \qquad (2)$$

Eq.(2) represents a system of nd_f non-linear scalar equations, where d_f is the number of degrees of freedom per atom. The solution of this system may be sought for instance by means of the Newton-Raphson method. Namely, we may use at the (s+1)th step of the iteration a truncated Taylor expansion of $\underset{\sim}{F}_i(\underset{\sim}{X},\underset{\sim}{Y})$ around the solution $\underset{\sim}{X}^{(s)}$ derived at the preceding step, thus obtaining

$$\underset{\sim}{F}_i(\underset{\sim}{X}^{(s)},\underset{\sim}{Y}) + \sum_{j=1}^{n}\underset{\sim}{M}_{ij}(\underset{\sim}{X}^{(s)},\underset{\sim}{Y})(\underset{\sim}{x}_j-\underset{\sim}{x}_j^{(s)}) = \underset{\sim}{0}, \qquad (3)$$

for i = 1,...,n, where

$$\underset{\sim}{M}_{ij}(\underset{\sim}{X},\underset{\sim}{Y}) \equiv \frac{\partial\underset{\sim}{F}_i}{\partial\underset{\sim}{x}_j}(\underset{\sim}{X},\underset{\sim}{Y}) = -\frac{\partial^2\Phi(\underset{\sim}{X},\underset{\sim}{Y})}{\partial\underset{\sim}{x}_i\partial\underset{\sim}{x}_j}. \qquad (4)$$

We thus arrive at the following iteration scheme

$$\sum_{j=1}^{n}\underset{\sim}{M}_{ij}(\underset{\sim}{X}^{(s)},\underset{\sim}{Y})(\underset{\sim}{x}_j^{(s+1)}-\underset{\sim}{x}_j^{(s)}) = -\underset{\sim}{F}_i(\underset{\sim}{X}^{(s)},\underset{\sim}{Y}), \qquad (5)$$

for i = 1,...,n and $s\geq1$, where $\underset{\sim}{X}^{(1)}$ denotes some initial guess of the atomic configuration. The iteration is continued until the changes in co-ordinates of all atoms between successive iterations and/or the magnitudes of all residual forces are within preset limits. Each system of linear algebraic equations (5) may be solved by using either direct or relaxation methods (see e.g. Sinclair 1971, Chang & Graham 1966).

The above atomistic calculation may be given a more explicit form when all atoms have the same nature (as in the case of dislocations and/or intrinsic point defects) and are assumed to interact via a central-force pair potential, say V. The potential energy of region I is then given by

$$\Phi = \sum_{i=1}^{n}\Big\{\frac{1}{2}\sum_{\substack{j=1\\ j\neq i}}^{n}V(r_{ij}) + \sum_{k=1}^{m}V(d_{ik})\Big\}, \qquad (6)$$

where $r_{ij} = |\underset{\sim}{x}_i - \underset{\sim}{x}_j|$, $d_{ik} = |\underset{\sim}{x}_i - \underset{\sim}{y}_k|$. Denote for conciseness

$$A(r) = V'(r)/r, \quad B(r) = A'(r)/r.$$

Then, introducing (6) into (1) yields

$$\underset{\sim}{F}_i = \sum_{\substack{j=1\\ j\neq i}}^{n}A(r_{ij})(\underset{\sim}{x}_j-\underset{\sim}{x}_i) + \sum_{k=1}^{m}A(d_{ik})(\underset{\sim}{y}_k-\underset{\sim}{x}_i). \qquad (7)$$

Next, since $\partial(a\underset{\sim}{v})/\partial\underset{\sim}{x} = \underset{\sim}{v}\,\partial a/\partial\underset{\sim}{x} + a\,\partial\underset{\sim}{v}/\partial\underset{\sim}{x}$, where $a(\underset{\sim}{x})$ is a scalar field and $\underset{\sim}{v}(\underset{\sim}{x})$ a vector field, both of which being supposed of class C^1 but otherwise arbitrary, it results from (4) and (7) that

$$\underset{\sim}{M}_{ip} = B(r_{ip})(\underset{\sim}{x}_p-\underset{\sim}{x}_i)(\underset{\sim}{x}_p-\underset{\sim}{x}_i) + A(r_{ip})\underset{\sim}{1},$$

for i,p = 1,...,n, $i\neq p$, and

$$\underset{\sim}{M}_{pp} = -\sum_{\substack{j=1\\ j\neq p}}^{n}\underset{\sim}{M}_{pj} - \sum_{k=1}^{m}\{B(d_{pk})(\underset{\sim}{x}_p-\underset{\sim}{y}_k)(\underset{\sim}{x}_p-\underset{\sim}{y}_k) + A(d_{pk})\underset{\sim}{1}\},$$

where $\underset{\sim}{1}$ is the unit tensor.

Once the provisional equilibrium configuration $\underset{\sim}{X}^{(e)}$ has been determined, we can calculate the residual forces acting on the fixed transition atoms $\underset{\sim}{y}_k$, k = 1,...,m. For instance, restricting ourselves again to the case of the central-force pair potential, we may write

$$\underset{\sim}{P}_k = \sum_i A(d_{ik}^{(e)})(\underset{\sim}{x}_i^{(e)} - \underset{\sim}{y}_k^{(e)}), \quad k = 1,\ldots,m, \qquad (8)$$

where $\underset{\sim}{x}_i$ denotes the position vector of any atom interacting with the transition atom $\underset{\sim}{y}_k$, while the superscript (e) indicates evaluation in the equilibrium configuration. Clearly, in order to find all $\underset{\sim}{P}_k$'s , we have to derive from the displacement field used for region II the positions of all atoms situated within a strip of width $2r_c$ around region I, where r_c denotes the cut-off radius of the interatomic potential (cf. also Sinclair 1971 and Sinclair, Gehlen, Hoagland & Hirth 1978).

Next, the configuration of the elastic region II has to be determined under the action of the forces $\underset{\sim}{P}_1,\ldots,\underset{\sim}{P}_m$ and taking into account the boundary conditions on the external surface of the crystal. This will result in new positions $\underset{\sim}{Y}$ of the transition atoms, eventually requiring a new atomistic calculation for region I, and so on, until all residual forces on transition and region-I atoms are within preset limits.

Other coupling conditions between regions I and II are also possible, for instance making them overlap (see e.g. Teodosiu 1982: 258). Finally, the solution of the coupled systems of equations corresponding to both regions may be found by a common strategy.

3 THE CONTINUUM REGION

Consider now region II of the crystal, which should be treated as an elastic continuum. We will adopt the Eulerian formulation, which proves to be more convenient when simulating dislocations (see Sinclair, Gehlen, Hoagland & Hirth 1978 and Teodosiu & Soós 1981). Consequently, all fields will be considered as depending on the positions of the particles in the imperfect crystal, i.e. in the deformed configuration.

Denote by Ω the spatial domain occupied by region II in the deformed configuration and let $\underset{\sim}{X}$, $\underset{\sim}{x}$ denote the position vectors of a current particle in the natural and deformed configurations, respectively. Let $\underset{\sim}{u}$ be the displacement field and denote by $\underset{\sim}{H}$ and $\underset{\sim}{h}$ its gradient with respect to $\underset{\sim}{X}$, respectively $\underset{\sim}{x}$. Then the deformation gradient $\underset{\sim}{F} = \partial\underset{\sim}{x}/\partial\underset{\sim}{X}$, its inverse $\underset{\sim}{F}^{-1} = \partial\underset{\sim}{X}/\partial\underset{\sim}{x}$, and the strain tensor $\underset{\sim}{E} = (1/2)(\underset{\sim}{F}^T\underset{\sim}{F} - \underset{\sim}{1})$ are given in terms of $\underset{\sim}{h}$ and $\underset{\sim}{H}$ by

$$\underset{\sim}{F} = \underset{\sim}{1} + \underset{\sim}{H}, \quad \underset{\sim}{F}^{-1} = \underset{\sim}{1} - \underset{\sim}{h}, \quad \underset{\sim}{E} = (1/2)(\underset{\sim}{H} + \underset{\sim}{H}^T + \underset{\sim}{H}^T\underset{\sim}{H}), \tag{9}$$

where $\underset{\sim}{1}$ denotes as before the unit tensor, while the superscript T denotes the transposition. Moreover, the obvious relation $\underset{\sim}{F}\underset{\sim}{F}^{-1} = \underset{\sim}{1}$ implies that

$$\underset{\sim}{H} = \underset{\sim}{h} + \underset{\sim}{H}\underset{\sim}{h}. \tag{10}$$

The Cauchy stress tensor $\underset{\sim}{T} = \underset{\sim}{T}^T$ is given by the constitutive equation of finite elasticity

$$\underset{\sim}{T} = (\det \underset{\sim}{F}^{-1})\, \underset{\sim}{F}\, \partial W(\underset{\sim}{E})/\partial\underset{\sim}{E}\, \underset{\sim}{F}^T, \tag{11}$$

where W is the strain-energy function. Assuming that W can be expanded in a power series of $\underset{\sim}{E}$ and using throughout rectangular Cartesian components and the summation convention for repeated indices, we may write

$$\partial W(\underset{\sim}{E})/\partial E_{kl} = c_{klmn}E_{mn} + (1/2)C_{klmnrs}E_{mn}E_{rs} + \dots, \tag{12}$$

where $\underset{\sim}{c}$ and $\underset{\sim}{C}$ are the tensors of second- and third-order elastic constants, respectively.

We shall assume in what follows that the quantity $\varepsilon = |\underset{\sim}{h}|$ is small against unity and neglect all terms of order $O(\varepsilon^3)$ or higher. It then follows from (9) and (10) that

$$\underset{\sim}{H} = \underset{\sim}{h} + \underset{\sim}{h}\underset{\sim}{h}, \quad \underset{\sim}{E} = (1/2)(\underset{\sim}{h} + \underset{\sim}{h}^T + \underset{\sim}{h}\underset{\sim}{h} + \underset{\sim}{h}^T\underset{\sim}{h}^T + \underset{\sim}{h}^T\underset{\sim}{h}),$$

while (11) and (12) lead after some intermediate calculation to the Eulerian form of the constitutive equation of second-order elasticity :

$$T_{ij} = (1 - h_{pp})\sigma_{ij} + h_{ip}\sigma_{pj} + h_{jp}\sigma_{pi} + c_{ijmn}h_{mp}h_{pn} + (1/2)(c_{ijmn}h_{pm}h_{pn} + C_{ijmnrs}h_{mn}h_{rs}), \tag{13}$$

where

$$h_{mn} = \partial u_m/\partial x_n, \quad \sigma_{ij} = c_{ijmn}h_{mn}. \tag{14}$$

Consider now the boundary-value problem for the elastic continuum. Denote by Ω the spatial domain occupied by region II in the deformed configuration, by Σ_o the separation surface between regions I and II, and by Σ the external surface of the crystal. Assume that the displacement vector $\underset{\sim}{u}$ is prescribed on a part Σ_u of Σ and that the stress vector $\underset{\sim}{t}$ is prescribed on the complementary part Σ_t of Σ, namely

$$\underset{\sim}{u} = \underset{\sim}{u}^* \text{ on } \Sigma_u, \quad \underset{\sim}{t} = \underset{\sim}{t}^* \text{ on } \Sigma_t, \tag{15}$$

where $\underset{\sim}{u}^*$ and $\underset{\sim}{t}^*$ are given functions of $\underset{\sim}{x}$. Since the elastic continuum is acted on by the surface forces $\underset{\sim}{t}^*$ on Σ_t and by the concentrated forces $\underset{\sim}{P}_1,\dots,\underset{\sim}{P}_m$ at the points $\underset{\sim}{y}_1,\dots,\underset{\sim}{y}_m$, the stress tensor $\underset{\sim}{T}$ has to satisfy the following equilibrium and traction conditions

$$\left.\begin{aligned} &\operatorname{div}\underset{\sim}{T} = \underset{\sim}{0} \text{ in } \Omega\backslash D, \quad D \equiv \{\underset{\sim}{y}_1,\dots,\underset{\sim}{y}_m\},\\ &\underset{\sim}{T}\,\underset{\sim}{n} = \underset{\sim}{t}^* \text{ on } \Sigma_t,\\ &\lim_{\eta\to 0}\int_{\Sigma_\eta(\underset{\sim}{y}_k)} \underset{\sim}{T}\,\underset{\sim}{n}\, d\Sigma = \underset{\sim}{P}_k, \quad k = 1,\dots,m, \end{aligned}\right\} \tag{16}$$

where $\Sigma_\eta(\underset{\sim}{y}_k)$ denotes the part contained in Ω of a sphere centred at $\underset{\sim}{y}_k$ and of radius η, while $\underset{\sim}{n}$ denotes the outward unit normal to Σ_t, respectively the inward unit normal to $\Sigma_\eta(\underset{\sim}{y}_k)$ (cf. also Turteltaub & Sternberg 1968 and Gurtin 1972: 179-183).

Eqs.(13) through (16) define a mixed boundary-value problem for the continuum part of the imperfect crystal when the displacements are continuous, for instance in the case of point defects. A slightly more complicated problem occurs, however, when considering also dislocations. In order to fix ideas, assume that the imperfect crystal contains a single dislocation loop L. Then the region I is generally taken as a thin tube of boundary Σ_o around the dislocation line L, while Ω becomes a doubly-connected region (fig.2). Let S be a smooth and two-sided cut rendering Ω simply-connected and hence allowing to define a (discontinuous) single-valued displacement field around the dislocation (see e.g. Teodosiu 1982: 37-39 and 105-109). Arbitrarily choose a positive sense on L and denote by $\underset{\sim}{\nu}$ the unit normal to S directed according to the right-hand rule with respect to the positive sense on L. Choose the face of S into which $\underset{\sim}{\nu}$ points as positive face S^+ and label the limiting values of all fields on S^+ and S^- by plus and minus signs, respectively. Then, the displacement field has to satisfy the jump condition

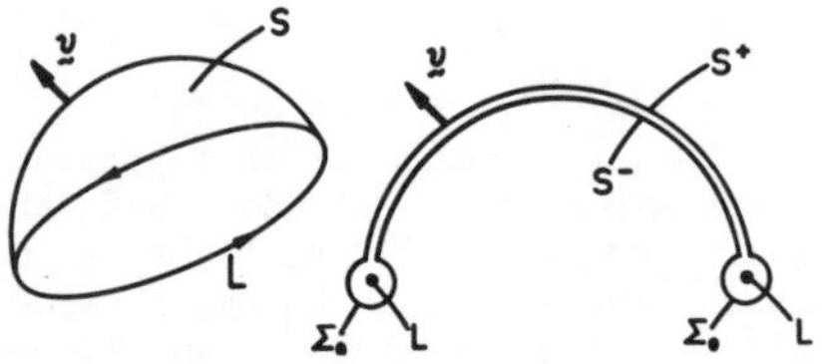

Fig. 2 Orientation of the cut barrier

$$\mathbf{u}^{+} - \mathbf{u}^{-} = \mathbf{b} \quad \text{on } S, \tag{17}$$

while $\mathbf{h}$ and $\mathbf{T}$ must be continuous through S, and hence

$$\mathbf{t}^{+} + \mathbf{t}^{-} = \mathbf{0} \quad \text{on } S. \tag{18}$$

For the subsequent finite element formulation of our problem it is useful to equivalently replace the static equations (16) and (18) by an adequate principle of virtual work. To this end we need some supplementary definitions. Let $\bar{\Omega} = \Omega \cup \Sigma \cup \Sigma_o$.

A kinematically admissible displacement field $\mathbf{u}$ is a vector field of class C^2 in $\bar{\Omega}\backslash S$ which satisfies the displacement boundary condition $(15)_1$, the jump condition (17), and such that its gradient $\mathbf{h}$ is continuous through S and its extension (by continuity) to $\bar{\Omega}$ is of class C^1. A virtual displacement field $\delta\mathbf{u}$ is a vector field of class C^2 in $\bar{\Omega}$ which satisfies the homogeneous conditions

$$\delta\mathbf{u} = \mathbf{0} \quad \text{on } \Sigma_u, \tag{19}$$

$$\delta\mathbf{u}^{+} - \delta\mathbf{u}^{-} = \mathbf{0} \quad \text{on } S. \tag{20}$$

A statically admissible stress field $\mathbf{T}$ is a symmetric second-order tensor field of class C^1 in $\bar{\Omega}\backslash S\backslash D$ which satisfies Eqs. (16) and (19). With these definitions it can be proved (see Appendix) that the following principle of virtual work holds.

THEOREM. A symmetric second-order tensor field $\mathbf{T}$ of class C^1 in $\bar{\Omega}\backslash S\backslash D$ is a statically admissible stress field if and only if the condition

$$\int_{\Omega} \mathbf{T}\cdot\delta\mathbf{h}\, d\Omega = \int_{\Sigma_t} \mathbf{t}^{*}\cdot\delta\mathbf{u}\, d\Sigma + \sum_{k=1}^{m} \mathbf{P}_k\cdot\delta\mathbf{u}_k \tag{21}$$

is satisfied for any virtual displacement field $\delta\mathbf{u}$, where $\delta\mathbf{h} = \partial(\delta\mathbf{u})/\partial\mathbf{x}$ and $\delta\mathbf{u}_k = \delta\mathbf{u}(\mathbf{y}_k)$.

Assume now that $\mathbf{T}$ is derived from a kinematically admissible displacement field via the constitutive equation (13). Then, the principle of virtual work provides the necessary and sufficient conditions for $\mathbf{T}$ to be a statically admissible stress field, and hence for $\mathbf{u}$ to be a solution of the mixed boundary-value problem (13)-(18).

4 FINITE ELEMENT DISCRETIZATION

Assume first that we have to solve a problem with continuous displacements, as in the case of point defects. We divide the region $\bar{\Omega}$ into finite elements and seek the displacement field and its gradient in the approximate forms

$$\mathbf{u} = \sum_{\alpha\in\bar{\Omega}} \mathbf{u}_\alpha \Phi_\alpha(\mathbf{x}), \quad \mathbf{h} = \sum_{\alpha\in\bar{\Omega}} \mathbf{u}_\alpha\, \mathrm{grad}\Phi_\alpha(\mathbf{x}), \tag{22}$$

where $\mathbf{u}_\alpha$ is the value of the displacement field at the global node α, while $\Phi_\alpha(\mathbf{x})$, $\alpha\in\bar{\Omega}$ denote the shape functions, which equal selected polynomial functions within each element and satisfy the conditions

$$\Phi_\alpha(\mathbf{x}_\beta) = \delta_{\alpha\beta}, \tag{23}$$

for any $\alpha,\beta\in\bar{\Omega}$, $\delta_{\alpha\beta}$ being the Kronecker delta (= 1 for $\alpha = \beta$, = 0 for $\alpha \neq \beta$). Similarly, using the same shape functions and considering (19), we shall take the virtual displacement field and its gradient in the form

$$\delta\mathbf{u} = \sum_{\alpha=1}^{N} (\delta\mathbf{u}_\alpha)\Phi_\alpha(\mathbf{x}), \quad \delta\mathbf{h} = \sum_{\alpha=1}^{N} (\delta\mathbf{u}_\alpha)\mathrm{grad}\Phi_\alpha(\mathbf{x}), \tag{24}$$

where $\delta\mathbf{u}_1,\ldots,\delta\mathbf{u}_N$ are the arbitrary nodal values of $\delta\mathbf{u}$, N being the number of global nodes in $\bar{\Omega}\backslash\Sigma_u$.

Introducing now (24) into (21) and taking into account that $\delta\mathbf{u}_\alpha$ are arbitrary, we obtain a system of N vectorial non-linear equations for the determination of the N unknown nodal displacement vectors

$$\mathbf{g}_\alpha(\mathbf{u}) = \mathbf{f}_\alpha(\mathbf{u}), \quad \alpha = 1,\ldots,N, \tag{25}$$

where

$$\mathbf{g}_\alpha(\mathbf{u}) = \int_{\Omega} \mathbf{T}\cdot\mathrm{grad}\Phi_\alpha\, d\Omega, \tag{26}$$

$$\mathbf{f}_\alpha(\mathbf{u}) = \int_{\Sigma_t} \mathbf{t}^{*}\, \Phi_\alpha\, d\Sigma + \sum_{k=1}^{m} \mathbf{P}_k\, \Phi_\alpha(\mathbf{y}_k). \tag{27}$$

By virtue of (23), the first term in the right-hand side of (27) contributes to $\mathbf{f}_\alpha$ only if the global node α belongs to Σ_t, while the last one contributes to $\mathbf{f}_\alpha$ only if the global node α and at least one transition atom belong to the same finite element. Clearly, system (25) contains Nd_f scalar equations and unknowns, where d_f is the number of degrees of freedom per node. The matrix form of this system for 3-D and 2-D problems could be easily derived, but this will not be done here.

System (25) may be solved iteratively by using various standard procedures. For instance, when using a direct iteration, we may first write the constitutive equation (13) as

$$T_{ij} = d_{ijmn}(\mathbf{u})\, h_{mn}, \tag{28}$$

where

$$d_{ijmn}(\underset{\sim}{u}) = (1-h_{pp})c_{ijmn}+\sigma_{nj}\delta_{im}+\sigma_{ni}\delta_{jm}+$$

$$+c_{ijms}h_{ns}+(1/2)(c_{ijsn}h_{ms}+C_{ijmnrs}h_{rs}).$$

Next, by introducing (28) into (25) and (26), and considering (15), we obtain

$$\sum_{\beta=1}^{N} \underset{\sim}{k}_{\alpha\beta}(\underset{\sim}{u})\,\underset{\sim}{u}_{\beta} = \hat{\underset{\sim}{f}}_{\alpha}(\underset{\sim}{u}),\ \alpha = 1,\ldots,N, \quad (29)$$

where

$$\{\underset{\sim}{k}_{\alpha\beta}(\underset{\sim}{u})\}_{ij} = \int_{\Omega} d_{ijmn}(\underset{\sim}{u})\frac{\partial\Phi_{\alpha}}{\partial x_j}\frac{\partial\Phi_{\beta}}{\partial x_n}\,d\Omega,\ \alpha,\beta\in\bar{\Omega}, \quad (30)$$

$$\hat{\underset{\sim}{f}}_{\alpha}(\underset{\sim}{u}) = \underset{\sim}{f}_{\alpha}(\underset{\sim}{u}) - \sum_{\beta\in\Sigma_u} \underset{\sim}{k}_{\alpha\beta}(\underset{\sim}{u})\,\underset{\sim}{u}^{*}_{\beta}. \quad (31)$$

We can now use the iteration scheme

$$\sum_{\beta=1}^{N} \underset{\sim}{k}_{\alpha\beta}(\underset{\sim}{u}^{(s)})\,\underset{\sim}{u}_{\beta}^{(s+1)} = \hat{\underset{\sim}{f}}_{\alpha}(\underset{\sim}{u}^{(s)}),\alpha=1,\ldots N, \quad (32)$$

where $s\geq 1$ is the step of the iteration, while $\underset{\sim}{u}^{(1)}$ denotes some initial guess of the displacement field.

Alternatively, we may solve (25) by employing the Newton-Raphson method, as in the case of the atomistic calculation. To this end, we use at the (s+1)th step of the iteration a truncated Taylor expansion of $\underset{\sim}{g}_{\alpha}(\underset{\sim}{u})$ around the solution $\underset{\sim}{u}^{(s)}$ derived at the preceding step, thus obtaining

$$\sum_{\beta=1}^{N} \bar{\underset{\sim}{k}}_{\alpha\beta}(\underset{\sim}{u}^{(s)})(\underset{\sim}{u}_{\beta}^{(s+1)}-\underset{\sim}{u}_{\beta}^{(s)}) = \bar{\underset{\sim}{f}}_{\alpha}(\underset{\sim}{u}^{(s)}),$$

$$\alpha = 1,\ldots,N, \quad (33)$$

for $s\geq 1$, where $\underset{\sim}{u}^{(1)}$ denotes some initial guess of the displacement field and

$$\bar{\underset{\sim}{k}}_{\alpha\beta}(\underset{\sim}{u}) \equiv \frac{\partial \underset{\sim}{g}_{\alpha}(\underset{\sim}{u})}{\partial \underset{\sim}{u}_{\beta}},\quad \alpha,\beta = 1,\ldots,N, \quad (34)$$

$$\bar{\underset{\sim}{f}}_{\alpha}(\underset{\sim}{u}) \equiv \underset{\sim}{f}_{\alpha}(\underset{\sim}{u}) - \underset{\sim}{g}_{\alpha}(\underset{\sim}{u}),\quad \alpha = 1,\ldots,N. \quad (35)$$

To derive a more explicit form of $\bar{\underset{\sim}{k}}_{\alpha\beta}$ we introduce (26) into (34) and apply the chain rule of differential calculus, considering also the second equation (22). We thus successively obtain, in Cartesian components,

$$\{\bar{\underset{\sim}{k}}_{\alpha\beta}(\underset{\sim}{u})\}_{im} = \frac{\partial g_{\alpha i}(\underset{\sim}{u})}{\partial u_{\beta m}} = \int_{\Omega}\frac{\partial T_{ij}}{\partial u_{\beta m}}\frac{\partial\Phi_{\alpha}}{\partial x_j}\,d\Omega$$

$$= \int_{\Omega}\frac{\partial T_{ij}}{\partial h_{pn}}\frac{\partial h_{pn}}{\partial u_{\beta m}}\frac{\partial\Phi_{\alpha}}{\partial x_j}\,d\Omega = \int_{\Omega}\bar{d}_{ijmn}(\underset{\sim}{u})\frac{\partial\Phi_{\alpha}}{\partial x_j}\frac{\partial\Phi_{\beta}}{\partial x_n}\,d\Omega, \quad (36)$$

where, by virtue of (13),

$$\bar{d}_{ijmn}(\underset{\sim}{u}) \equiv \partial T_{ij}/\partial h_{mn} = c_{ijmn}(1-h_{pp})-\sigma_{ij}\delta_{mn}+$$

$$\sigma_{in}\delta_{jm}+\sigma_{nj}\delta_{im}+c_{sjmn}h_{is}+c_{ismn}h_{js}+c_{ijsn}h_{ms}+$$

$$c_{ijms}h_{ns}+c_{ijsn}h_{sm}+C_{ijmnrs}h_{rs}.$$

For either of these schemes, the iteration is continued until the changes in magnitude of all nodal displacements and/or the magnitudes of the unequilibrated forces are within preset limits. After each iteration step the positions of the global nodes and of the transition atoms must be actualized by using the obvious relations

$$\underset{\sim}{x}_{\alpha}^{(s+1)} = \underset{\sim}{x}_{\alpha}^{(s)} + \underset{\sim}{u}_{\alpha}^{(s+1)} - \underset{\sim}{u}_{\alpha}^{(s)},\ \alpha = 1,\ldots,N,$$

$$\underset{\sim}{y}_{k}^{(s+1)} = \underset{\sim}{y}_{k}^{(s)} + \underset{\sim}{u}_{k}^{(s+1)} - \underset{\sim}{u}_{k}^{(s)},\ k = 1,\ldots,m.$$

Clearly, the whole computation is easier to survey when the transition atoms themselves are taken as nodes of the finite element mesh, and this alternative should be preferred whenever the range of the interatomic potential is not too large.

Let us consider finally the case of a single dislocation loop. The fields $\underset{\sim}{u}$ and $\delta\underset{\sim}{u}$ have now to satisfy conditions (17), respectively (20), on S. Consider that S^{+} and S^{-} are boundaries of Ω and denote by $\beta\in S^{+}$ and $\gamma\in S^{-}$ two congruent global nodes on S. Since $\underset{\sim}{u}_{\gamma} = \underset{\sim}{u}_{\beta} - \underset{\sim}{b}$, we may eliminate the unknown $\underset{\sim}{u}_{\gamma}$ by simply adding $\underset{\sim}{k}_{\alpha\gamma}$ to $\underset{\sim}{k}_{\alpha\beta}$ and $\underset{\sim}{k}_{\alpha\gamma}\underset{\sim}{b}$ to $\hat{\underset{\sim}{f}}_{\alpha}$ for each α in (32). On the other hand, since $\delta\underset{\sim}{u}_{\gamma} = \delta\underset{\sim}{u}_{\beta}$, these virtual displacement vectors are no longer independent, and we have to sum up equations β and γ in (32). Proceeding in the same way with all pairs of congruent nodes on S, it is easily seen that the number of vectorial unknowns and equations of system (32) diminishes by exactly the number of pairs of congruent nodes. Similar considerations hold, of course, for system (33). Alternatively, one may use special techniques that allow to avoid the assembly of equations corresponding to nodes on repeating boundaries and lead directly to the reduced system of equations (see Zienkiewicz 1977: 686-688 and Teodosiu, Soós & Roşu 1983).

5 POSSIBILITY OF USING ENGINEERING CODES

Although non-linear engineering codes have not yet been designed for treating specific problems of crystal imperfections, it is of course tempting to try adapting them to such applications, because they generally offer the user standard facilities for data pre- and postprocessing, as well as very effective routines for iteratively solving large systems of linear and non-linear equations. This possibility has been recently investigated on the lines of this paper by S.Kohlhoff, a graduate student of the University of Stuttgart, with the non-linear code LARSTRAN, which was developed at the Institute for Computer Applications in Stuttgart. It came out that the basic changes to be implemented are: (i)the consideration of the two faces of S as a repeating boundary (coupled degrees of freedom), after imposing the rigid displacement jump condition, (ii) the treatment of the atomic array in region I as a structure of pinned rods with non-linear elastic properties and (iii) the coupling of this structure with the finite element part of the model at the sites of the transition atoms, which are again considered as a repeating boundary.

Preliminary results obtained in the case of a straight dislocation lying in the axis of a non-linear anisotropic elastic cylinder show that the computing time necessary for the finite element treatment of the continuum region is comparable with that required to numerically handle approximate non-linear solutions obtained for this region (see e.g. Seeger, Teodosiu & Petrasch 1975 and Petrasch & Belzner 1976) and is at any rate a small fraction of that spent anyhow for the atomistic calculation. On the other hand, the finite element approach has, of course, the advantage of a higher flexibility in modelling complex configurations and boundary conditions, without previous knowledge of any analytical solutions.

AKNOWLEDGEMENT

The author expresses his gratitude to Professor Seeger for many stimulating discutions and for his criticism of the manuscript.

REFERENCES

Chang, E. & L.T. Graham 1966. Edge dislocation core structure and the Peierls barrier in body-centred cubic iron. phys. stat. sol. 18: 99-103.

Gurtin, M.E. 1972. The Linear Theory of Elasticity. In S. Flügge & C. Truesdell (eds.), Handbuch der Physik, vol. VIa/2, p.1-295. Berlin, Heidelberg, New York: Springer.

Petrasch, P. & V. Belzner 1976. Elasto-atomistic coupling procedure to determine the core configuration of edge dislocations in ionic crystals. J. Physique 37, Colloque C7: 553-556.

Seeger, A., C. Teodosiu & P. Petrasch 1975. Second-order effects in the anisotropic elastic field of a straight edge dislocation. phys. stat. sol. (b) 67: 207-224.

Sinclair, J.E. 1971. Improved atomistic model of a bcc dislocation core. J. Appl. Phys. 42: 5321-5329.

Sinclair, J.E., P.C. Gehlen, R.G. Hoagland & J.P. Hirth 1978. Flexible boundary conditions and nonlinear geometric effects in atomic dislocation modelling. J. Appl. Phys. 49: 3890-3897.

Teodosiu, C. & E. Soós 1981. Non-linear elastic models of single dislocations. Rev. Roum. Sci. Techn.-Méc. Appl. 26: 731-745.

Teodosiu, C. 1982. Elastic Models of Crystal Defects. Berlin, Heidelberg, New York: Springer.

Teodosiu, C., E. Soós & I. Roşu 1983. A finite element model of the hot working of axisymmetric products. I. Finite element formulation and algorithm of time integration. Rev. Roum. Sci. Techn.-Méc. Appl. 28: 575-601.

Turteltaub, M.J. & E. Sternberg 1968. On concentrated loads and Green's functions in elastostatics. Arch. Rational Mech. Anal. 29: 193-240.

Zienkiewicz, O.C. 1977. The Finite Element Method. London, New York, St. Louis: Mc Graw-Hill.

APPENDIX

In order to prove the theorem stated in Sect. 3, we first establish a preliminary relation. Let Ω_η denote the domain obtained from $\bar{\Omega}\setminus S$ by eliminating disjoint balls centred at the points $\underset{\sim}{y}_k$ and of a sufficiently small radius η, bounded by the surfaces $\Sigma_\eta(\underset{\sim}{y}_k)$ defined in the main text. Since $\underset{\sim}{T}$ is assumed of class C^1 in Ω_η, we may apply the divergence theorem and, by taking into account Eqs. (19), (20), and the symmetry of $\underset{\sim}{T}$, we obtain

$$\int_{\Omega_\eta} (\mathrm{div}\underset{\sim}{T})\cdot\delta\underset{\sim}{u}\, d\Omega = \int_{\Omega_\eta} \{\mathrm{div}(\underset{\sim}{T}^T\underset{\sim}{u}) - \underset{\sim}{T}\cdot\mathrm{grad}\delta\underset{\sim}{u}\}\, d\Omega$$

$$= -\int_{\Omega_\eta} \underset{\sim}{T}\cdot\mathrm{grad}\delta\underset{\sim}{u}\, d\Omega + \int_{\Sigma_t} \underset{\sim}{t}\cdot\delta\underset{\sim}{u}\, d\Sigma + \int_S (\underset{\sim}{t}^+ + \underset{\sim}{t}^-)\cdot\delta\underset{\sim}{u}\, dS$$

$$+\sum_{k=1}^{m}\int_{\Sigma_\eta(\underset{\sim}{y}_k)} \underset{\sim}{t}\cdot\delta\underset{\sim}{u}\, d\Sigma$$

for any virtual displacement field $\delta\underset{\sim}{u}$. On the other hand, we may write

$$\int_{\Sigma_\eta(\underset{\sim}{y}_k)} \underset{\sim}{t}\cdot\delta\underset{\sim}{u}\, d\Sigma = \int_{\Sigma_\eta(\underset{\sim}{y}_k)} \underset{\sim}{t}\cdot(\delta\underset{\sim}{u} - \delta\underset{\sim}{u}_k)\, d\Sigma + \left(\int_{\Sigma_\eta(\underset{\sim}{y}_k)} \underset{\sim}{t}\, d\Sigma\right)\cdot\delta\underset{\sim}{u}_k .$$

and the first integral in the right-hand side tends to 0 as $\eta\to 0$, by virtue of the continuity of $\delta\underset{\sim}{u}$. Letting now $\eta\to 0$, we obtain from the last two relations the desired preliminary result:

$$\int_\Omega (\mathrm{div}\underset{\sim}{T})\cdot\delta\underset{\sim}{u}\, d\Omega = -\int_\Omega \underset{\sim}{T}\cdot\mathrm{grad}\delta\underset{\sim}{u}\, d\Omega + \int_{\Sigma_t} \underset{\sim}{t}\cdot\delta\underset{\sim}{u}\, d\Sigma$$

$$+\int_S (\underset{\sim}{t}^+ + \underset{\sim}{t}^-)\cdot\delta\underset{\sim}{u}\, dS + \sum_{k=1}^{m}\left(\lim_{\eta\to 0}\int_{\Sigma_\eta(\underset{\sim}{y}_k)} \underset{\sim}{t}\, d\Sigma\right)\cdot\delta\underset{\sim}{u}_k \quad (A1)$$

First assume that $\underset{\sim}{T}$ is statically admissible, i.e. it satisfies Eqs.(16) and (18). It then follows from (A1) that Eq.(21) is satisfied for any virtual displacement field $\delta\underset{\sim}{u}$. Conversely, assume that Eq.(21) is fulfilled for any virtual displacement field $\delta\underset{\sim}{u}$. Then, introducing (21) into the right-hand side of (A1) and rearranging terms yields

$$\int_{\Omega}(\mathrm{div}\underset{\sim}{T})\cdot\delta\underset{\sim}{u}\,d\Omega - \int_{\Sigma_t}(\underset{\sim}{t}-\underset{\sim}{t}^{*})\cdot\delta\underset{\sim}{u}\,d\Sigma - \int_{S}(\underset{\sim}{t}^{+}+\underset{\sim}{t}^{-})\cdot\delta\underset{\sim}{u}\,dS$$

$$- \sum_{k=1}^{m}\left(\lim_{\eta\to 0}\int_{\Sigma_{\eta}(\underset{\sim}{y}_k)}\underset{\sim}{t}\,d\Sigma - \underset{\sim}{p}_k\right)\cdot\delta\underset{\sim}{u}_k = 0.$$

Since $\delta\underset{\sim}{u}$ is arbitrary in Ω, on Σ_t, on S and at the $\underset{\sim}{y}_k$'s, it results that Eqs.(16) and (18) are satisfied, and hence $\underset{\sim}{T}$ is statically admissible, which completes the proof.

5th International Symposium on Continuum Models of Discrete Systems / Nottingham / 14-20 July 1985

Continuum description of mechanical behaviour for materials with heterogeneous oriented structure

A.Litewka
Technical University of Poznań, Poland

ABSTRACT; The aim of this note is to describe the overall mechanical response for the damaged solid with regularly distributed cracks. The yield criterion for such a material is formulated as a scalar function including the damage variable in the form of the symmetric second rank tensor. The theoretical results concerning the plastic constants are verified experimentally employing the models with simulated oriented damage in the form of narrow rectangular slots arranged in the square pattern. Two different orientations of the openings were considered: in the first case the longitudinal axes of the slots were parallel to the pitch direction and in the second one the cracks were arranged diagonally. In both arrangements the overall mechanical response is similar to that observed for orthotropic solid but in the limit case when the rectangular slots are reduced to square or circular holes the tetragonal or hexagonal symmetry is obtained. These two types of the symmetry were also considered and the appropriate yield criteria were formulated.

1 INTRODUCTION

Development of modern technology creates new fields of interest for the mechanics of solid and one of them is a continuum theory of the materials with heterogeneous oriented structure. When describing the overall mechanical response of porous and perforated materials and also damaged solids with many uniformly distributed cracks the notion of the equivalent material is utilized (O'Donnell, Langer, 1962, Porowski, O Donnell, 1974, Litewka, Rogalska, 1983, Litewka, Sawczuk, 1984). The constitutive equations for such materials should account for the stiffness and strength reduction of the equivalent material due to development of cracks and voids and also for the specific symmetry of the material structure. These equations are derived employing the theory of tensor functions representations and the theory of scalar invariants (Spencer, 1971 Boehler, Raclin, 1977).

To illustrate the specific features of the constitutive equation formulation for the heterogeneous material its initial plastic flow will be discussed in the paper. In particular the aim of this note is to describe the modification of the effective yield surface for the orthotropically damaged solid with regularly distributed rectilinear cracks. According to Vakulenko and Kachanov (1971) the current state of the damaged material is described by the damage variable assumed in the form of the symmetric second rank tensor. Two other specific cases of the voids distribution discussed in the paper lead to tetragonal and hexagonal symmetries of the material which cannot be described by the damage tensor assumed. Some results concerning the yield criteria formulation for the materials possessing such higher symmetries are also presented.

2 ORTHOTROPICALLY DAMAGED MATERIAL

2.1. Yield criterion

The considerations presented in this section concern the cracked material possessing three mutually perpendicular planes of the symmetry associated with the axes of Cartesian coordinate system x_1, x_2, x_3. In the case of such a symmetry the damage variable can be assumed in the form of the symmetric second rank tensor (Vakulenko, Kachanov, 1971, Murakami, Imaizumi, 1982)

$$\underset{\sim}{D} = D_1 \underset{\sim}{n}_1 \otimes \underset{\sim}{n}_1 + D_2 \underset{\sim}{n}_2 \otimes \underset{\sim}{n}_2 + D_3 \underset{\sim}{n}_3 \otimes \underset{\sim}{n}_3 \qquad (1)$$

where D_1, D_2, D_3 are the principal values of the damage tensor, corresponding to the planes of the symmetry defined by the unit normal vectors $\underset{\sim}{n}_1$, $\underset{\sim}{n}_2$, $\underset{\sim}{n}_3$. As was proposed by Litewka (1985) the principal values D_1, D_2, D_3 can be calculated as follows

$$D_i = \frac{S_{ci}}{S_{Li}}, \quad i = 1, 2, 3 \qquad (2)$$

where S_{ci} is a crack area and S_{Li} is a net area of a material remained on the plane normal to the coordinate axis x_i.

The yield criterion for the material discussed can be formulated as an isotropic scalar function of the stress and damage tensor. According to the theory of invariants (Spencer, 1971) such a criterion has a form of polynomial function of the scalar invariants derived for the tensors involved. In the case of two symmetric tensors the irreducible set of invariants consists of ten elements. However, only some of them must be included in the final form of the yield criterion. Some information concerning this problem can be furnished by the analysis of the strain energy calculated for the equivalent material having the same mechanical properties as the cracked material discussed. To this end the elastic stress-strain relation derived by Litewka (1985) was employed

$$\underset{\sim}{E} = (A\, tr\underset{\sim}{T} + \beta\, tr\underset{\sim\sim}{DT})\, \underset{\sim}{I} + 2B\, \underset{\sim}{T} + 2\gamma (\underset{\sim\sim}{TD} + \underset{\sim\sim}{DT}) + \beta\, \underset{\sim}{D}\, tr\underset{\sim}{T} \qquad (3)$$

where $\underset{\sim}{E}$, $\underset{\sim}{T}$ and $\underset{\sim}{I}$ are the strain, stress and unit tensors respectively. The constants β and γ are determined experimentally or theoretically for a given arrangement and dimensions of the cracks. Two other constants A and B are expressed by the Young modulus E and Poisson ratio ν of the matrix material

$$A = -\nu/E, \quad B = (1 + \nu)/2E$$

Taking into account the equation (3) the elastic strain energy calculated for the homogenized equivalent material has a form

$$\Phi_e = (A/2 + B/3)\, tr^2\underset{\sim}{T} + B\, tr\underset{\sim}{S}^2 + \beta\, tr\underset{\sim}{T}\, tr\underset{\sim\sim}{DT} + 2\gamma\, tr\underset{\sim\sim}{DT}^2 \qquad (4)$$

It is reasonable to assume that a yield criterion for the cracked solid has the form similar to that obtained for the elastic strain energy and expressed by the equation (4) containing only four invariants. Taking into consideration the values β and γ determined in (Litewka, 1985)

$$\beta = 0$$

$$\gamma = \frac{1}{4D_2}\left(\frac{1}{E_2} - \frac{1}{E}\right)$$

where E_2 is the effective Young modulus of the damaged material for the uniaxial loading in the direction of x_2 axis, the term containing $tr\underset{\sim\sim}{DT}$ disapears. Therefore, only three invariants $tr\underset{\sim}{T}$, $tr\underset{\sim}{S}^2$ and $tr\, \underset{\sim\sim}{DT}^2$ are included in the yield criterion proposed in the form

$$C_1\, tr^2\underset{\sim}{T} + C_2\, tr\underset{\sim}{S}^2 + C_3\, tr\underset{\sim\sim}{DT}^2 - \sigma_o^2 = 0 \qquad (5)$$

where C_1, C_2 and C_3 are unknown constants and σ_o is the uniaxial yield stress for the matrix material.

The validity of the yield criterion (5) and the assumed definition of the damage variable (1) and (2) can be verified after calculation the constants C_1, C_2 and C_3. This can be done specifying the equation (5) for the prescribed simple stress states like an uniaxial tension in the direction of x_1 and x_2 axes and for a biaxial uniform tension. In the result the following set of linear equations is obtained

$$\begin{aligned} C_1 + 2C_2/3 + D_1C_3 &= (\sigma_o/\sigma_{1o})^2 \\ C_1 + 2C_2/3 + D_2C_3 &= (\sigma_o/\sigma_{2o})^2 \\ 4C_1 + 2C_2/3 + (D_1 + D_2)C_3 &= (\sigma_o/T_o)^2 \end{aligned} \qquad (6)$$

where σ_{1o}, σ_{2o} and T_o are the respective yield stresses for the cracked solid uniaxially loaded in the direction of x_1 and x_2 axes and for the biaxial uniform tension. These yield stresses required to calcualate C_1, C_2 and C_3 from (6) can be measured experimentally or determined from the theoretical considerations.

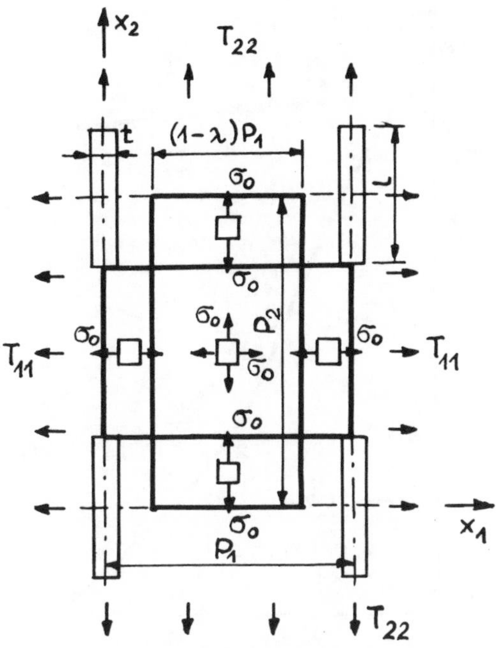

Fig. 1 Statically admissible discontinuous fields of the stresses for biaxial loading of the cracked solid.

2.2. Determination of the yield stresses

The alternative method of the yield function formulation consists in the comparison of the strain energy (4) calculated for the equivalent homogeneous material with that accumulated in the matrix material. However, this method cannot be used without some simplifying assumptions concerning the stress distribution within the unit cell of the material structure (see e.g. Markov, 1981) . Because of the complexity of the elastic-plastic stress state in the matrix material of the solid with cracks such a comparison of the strain energy can be done effectively only for some selected types of the loading. Considering for example the uniaxial loading in the direction parallel and perpendicular to the longitudinal axis of the cracks and the biaxial uniform loading the statically admissible discontinuous fields of the stress can be constructed. The example of such a field in the case of the biaxial uniform loading is shown in fig. 1. The elastic strain energy estimated for the matrix material taking into account these fields of the stress was equated to that calcu-

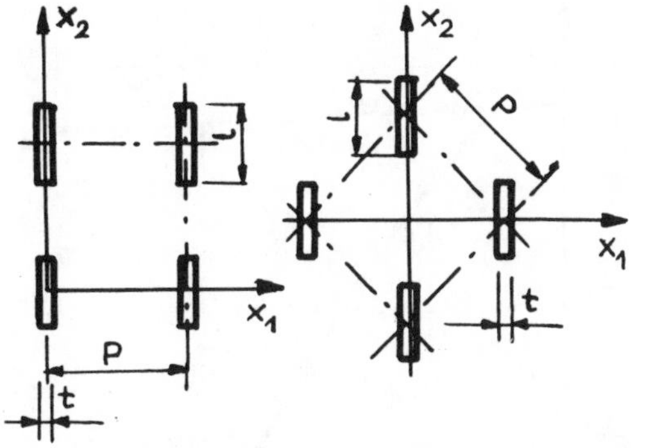

Fig. 2 Cracks arrangements in the models simulating the damaged material.

lated from (4). This made it possible to determine the yield stress of the cracked material T_o in the form

$$T_o = (1-\lambda)\left\{2(1-\tau)\left[1-\nu(1-\lambda)\right]\right\}^{1/2} \left[2(1-\nu)(1-\tau)(1-\lambda) + \lambda(\lambda+\tau-2\lambda\tau)\right]^{-1/2} \sigma_o \tag{7}$$

Similar but much more simple fields of the stresses can be constructed for the uniaxial loading and finally the uniaxial yield stresses for loading in the directions of x_1 and x_2 axes are determined

$$\sigma_{1o} = \frac{1-\lambda}{\sqrt{1-\lambda+\lambda^2}}\,\sigma_o \qquad \sigma_{2o} = \frac{1-\tau}{\sqrt{1-\tau+\lambda\tau}}\,\sigma_o \tag{8}$$

where $\lambda = l/P_2$ and $\tau = t/P_1$ are the dimensionless length and width of the crack.

2.3. Experimental verification

The validity of the yield criterion (5) was verified experimentally employing the uniaxially loaded models simulating the cracked materials. To this end the rectangular openings arranged as shown in fig. 2 were cut out in the flat specimens made of the aluminium alloy sheet metals. The detailed description of the models

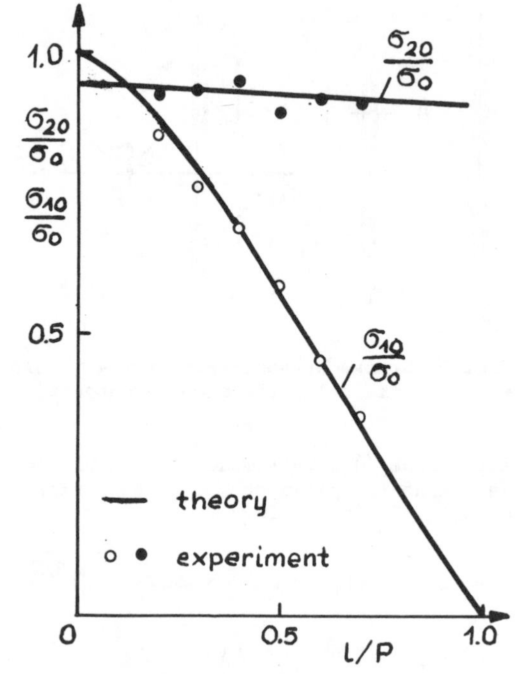

Fig. 3 Uniaxial yield stresses for the damaged solid with cracks arranged in the pitch direction.

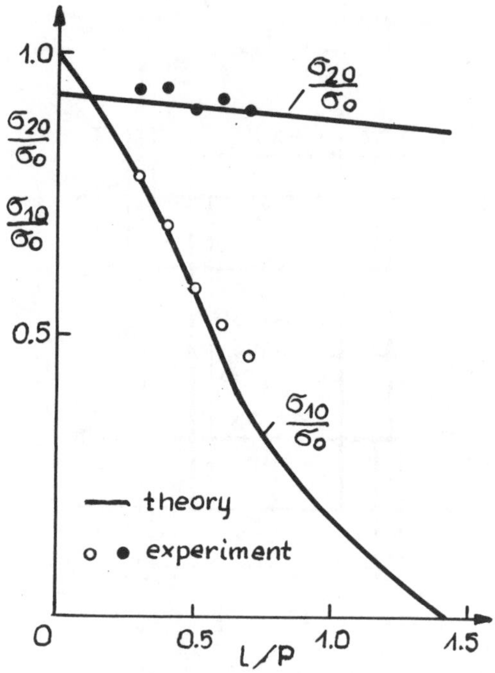

Fig. 4 Uniaxial yield stresses for the damaged material with diagonal cracks arrangement.

preparation and the experimental technique used can be found in the previous papers (Litewka, Stanisławska, 1983 and Litewka, Sawczuk, 1984).

The uniaxial yield stresses measured experimentally for the models with crack arranged in the pitch direction loaded in the direction parallel and perpendicular to the longitudinal cracks axes were compared in fig. 3 with the theoretical values calculated from equations (8). Similar comparison concerning the diagonal cracks arrangement is shown in fig. 4. To calculate the theoretical values of the appropriate uniaxial yield stresses required to determine the constants C_1, C_2 and C_3 for the material with diagonal crack arrangement the analysis similar to that presented in the section 2.2 was performed.

3 MATERIALS WITH HIGHER SYMMETRIES

3.1. Tetragonal symmetry

This type of the symmetry encountered in the perforated plates with square penet-

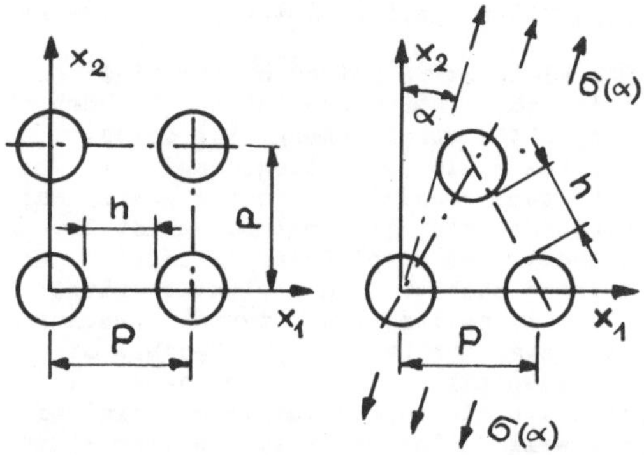

Fig. 5 Square and triangular penetration patterns.

ration pattern (fig. 5) can be also obtained as a limit case of the damaged material shown in fig. 1 when $P_1 = P_2$ and $t = 1$. The yield criterion accounting for the specific material symmetry of the square penetration pattern was formulated by Markov (1981) and Litewka and Rogalska (1983). Here cannot be employed the method used succesfully in the case of the orthorопic materials where the

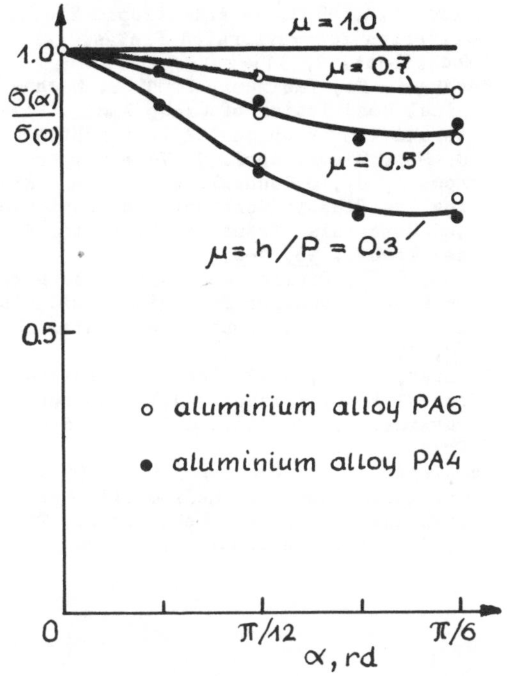

Fig. 6 Dependence of the uniaxial yield stress for perforated material with triangular penetration pattern on the loading orientation.

symmetry is fully described by the independent variable assumed in the form of the symmetric second rank tensor. Now the suitable variable entering the yield criterion should have the form of the fourth order tensor. To avoid the mathematical problems expected when deriving the representation of the scalar function containing the fourth order tensor as an independent variable the alternative method is used. It seems that the most conveniant way of the yield criterion formulation is to apply the results obtained by Smith and Rivlin (1958) concerning the scalar invariants of the single symmetric second rank tensor determined under the symmetry groups corresponding to the various crystal classes.

The specific symmetry of the material with the square penetration pattern corresponds to the ditetragonal-dipyramidal crystal class. The appropriate yield criterion derived as a polynomial function of the scalar invariants determined by Smith and Rivlin (1958) has a form

$$A\,tr^2\underset{\sim}{T} + tr\underset{\sim}{S}^2 + B\,T_{12}^2 + C\,T_{11}T_{22} + D\,T_{33}^2 - \frac{2}{3}\sigma_0^2 = 0 \qquad (9)$$

where A, B, C and D are the material constants measured experimentally (see Litewka, Sawczuk, 1981 and Litewka and Rogalska, 1983).

3.2. Hexagonal symmetry

Usually it is assumed that the mechanical properties of the material possessing the hexagonal symmetry like for example the perforated plate with the triangular penetration pattern shown in fig. 5 are independent on the loading orientation (see e.g. O Donnell, Langer, 1962). However, the experiments do not corroborate this simplifying assumption. To this end the specimens of the perforated material with the triangular array of the circular holes were subjected to the uniaxial loading in various directions with respect to the symmetry axes of the material structure. The results are shown in fig. 6 where $\sigma(o)$ is the uniaxial yield stress measured for the loading in x_2 direction and $\sigma(\alpha)$ is that determined for loading inclined at the angle α with respect to the x_2 axis. It is seen from fig. 6 that for the higher density of holes determined by the ligament efficiency $\mu = h/P$ the directional properties of the material cannot be neglected. It means that the yield criterion formulated for such materials should account for the specific symmetry of the material structure. This problems seems to be mathematically solved and the appropriate form of the yield criterion could be found as a polynomial function of the scalar invariants determined by Smith and Rivlin, (1958). The symmetry of the perforated material with the triangular penetration pattern corresponds to the dihexagonal-dipyramidal crystal class and in this case the most important invariant responsible for the material symmetry has the form

$$J = T_{11}\left[(T_{11} + 3T_{22})^2 - 12T_{12}^2\right]$$

However, the attempts to formulate the yield criterion as a quadratic polynomial function in the form

$$f\,(tr\underset{\sim}{T},\ tr\underset{\sim}{S}^2,\ J) - \sigma_0^2 = 0 \qquad (10)$$

were not succesfull. Several different

forms of the function (10) were considered and in each case the yield surface did not fulfilled the requirement of the convexity. These disturbances were caused by the invariant J which cannot be omitted because it is the only term in the function (10) accounting for the material symmetry.

4 CONCLUSIONS

The symmetry of the internal structure of the heterogeneous materials should be accounted for when formulating the constitutive equations of the mechanics. In particular it concerns the cracked and perforated solids. According to the specific symmetry of the material considered various methods of the constitutive equation formulation can be used. The yield criterion for the materials possessing three mutually perpendicular planes of the symmetry like for example orthotropically damaged solid can be derived employing the symmetric second rank damage tensor. In the case of the tetragonal and hexagonal symmetry the yield criterion should be derived as a scalar function which is form invariant under the appropriate symmetry group.

REFERENCES

Boehler, J.P., Raclin, J. 1977. Représentations irréducibles des fonctions tensorielles non-polynomiales de deux tenseurs symétriques dans quelques cas d'anisotropie. Arch. Mech. Stos. 29, 431-444.

O'Donnell, W.J., Langer, B.F. 1962. Design of Perforated Plates. Trans. ASME, J. Engng Ind. 84, 307-320.

Litewka, A. 1985. Effective Material Constants for Orthotropically Damaged Solid. Arch. Mech. Stos., 37, in press.

Litewka, A., Rogalska, E. 1983. Analysis of Yield Criteria for Perforated Plates. Res Mech., 9, 207-218.

Litewka, A., Sawczuk, A. 1981. A Yield Criterion for Perforated Sheets. Ing.-Arch. 50, 393-400.

Litewka, A., Sawczuk, A. 1984. Experimental Evaluation of Overall Anisotropic Material Response on Continuous Damage. In G.J. Dvorak, R.T. Shield (ed.), Mechanics of Material Behavior, the Daniel C. Anniversary Volume, p. 239-252, Amsterdam : Elsevier.

Litewka, A., Stanisławska, J. 1983. Experimental Simulation of Anisotropic Damage. Mech. Teor. Stos. 21, 361-370.

Markov, K.Z. 1981. An Anisotropic Yield Criterion for Perforated Plates. Res Mech. Lett. 1, 315-318.

Murakami, S., Imaizumi, T. 1982. Mechanical Description of Creep Damage State and Its Experimental Verification. J. Méc. Théor. Appl. 1, 743-761.

Porowski, J., O Donnell, W.J. 1974. Effective Plastic Constants for Perforated Materials. Trans. ASME, Press. Vessel Techn., 96, 1-8.

Smith, G.H., Rivlin, R.S. 1958. The Strain-Energy Function for Anisotropic Elastic Materials. Trans. Amer. Math. Soc. 88, 175-193.

Spencer, A.J.M. 1971. Theory of invariants. In C. Eringen (ed.) Continuum Physics, p. 239-353, London: Academic Press.

Vakulenko, A.A., Kachanov, M.L. 1971. Continuum Theory of Medium with Cracks. Izv. A.N. S.S.S.R., Mekh. Tverd. Tela, 6, 159-166, (in Russian).

5th International Symposium on Continuum Models of Discrete Systems / Nottingham / 14-20 July 1985

On the patterning of microstructures and deformation

E.C.Aifantis
MM Program, Michigan Technological University, Houghton, USA

ABSTRACT: We discuss a method for the analysis of spatial patterns of defects, microstructures and deformation occuring during mechanical loading. We take the point of view that such patterns is the result of instabilities occuring in partial differential equations modelling the evolution of these quantities. The approach is thus similar to that used in other related problems of self-organization or synergetics in chemical and biological systems driven far from thermodynamic equilibrium. For simplicity, we confine attention to one dimension only and consider, in particular, the stability of the periodic dislocation structures of persistent slip bands observed during cycling straining of single monocrystals and the development of finite size localized zones or bands developed during simple shearing of infinite blocks.

1. INTRODUCTION

This communication contains a brief account of recent developments concerned with the prediction and modelling of spatially varying structures commonly observed in various materials science problems. In particular, the tendency of defects to form colonies during deformation is discussed and a method of analysis is presented. We take the point of view that such ordered structures occur, in far from thermodynamic equilibrium conditions, as a result of dynamical instabilities arising in the partial differential equations modelling the motion and generation of defects.

Even though this approach, often known as self-organization or synergetics, has extensively been applied to the study of chemical and biological systems, not much attention has been given to related phenomena of materials science usually induced under the action of an applied stress. For example, the ordering of vacancies and voids has been addressed in [1],[2] and the patterning of dislocation species has been recently discussed in [3]-[5]. A brief introductory review pertaining to these problems is given in [6] where the related problem of localization of deformation in one [7] and two [8] dimensions is treated from the point of view of providing shear bands of finite widths.

Since this approach is not too familiar among researchers in mechanics, it was thought to be instructive to concentrate in one-dimensional situations only and discuss, in particular, two typical and rather elucidative examples.

The first example is concerned with the prediction of periodic solutions for the dislocation distribution within a persistent slip band (PSB) during fatigue. It is shown that the wavelength of these layered dislocation structures is a material property, in accordance with experimental observations. The qualitative results obtained in this example may be thought of as the beginning towards a three-dimensional quantitative theory for the formation and stability of dislocation cell walls. Such a theory may be particularly useful for determining the strength of materials in cases where the overall mechanical behavior is determined by the dislocation network rather than the properties of crystals surrounding it [9].

The second example is concerned with illustrating the possibility of obtaining finite size shear band widths. This results from the non-monotonicity of the stress-strain graph and the stabilizing role of higher-order strain gradients. Again, the qualitative findings obtained in this example may be thought of as the beginning towards a quantitative three-dimensional theory of localization of deformation which in addition to the orientation of the band it will also predict its width, thus providing the extra divident of a correct post-localization treatment of deformation.

2. PERIODIC DISLOCATION DISTRIBUTIONS IN PSBs

In this section we consider the development and stability of the periodic layerlike dislocation structures observed in persistent slip bands during cycling loading. It has been experimentally noted during fatigue studies of monocrystals and polycrystals [10],[11] that two distinct types of dislocation structures occur: rodlike veins and layerlike persistent slip bands (PSBs).

The first PSB occurs when the stress reaches a certain critical value which subsequently remains constant until the specimen is filled up with PSBs. Each PSB consists of groups of sequential primary slip planes (say x-y planes) extended over a small finite distance in the perpendicular direction z. The structure of PSBs is not uniform but it consists instead of alternating layers of rich and poor dislocation regions characterized by an intrinsic wavelength and a wavevector parallel to the applied resolved shear stress. We thus have within each PSB a periodic dislocation distribution in the x direction (slip direction) with the dislocation density remaining uniform in the y and z directions. The physical submechanism for the appearance of such structures is the formation of dislocation dipoles of a density varying uniformly in the directions of y and z [10],[11]. The PSBs are surrounded by rodlike vein structures consisting of dense dislocation dipoles or multipoles which, however, do not exhibit the one-dimensional order of PSBs that are of exclusive concern here.

With the above assumptions we have reduced the problem to a one-dimensional one with the slip direction x being the only direction along which a variation of properties exist. In this connection, and without loss of generality, we may view the coordinate system as coinciding with the crystallographic axes of the specimen. The dislocation population may then effectively be considered as composed of infinite straight parallel edge dislocations that can glide, interact, be generated, and annihilated along their slip plane. Moreover, we distinguish between slow moving or immobile and fast moving or mobile dislocations as this distinction is essential in revealing the essence of the kinetic character of dislocation reactions.

The starting point of the analysis is the conservation equation for the effective mass of dislocation species in the form

$$\partial_t \rho_\kappa + \nabla_x j_\kappa = \hat{c}_\kappa \quad , \tag{1}$$

where ρ denotes density, j flux, $\hat{c}$ production/annihilation and the index κ takes the values 1 and 2 for immobile and mobile dislocation populations respectively. A justification for equation (1) can be found, for example, in [6] where the concept of normal (associated with the lattice) and excited (associated with the dislocations) states is utilized. For the present case of parallel straight dislocations, relation (1) is also consistent with the theory for continuous distributions of dislocations with the exception that the generation term $\hat{c}_\kappa$ is included here.

The second conservation law to be utilized is that of linear momentum for the dislocated state of the form

$$\nabla_x T_\kappa = \hat{f}_\kappa \quad , \tag{2}$$

where T_κ denotes the stress associated with the dislocation species and $\hat{f}_\kappa$ the force generated by the exchange of momentum between dislocated and lattice states. Specifically, T_κ measures the interaction effects between dislocations of the same type and $\hat{f}_\kappa$ the interaction effects between a particular type of dislocations with the others as well as the lattice.

Conservation laws similar to (1) and (2) also hold for the lattice state (ρ_L, j_L, T_L, $\hat{f}_L$, $\hat{c}_L$), but they will not be listed explicitly as they are not used directly here. It is only useful to note that addition of $(2)_{1,2}$ and a corresponding equation for the lattice state together with the definition

$$S = T_L + \sum_{\kappa=1}^{2} T_\kappa \quad , \tag{3}$$

for the total stress S, and the requirement of total momentum balance

$$\hat{f}_L + \sum_{k=1}^{2} \hat{f}_\kappa = 0 \quad , \tag{4}$$

leads to the usual equilibrium equation

$$\nabla_x S = 0 . \qquad (5)$$

In this connection, we note that inertia effects associated with the lattice and dislocated states are not included.

The above set of balance equations must be supplemented with constitutive relations. For the present case of one-dimensional slip we make the following assumptions for the stresses and forces associated with the dislocated states

$$T_\kappa = \pi_\kappa(\rho_\kappa) , \quad \hat{f}_\kappa = \alpha_\kappa(\rho_\kappa) + \beta_\kappa(\rho_\kappa) j_\kappa - \gamma_\kappa(\rho_\kappa) T_L , \qquad (6)$$

where $\pi_\kappa(\rho_\kappa)$ is the interaction stress between dislocations, $\alpha_\kappa(\rho_\kappa)$ is a threshold friction-like resistance acting on the dislocation state as a result of the discrete nature of the lattice, $\beta_\kappa(\rho_\kappa) j_\kappa$ is the drag force exerted on the mobile dislocation state, and $\gamma_\kappa(\rho_\kappa) T_L$ is a continuum generalization of the Peach-Koehler force arising in the theory of individual dislocations. Even though the type of functional and sign dependence of the above coefficients may be modified according to particular situations and the time and space scales for which the theory is being applied, the structure of relations (6) remains generally valid.

Next, we insert (6) into (2) and adopt a few mild linearity assumptions for the coefficients appearing in the stress and force relations (6). We also make use of certain equilibrium conditions satisfied by these coefficients as a result of (3) and (5) and the definition of yielding. Then, by utilizing the conservation of effective mass for the dislocation species (1), we obtain the following reaction-diffusion equations for the immobile and mobile dislocation populations ρ_1 and ρ_2

$$\left.\begin{aligned} \partial_t \rho_1 &= D_1 \nabla_{xx}^2 \rho_1 + g(\rho_1) - h(\rho_1, \rho_2) , \\ \partial_t \rho_2 &= D_2 \nabla_{xx}^2 \rho_2 + h(\rho_1, \rho_2) . \end{aligned}\right\} \qquad (7)$$

The coefficient D_1 is the diffusivity associated with the immobile state and measures random-like effects such as interaction with vacancies, thermal events, bowing movements due to local internal stresses, etc. Unlike to the Brownian-like nature of D_1, the diffusivity D_2 models the drift-like motion of dislocations liberated by the applied stress.

The source term $g(\rho_1)$ is associated with the immobile state and is not assigned any specific form. The source term $h(\rho_1,\rho_2)$, however, is modelling interaction effects between immobile and mobile states and is assumed here to have the following form

$$h(\rho_1, \rho_2) = b\rho_1 - c\rho_2\rho_1^2 \qquad (8)$$

The coefficient b measures the rate with which immobile dislocations break free when the applied stress exceeds a certain threshold. Finally, the coefficent c measures the pinning rate of freed dislocations by immobile dipoles.

Next, we introduce scaled quantities by the relations

$$\rho_1 \to \sqrt{c}\rho_1 , \quad \rho_2 \to \sqrt{c}\rho_2 , \quad g(\rho_1) \to \sqrt{c}\, g\left(\frac{\rho_1}{\sqrt{c}}\right) \qquad (9)$$

and then (7) with the aid of (8) gives the final system of coupled reaction-diffusion equations

$$\left.\begin{aligned} \partial_t \rho_1 &= D_1 \nabla_{xx}^2 \rho_1 + g(\rho_1) - b\rho_1 + \rho_2 \rho_1^2 , \\ \partial_t \rho_2 &= D_2 \nabla_{xx}^2 \rho_2 + b\rho_1 - \rho_2 \rho_1^2 . \end{aligned}\right\} \qquad (10)$$

The homogeneous steady state solution of (10) is given by

$$g(\rho_1^{o}) = 0 , \quad \rho_1^{o} \rho_2^{o} = b , \qquad (11)$$

and small pertrubations from it of the form

$$\tilde{\rho}_1 = \rho_1 - \rho_1^{o} , \quad \tilde{\rho}_2 = \rho_2 - \rho_2^{o} , \qquad (12)$$

are shown to satisfy the following matrix equation

$$\partial_t \begin{pmatrix} \rho_1 \\ \rho_2 \end{pmatrix} = \begin{bmatrix} D_1 \nabla_{xx}^2 + b + g'(\rho_1^{o}) & \rho_1^{o2} \\ -b & D_2 \nabla_{xx}^2 - \rho_1^{o2} \end{bmatrix} \times \begin{pmatrix} \rho_1 \\ \rho_2 \end{pmatrix} \qquad (13)$$

where the bars ~ were dropped for convenience. Taking the Fourier transform in (13) defined as usual by

$$\rho_q \sim \int_{-\infty}^{\infty} \rho(x)e^{iqx}dx \quad , \qquad (14)$$

and setting $g'(\rho_1^o) = -a(<0$ for the stability of homogeneous states), we can show that in Fourier space, (13) becomes

$$\partial_t \begin{pmatrix} \rho_{1q} \\ \rho_{2q} \end{pmatrix} = \begin{bmatrix} b-a-q^2D_1 & \rho_1^{o2} \\ -b & -\rho_1^{o2}-q^2D_2 \end{bmatrix} \begin{pmatrix} \rho_{1q} \\ \rho_{2q} \end{pmatrix}, \qquad (15)$$

The stability of (15) is determined by the characteristic equation

$$\omega^2 + \beta\omega + \gamma = 0 \;, \qquad (16)$$

where

$$\beta=-\mathrm{tr}[\;]=q^2(D_1+D_2)+(a-b+\rho_1^{o2}),$$
$$\gamma=\det[\;]=q^4D_1D_2+q^2[D_2(a-b)+D_1\rho_1^{o2}] + a\rho_1^{o2} \;. \qquad (17)$$

The homogeneous solution (ρ_1^o,ρ_2^o) becomes unstable when the real part of at least one of the roots of (17) vanishes. Indeed, there are two possible types of instabilities: (a) A Hopf bifurcation leading to temporal oscillations occurs for

$$q=0; \quad b \geq b_c^o \equiv a+\rho_1^{o2} \;. \qquad (18)$$

(b) A Turing instability leading to spatially periodic solutions occurs for

$$q^2 = q_c^2 = \left[\frac{a\rho_1^{o2}}{D_1D_2}\right]^{1/2} ;$$
$$b \geq b_c \equiv \left[a^{1/2}+\rho_1^{o}\left(\frac{D_1}{D_2}\right)^{1/2}\right]^2 , \qquad (19)$$

The first type of instability has experimentally been observed in the form of strain bursts for slow increases of the stress amplitude [12]. The second type of instability has been experimentally observed in the form of layered ladder-like dislocation structures of PSBs for sudden impositions of stress amplitude [10]. The quantity q_c denotes the wavenumber corresponding to the preferred wavelength $\lambda_c = 2\pi/q_c$.

It also turns out by comparing (18) and (19) that the patterning instability is reached before the temporal oscillations when $b_c < b_c^o$, that is when

$$\frac{D_1}{D_2} < \frac{a}{\rho_1^{o2}}\left[\left(1+\frac{\rho_1^{o2}}{a}\right)^{1/2}-1\right]^2 . \qquad (20)$$

Since it could be argued that in most cases $D_1 \ll D_2$, it may be expected that PSB structures should form before the occurence of strain bursts.

In concluding this section we provide another justification of the diffusive nature of the mobile dislocated state independently of the arguments embodied in (6) and (2). To this end we only utilize the balance equations of effective mass for the mobile dislocation species where, however, a further distinction between positive and negative dislocations is also made. The details of this procedure can be found in [13] where this problem is considered within a more general setting. Based on the above distinction together with expression (8) for the mobilization rate, we can begin with the following balance equations

$$\left.\begin{aligned} \partial_t\rho_2^+ + \nabla_x j_2^+ &= \frac{b}{2}\rho_1 - \frac{c}{2}\rho_2^+\rho_1^2 \;, \\ \partial_t\rho_2^- + \nabla_x j_2^- &= \frac{b}{2}\rho_1 - \frac{c}{2}\rho_2^-\rho_1^2 \;, \end{aligned}\right\} \qquad (21)$$

In writing (21) we have assumed that positive and negative dislocations have the same probability to be produced by the applied stress and captured by immobile dipoles. We also assume that positive and negative dislocations will travel with equal but positive speed $j_2^\pm = \rho_2^+ v = -\rho_2^- v$, so that (21) can be rewritten in terms of the sum $\rho_2=\rho_2^+ + \rho_2^-$ and the difference $\delta=\rho_2^+ - \rho_2^-$ as follows

$$\left.\begin{aligned} \partial_t\rho_2 &= -v\nabla_x\delta + b\rho_1 - c\rho_2\rho_1^2 \;, \\ \partial_t\delta &= -v\nabla_x\rho_2 - c\rho_1^2\delta \;. \end{aligned}\right\} \qquad (22)$$

Next by assuming that both ρ_2 and δ vary slowly over the period of one cycle and that the velocity v has the form $v=v_a\cos\omega t$ due to cyclic loading (with v_a denoting amplitude and ω frequency not necessarily identified with the test frequency) we can integrate $(22)_2$ to obtain an explicit solution for $\delta(t)$ which when introduced in $(22)_1$ gives

$$\partial_t\rho_2 = D_2\nabla_{xx}^2\rho_2 + b\rho_1 - c\rho_2\rho_1^2 \ , \qquad (23)$$

with

$$D_2 = \frac{v_a^2}{2}\,\frac{c\rho_1^{o2}}{c^2\rho_1^{o4}+\omega^2} \ , \qquad (24)$$

that is, an explicit formula for the diffusivity of the mobile state.

3. PATTERNING OF DEFORMATION

In this section we illustrate briefly the possibility for the deformation to localize in the form of patterned solutions previously discussed for dislocation distributions. Before we proceed with an explicit example pertaining to shear band formation it is instructive to make the following qualitative remark. Roughly speaking, let us assume that the increment of the plastic strain ε_p is proportional to the increment of the mobile dislocation density ρ_2 and that the increment of the internal strain ε_i is proportional to the density of the immobile dislocation density ρ_1. It then follows that plastic and internal strains will obey coupled partial differential equations of the reaction-diffusion type as it was the case for the dislocation densities in (10). This observation sets up the frame for studying the patterning of deformation. Below we consider such an example of a patterned solution for the deformation by taking a fresh look to the problem of simple shear.

First let us recall from the previous section the stress relation

$$S = T_L + T_D \ , \qquad (25)$$

where T_L is the lattice stress and T_D an average dislocation stress including all possible types of dislocations. We assume conditions of static equilibrium such that

$$\nabla_z S = 0 \ . \qquad (26)$$

The coordinate notation here is slightly different than that employed in the previous section. In particular, we assume that all properties vary now uniformly in the x and y directions and allow a possible non-uniformity to develop in the z direction only. Thus, instead of looking for possible patterning on the slip plane, as was the case with the PSBs, we now examine the possibility of strain inhomogeneities in the direction perpendicular to it. We neglect rotation effects and assume that the lattice stress T_L relates to the shear strain γ with a non-monotone graph

$$T_L = \tau(\gamma) \ , \qquad (27)$$

where $\tau(\gamma)$ may be of sinusoidal type or a generalization of it. The stress T_D is assumed to be linear in the dislocation density ρ, that is

$$T_D = \mu\rho \ , \qquad (28)$$

with μ being a constant whose sign does not affect the argument to follow. Finally, we consider steady-states for the evolution of the average dislocation density ρ. We thus employ equation (1) with the time-dependent term $\partial_t\rho$ neglected, the flux term $\nabla_x j$ being of a diffusive form, and the source term $\hat{c}$ given by a linear function of ρ with the coefficient of linearity possibly dependent on the strain γ, that is

$$\rho_{xx} = \lambda(\gamma)\rho \ . \qquad (29)$$

On combining equations (25)-(29) we can eliminate the internal-like variable ρ in favor of higher order gradients in the macroscopic strain γ. Specifically, we have

$$S = \tau(\gamma) - a(\gamma)\gamma_{zz} - b(\gamma)\gamma_z^2 \ , \qquad (30)$$

with $a=d\tau/\lambda d\gamma$ and $b=d^2\tau/\lambda d\gamma^2$. On inserting the gradient-dependent constitutive equation (30) into the equilibrium equation (26) we obtain the following nonlinear differential equation for γ

$$a(\gamma)\gamma_{zz} + b(\gamma)\gamma_z^2 = \tau(\gamma) - \tau_o \ , \qquad (31)$$

with τ_o denoting the externally applied shear stress, that is the stress at $z=\pm\infty$. The solution of (31) for the boundary condition $\gamma(\pm\infty)=\gamma_o$ turns out to be a bell-shape symmetric graph with a maximum γ_* occuring at a point $z=z_*$.

The details of this construction can be found, for example, in [6],[7] but for the completeness of the presentation we give below the main results. Thus if $\bar{z}$ is an arbitrary point the solution reads

$$z = \bar{z} - \int_{\gamma(\bar{z})}^{\gamma(z)} \frac{d\gamma}{\sqrt{2F(\gamma)/G(\gamma)}} \ , \qquad (32)$$

where

$$F(\gamma) = \int_{\gamma_0}^{\gamma} (\tau-\tau_0)E(\gamma)d\gamma \ ,$$

$$G(\gamma) = aE(\gamma) \ , \quad E(\gamma) = \frac{1}{a} e^{2\int \frac{b}{a} d\gamma} \ . \quad (33)$$

Existence of solution is guaranteed when the area condition

$$\int_{\gamma_0}^{\gamma_*} (\tau-\tau_0)E(\gamma)d\gamma=0 \ , \quad (34)$$

is satisfied. This condition suggests that localization occurs at stress levels lower than those defined by the condition $d\tau/d\gamma = 0$, in agreement with the experimental observations. Moreover, (32) gives an estimate of the shear band width and thus the present approach removes difficulties associated with earlier analyses of localization and post-localization behavior where, among other things, the solutions were dependent on the mesh size. A more elaborate discussion on this problem can be found in references [6],[7] and [8].

REFERENCES

1. G. Martin, Long Range Periodic Decomposition of Irradiated Solid Solutions, Phys. Rev. Lett. 50, 250-252, 1983.
2. K. Krishan, Void Ordering in Metals During Irradiation, Phil. Mag. A 45, 401-417, 1982.
3. D. Walgraef and E.C. Aifantis, Mechanics of Microstructures, MM Reports No. 4 & 7, Michigan Technological University, 1984.
4. D. Walgraef and E.C. Aifantis, On the Formation and Stability of Dislocation Patterns I, II, III, Int. J. Engng. Sci. (in press).
5. D. Walgraef and E.C. Aifantis, Dislocation Patterning in Fatigued Metals as a Result of Dynamical Instabilities, J. Appl. Phys. 58, 688-691, 1985.
6. E.C. Aifantis, Continuum Models for Dislocated States and Media with Microstructures, in: The Mechanica of Dislocations, Eds. E.C. Aifantis and J.P. Hirth, pp. 127-146, ASM, Metals Park, 1985.
7. E.C. Aifantis, On the Microstructural Origin of Certain Inelastic Models, Transactions of ASME, J. Engng. Mat. Tech. 106, 326-330, 1984.
8. N. Triantafyllidis and E.C. Aifantis, A Gradient Approach to Localization of Deformation I. Hyperelastic Materials, J. of Elasticity (in press).
9. D.C. Drucker, Material Response and Continuum Relations; or From Microscales to Macroscales, Transactions of ASME, J. Engng. Mat. Tech. 106, 286-289, 1984.
10. H. Mughrabi, Cyclic Plasticity of Matrix and Persistent Slip Bands in Fatigued Metals, in: Continuum Models for Discrete Systems 4, Eds. O. Brulin and R.K.T. Hsieh, 241-257, North Holland, 1981.
11. T. Tabata, H. Fujita, M. Hiraoka and K. Onishi, Dislocation Behavior and the Formation of Persistent Slip Bands in Fatigued Copper Single Crystals Observed by High-Voltage Electron Microscopy, Phil. Mag. A47, 841-857, 1983.
12. P. Neumann, Strain Bursts and Coarse Slip During Cycling Deformation, Z. Metallkde. 59, 927-934, 1968.
13. D. Walgraef and E.C. Aifantis, forthcoming.

5th International Symposium on Continuum Models of Discrete Systems / Nottingham / 14-20 July 1985

Fractals

D.S.Broomhead
Royal Signals and Radar Establishment, Great Malvern, UK

ABSTRACT: The idea of a fractal as a mathematical object and as a model of natural phenomena is introduced by way of simple examples. The characterization of fractals both mathematically and experimentally is then considered. Finally a model exemplifying the generation of fractal geometries in nature is discussed.

1 INTRODUCTION

The word "fractal" was coined by Mandelbrot (see Mandelbrot 1982), an act which has subsequently given voice to many. The present talk is intended to motivate interest in, rather than to review exhaustively, this large area of research.

Let us motivate the motivation by looking at a classic piece of work by Lovejoy (1982) on the morphology of equatorial cloud formations. Fig. 1 (taken from this reference) summarizes the results of studies of cloud images obtained using satellite-born infra-red and radar techniques. The images were divided into resolution cells, or "pixels", whose size was determined by the resolution limit of the imaging technique employed. The area, A, and the perimeter, P, of each cloud image were then estimated by counting, respectively, the number of pixels covering the whole image and the boundary of the image. The log-log plot of A versus P clearly indicates a power law relationship. This is not of itself surprising, since one expects generally for closed curves drawn on a surface that $P \sim \sqrt{A}$. The surprise is that this is not the power law implied by Fig. 1. For cloud images one finds $P \sim \sqrt{A}^{d_f}$ where $d_f \sim 1.35$.

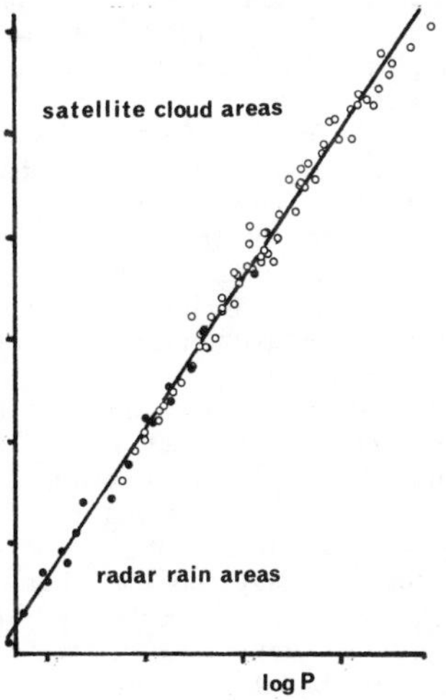

Fig. 1
The perimeter-area scaling of cloud images

The conclusion to be drawn from this work is that there is something unusual about the perimeter of clouds. Fig. 2 shows another shape, a Koch island as illustrated in Mandelbrot (1982), which, although a manifestly poor model of a cloud, does have again an unusual perimeter. This comment can be made more precise by considering the relationship between the area and the perimeter of the Koch island. Consider the recursive generation process indicated in Fig. 2. Initially one begins with a square of side L and rearranges its perimeter as shown. The resulting figure, which can be thought of as being constructed of squares of side L/6, has the same area; its perimeter, however, has increased. The recursion is to treat the unshared edges of the smaller squares in the same way as the edges of the original large square. One now assumes that the observation of the resulting object has limited resolution. For simplicity let us say that the smallest resolvable length is $\delta l = L/6^n$, where n is a

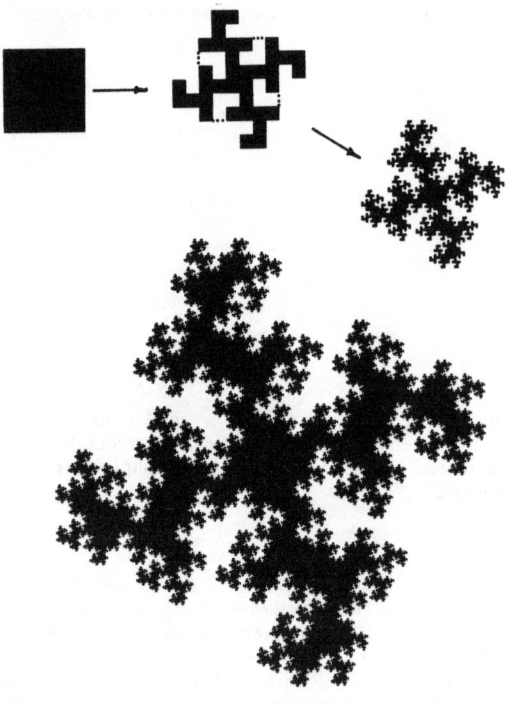

Fig. 2
A Koch island

positive integer. In this case the Koch island is indistinguishable from the nth iterate of the generation procedure, where $n=\ln(L/\delta l)/\ln 6$. In particular, P, appears to consist of 18^n pixels and is estimated to be $P=18^n\delta l = \delta l(L/\delta l)^{\ln 18/\ln 6}$. Since $A=L^2$, one obtains a scaling law analogous to that found in Fig. 1: $P \sim \sqrt{A}^{d_f}$, where $d_f=\ln 18/\ln 6 \simeq 1.61$.

Despite the manifest differences, therefore, both the Koch island and the clouds in Lovejoy's study have perimeters which appear to be too large for the area that they enclose - one is reminded of the skin of a dried fruit such as a prune. In consequence, rather than appearing as a clear boundary between inside and outside the perimeter is strongly folded to the extent that it appears to form a diffuse "boundary layer". The following section will contain useful generalisations of the concept of dimension which can be used to characterize this kind of property. It will be shown that the quantity d_f used above may be thought of as the "dimension" of the perimeter. Intuitively one thinks of a boundary curve as being a 1-dimensional object. The fact that, for both the Koch island and the cloud images, $d_f>1$ is interpreted as being indicative of the tendency of their perimeters to fill a 2-dimensional region.An object for which a generalised dimension such as d_f exceeds the intuitive dimension (defined precisely in the next section) is termed a fractal. It will emerge that this property results from the fractal having structure on all length scales. As such fractals make good models for many naturally occurring objects - cloud formations, for example. Fractal models have several nice features - in particular, as the calculation of d_f for the Koch island shows, they have symmetry. Thus, in the same way that a perfect lattice can be used as a basic model of a crystal and is
useful because it transforms in a simple way under the action of a suitable space group, so a fractal usefully models multiscaled objects in nature because it has simple properties under the action of a scaling group.

2 DIMENSION, METRIC AND MEASURE

The previous section referred to an "intuitive" notion of dimension. A more precise statement of this basic idea will now be given. The objective is to define the dimension of a set of points, Φ, found in a suitable d_u-dimensional space, U. Assuming that the latter has the structure of a linear vector space one can assert that its dimension is the largest number of mutually independent vectors that it can support. This is, however, too restrictive a definition to be applied to the general set Φ.

Topolgical dimension, d_T, requires only that continuity has meaning on the set. The definition is a recursive one: the topological dimension of Φ is $d_T=1+d_T'$ where d_T' is the topological dimension of a set whose removal would divide Φ. A point is taken to have $d_T=0$. It follows from this definition that a line has $d_T=1$ since it is divided by the removal of a point. Similarly a surface, since it is divided by the removal of a line, has $d_T=2$. These examples illustrate that d_T coincides with an intuitively reasonable idea of dimension - in particular note that it is always an integer.

The topological dimension of Φ is an intrinsic property of the set itself. It has an upper bound d_u but, apart from this owes nothing to U. In consequence, it is completely insensitive to any fractal character that Φ may possess. For example, consider the recursive process by which the Koch island is generated from a square. Since this may be thought of as a sequence of continuity preserving distortions, it has no effect on the topology of the square. The topological dimension of the perimeter of the

Koch island is therefore just that of the perimeter of the square i.e. $d_T=1$. The fractal structure of the perimeter of the Koch island is to do with the separations and relative positions of all points in the set. It follows that the definition of a dimension which takes account of this structure must employ metric properties on the space U. With this in mind consider the following generalisation of dimension.

Hausdorff dimension, d_H, is defined in terms of coverings of Φ with sets of d_u-dimensional cubes (for a good survey see Farmer, Ott, Yorke 1983). Initially the sizes of the cubes $\{\epsilon_i\}$ are allowed to vary subject to the constraint that all $\epsilon_i \leqslant \epsilon$, where ϵ is an arbitary constant. The quantity:

$$l_d(\epsilon) = \inf \sum_i \epsilon_i{}^d$$

($d \geqslant 0$ is a real parameter), is defined to have the form of a measure of the volume of the whole set. The summation is taken over a given choice of cover, while the infimum is taken over all possible covers subject to the constraint on the $\{\epsilon_i\}$. The effect of the infimum is to select a cover which is most efficient in the sense that it uses the smallest possible volume to cover Φ. Taking the limit

$$l_d = \lim_{\epsilon \to 0} l_d(\epsilon)$$

essentially gives the volume of the set assuming that the volume of an ϵ_i-cube is $\epsilon_i{}^d$. Hausdorff proved the existence of a critical value of d ($=d_H$) such that: for $d>d_H$, $l_d=0$ and for $d<d_H$, $l_d=\infty$. The quantity d_H is the Hausdorff dimension of Φ, while l_{d_H} is its Hausdorff measure or volume.

As an example consider Φ to be the perimeter of the Koch island. Having chosen $\epsilon=L/6^n$, where n is an arbitrary positive integer, a particular choice of cover is to take 18^n identical cubes of size $\epsilon_i=\epsilon$. In this case $\Sigma\epsilon_i{}^d=18^n(L/6^n)^d$. A new cover may be generated by replacing one the cubes with a set of smaller cubes, say of size $L/6^{n+1}$. This requires 18 of the smaller cubes and causes the value of the summation to change by an amount: $(18-6^d)(L/6^{n+1})^d$. There are two cases to be dealt with:

1.For $18>6^d$ the effect of changing the cover is to increase the value of the sum. Therefore, the infimum corresponds to the original choice. thus $l_d(\epsilon)=18^n(L/6^n)^d$. It follows that

$$l_d = \lim_{n\to\infty} L^d(18/6^d)^n = \infty$$

2.For $18<6^d$ the value of the summation decreases with the $\{\epsilon_i\}$. Therefore $l_d(\epsilon)=0$ which implies $l_d=0$.

The Hausdorff dimension of the perimeter of the Koch island is the critical value of d which separates these two cases: $d_H=\ln 18/\ln 6$. Note that the scaling arguments of the previous section also gave this value for the exponent d_f.

It is an easy matter to show that the Hausdorff and topological dimensions coincide for smooth objects such as surfaces and lines. The above example shows that this is not necessarily the case, indeed d_H need not be an integer. Intuitively, the value of d_H characterises the way in which the set Φ fills the space U. Thus $1<d_H<2$ suggests that the perimeter of the Koch island is a curve which is in some sense tending to fill an area of the plane.

Mandelbrot used this fundamental distinction between Hausdorff and topological dimension to **define** a fractal as a set for which $d_H>d_T$. Unfortunately, it is often a difficult task to calculate the Hausdorff dimension of a set, generally because of the need to obtain the infimum over the range of possible covers. Consequently, other related but more convenient definitions of dimension are often employed.

Capacity, d_c, is also defined via a cover of Φ using d_u-dimensional cubes. Here, however, the cover is uniform, all cubes having the size ϵ.

$$d_c = \lim_{\epsilon\to 0} \frac{\ln N(\epsilon)}{\ln \epsilon^{-1}} \quad ,$$

where $N(\epsilon)$ is the number of cubes required to cover Φ. Since the uniform cover is not necessarily the most efficient, d_c is strictly an upper bound to d_H.

Returning to the example of the Koch island perimeter, if one sets $\epsilon=L/6^n$ then the recursive generation algorithm implies $N(\epsilon)=18^n$, from which follows:

$$d_c = \lim_{n\to\infty} \frac{\ln 18^n}{\ln 6^n} = \frac{\ln 18}{\ln 6} \quad .$$

As in this case it is often true that $d_H=d_c$, for this reason, when the context is not clear (see the introduction) the general term "fractal dimension", d_f, has been employed. It should be noted, however that this terminology also appears in the literature with more specific meanings. An example of a set for which $d_c \neq d_H$ is $\Phi=\{1, 1/2, 1/3, 1/4,....\} \subset [0,1]$. It is an instructive exercise to show that $d_H=0$ while $d_c=1/2$, since the effect of the infimum becomes clear in the analysis.

Pointwise dimension, d_p, is closely related to capacity, although its definition uses a new generalisation - a probability measure $d\mu(x)$ defined on the set Φ. There are a variety of contexts in which this is useful - the most obvious is to think of the measure as giving the mass density of the material from which the fractal has been formed. Thus if one writes:

$$\mu(B_\epsilon(x)) = \int_{B_\epsilon(x)} d\mu(x') \quad ,$$

as the total probability contained in the ball $B_\epsilon(x)=\{x_i \mid \|x-x_i\|<\epsilon, x_i \in \Phi\}$, it may also be interpreted as the mass of a ball of radius ϵ cut from the fractal at a point x. The definition proceeds as follows. If $d_p(x)$ defined by:

$$d_p(x) = \lim_{\epsilon\to 0} \frac{\ln \mu(B_\epsilon(x))}{\ln \epsilon} \quad ,$$

is independent of x for almost all x with respect to $\mu(x)$, then $d_p=d_p(x)$ is defined as the pointwise dimension of Φ (with respect to $\mu(x)$).

There is a definition of fractal dimension, commonly used in work on the statistical mechanics of aggregation and growth, which is similar to the above with the interpretation of $d\mu(x)$ as a mass density. In this case the scaling relation between the radius of gyration, R, and the mass, M, of an aggregating cluster is written $M \sim R^{d_f}$. For small radii the definition of pointwise dimension leads to the same form: $\mu(B_\epsilon(x)) \sim \epsilon^{d_p(x)}$, and may be thought of as an idealisation of the mass scaling exponent. In practice, however, one cannot allow $R\to 0$, nor can the x-independence of $d_p(x)$ be established.

A survey of the statistical mechanics literature reveals a bewildering variety of fractal dimensions (see contributions to Family, Landau 1984 and Stanley, Ostrowsky 1985). Much can be done to unify these conceptually, by considering them as pointwise dimensions associated with appropriate measures. For example, a measure corresponding to the density of material being transported through a percolating cluster will give a dimension associated with its, so-called, backbone. Similarly, a measure giving the flux of new material onto a growing surface will give a fractal dimension for the set of active growth sites.

By way of a finale to this section we shall now introduce an uncountable number of dimensions! This will be referred to as the **Hentschel-Procaccia continuum**, $\{d_q \mid q \in \mathbb{R}\}$ (Hentschel, Procaccia 1983) and is defined as follows:

$$d_q = \frac{1}{q-1} \lim_{\epsilon\to 0} \left[\frac{\ln\left[\int d\mu(x)\ \mu(B_\epsilon(x))^{q-1}\right]}{\ln \epsilon} \right] ,$$

where q is any real number. The measure of the ϵ-ball, $\mu(B_\epsilon(x))$, is again used, in contrast with the pointwise dimension, however, it (or some power of it) is averaged over the measure $d\mu(x)$. There is, therefore, no analogue of the requirement that $d_p(x)$ be constant almost everywhere. Indeed, if the fractal is homogeneous in the sense that $\mu(B_\epsilon(x))=\mu(B_\epsilon)$, a constant, almost everywhere, then the $\{d_q\}$ reduce to a single quantity:

$$\lim_{\epsilon\to 0} \left[\ln\mu(B_\epsilon)/\ln\epsilon\right] ,$$

which is, in fact, the pointwise dimension.

The nature of the information carried by the $\{d_q\}$ can be illustrated with some particular examples. For integer values of $q>2$, d_q gives the scaling with radius of the probability of finding q points within ϵ-balls constructed on the fractal. In particular, d_2 gives the scaling at small separations of the two point correlation function of points in Φ with respect to $d\mu(x)$. d_2 is often referred to as the **correlation dimension** and is particularly useful since experimental probes often provide information about the pairwise correlation function.

Other cases of interest are the limit $q\to 0$, wich gives the capacity, d_c, defined earlier, and the limit $q\to 1$ which yields the **information dimension**, d_s, defined by:

$$d_s = \lim_{\epsilon\to 0} \left[\int d\mu(x) \ln\mu(B_\epsilon(x))/\ln\epsilon\right] .$$

The latter has been discussed in some detail recently (Farmer 1982) and may be thought of as the exponent for the scaling of the information content of a measurement made on a fractal with the resolution of the measurement.

The set $\{d_q\}$ forms an ordered sequence since it can be shown that $q'>q$ implies that $d_{q'} \leq d_q$. The equality holds in the case of a homogeneous fractal where, as we have discussed, the whole heirachy reduces to a single quantity. Any deviation from this degenerate situation is a measure of the degree to which the scaling properties differ from place to place within the fractal set.

From the point of view of the following discussion the dimensions described in this section will be sufficient. It is probably not true to say that they characterise all the interesting properties of fractals. One can certainly think of other quantities which are potentially useful. For example there is the "thickness" of fractal sets defined by Newhouse (see discussion in Guckenheimer, Holmes 1983) which characterises the degree to which such sets can occupy the same volume without intersecting each other.

3 THE PHYSICAL PROPERTIES OF FRACTAL OBJECTS

The scattering of short wavelength radiation from fractals and "subfractals" (which are objects with a fractal surface gradient) gives rise to scintillation effects which are commonly observed in nature. The glittering of a disturbed water surface, the twinkling of starlight passing through the turbulent atmosphere (Berry 1977, Berry 1979, Jakeman 1982) are examples. In fact our ability to perceive **texture** visually is indicative of correlations in the scattering medium which extend over many length scales (E. Jakeman: private communication), and points to the ubiquity of fractals in nature.

The general question of what physical properties are implied by an object having a fractal geometry is as large as it is fascinating. In keeping with the basic philosophy of this article we shall be content with a limited approach. The discussion to follow will be concerned with some experimental methods aimed at providing evidence for the existence of fractal structures and obtaining estimates of fractal dimensions. Broader physical implications of the effect of a fractal structure on mechanical properties (see for example:Turcotte, Smalley, Sola 1985, Alexander, Orbach 1982) or on transport properties such as conductivities or diffusion (see various articles in Family, Landau 1984 and Stanley, Ostrowskii 1985) will not be dealt with here.

In many cases it is possible to make observations on a system which give directly a characteristic scaling law of a fractal. Observations which deal directly with an image of the fractal, such as Lovejoy's study, are most direct and least vulnerable to ambiguous interpretation. An example of this approach is work done on metallic films sputtered onto inert substrates (Voss, Laibowitz, Alessandrini 1982, Kapitulnik, Deutscher 1982). This work employed digital image processing techniques applied to electron micrographs to extract not only the scaling relations but also impressive visualisations of the fractals themselves.

It was remarked in section 2 that scaling laws which are easy to measure are the mass-radius relation, giving something like the pointwise dimension, d_p, and the small scale behaviour of the particle-particle correlation function, which gives d_2, the correlation dimension. In a recent experiment Schaefer, Martin, Wiltzius, Cannell (1984) demonstrated that using light scattering techniques both can be obtained independently for the same system. Light scattering measures the scattered intensity as a function of angle of scatter, this relates directly to the structure factor S(k), where k is the magnitude of the scattering vector. S(k) is the Fourier transform of the pair distribution function, $g_2(R)$. One loosely interprets k as being the inverse separation of the scatterers contributing to the observed intensity. The experiment of Schaefer et al was concerned with the coagulation of colloidal suspensions of silica spheres. The observed S(k) has several distinct regions:

1. At small k the separation of the scatterers is large enough for the dominant contribution to the scattering to arise from scatterers on distinct aggregates within the fluid. The experiment sees the system as a suspension of aggregate particles which when sufficiently dilute can be modelled as a perfect gas. In this limit leading terms in the expansion of the structure factor give the following Lorenzian form:

$$S^{-1}(k) \simeq [cM_w]^{-1}[1 + k^2\langle R\rangle^2/3 + \dots]$$

where c is the solution concentration by weight, M_w is the weight averaged molecular weight of the aggregates while $\langle R\rangle$ is the mean aggregate radius. thus a study of the behaviour of S(k) at small k gives both the mass and radius of aggregates, which for a range of coagulation times can be plotted on a log-log plot to give $M_w \sim \langle R\rangle^{d_f}$. For their system Schaefer et al find $d_f \simeq 2.12$. In three dimensions one expects $M \sim R^3$ for solid bodies, which suggests that these aggregates have rather an open structure.

2. The second region of interest arises when the separation of the scatterers is such that scattering is dominated by scatterers on different spheres within the same aggregate. For fractal aggregates at small interparticle separations the pair distribution function scales as $g_2(R) \sim R^{d_2}$, which implies that $S(k) \sim k^{-d_2}$ when the scatterers come from the same fractal. Schaefer et al, give a log-log plot of S(k) in this region. The data used was from both light scattering and small angle X-ray scattering. The plot shows a linear region extending over about two orders of magnitude in k which has a slope $-d_2 = -2.12 \pm 0.05$. Note that this value agrees with the value obtained for d_f using the small k limit.

These observations lend confidence to the interpretation that the aggregates are (homogeneously) fractal and provide more evidence for the sponaneous generation of fractals in nature.

4 A FRACTAL GROWTH MODEL

We have commented on the ubiquity of fractals in nature, but have said nothing of the reasons for this. The present section is concerned with a particular model for an aggregation process which suggests a general (though not, of course, unique) criterion responsible for the growth of fractals. The model in question is called Diffusion Limited Aggregation (DLA) and has generated considerable interest in the literature since its description by Witten, Sander (1981). The form of this section owes much to Ball (1985).

The model describes the slow growth of a single aggregate which is seeded by a single fixed particle placed at the centre of a large sphere (or, in 2-dimensions, a circle). The sphere is an isotropic source of material which is able to diffuse and adhere to the aggregate. Mathematically, one has a diffusion problem:

$$\partial_t u(r,t) = D\nabla^2 u(r,t) \qquad ,$$

(where D is the diffusion constant and u(r,t) is the concentration field of the material) subject to the following boundary conditions:

1. $u(r,t)=0$ as the surface of the aggregate is approached from the outside
2. $u(r,t)=u_{source}$ at large radius
3. $\rho\langle v\rangle=D\nabla u(r,t)$ at the growing surface, where ρ is the density of the aggregate, $\langle v\rangle$ is the mean velocity of the surface, and ∇u is the concentration gradient as the aggregate surface is approached from the outside.

The important feature of this model is that the growing interface is unstable. To see this, imagine that the interface is initially planar but becomes distorted by a fluctuation which generates a small localised peak. The distortion generates a locally large value of ∇u and hence, in this region of the aggregate surface, the growing interface will advance more rapidly, thereby amplifying the original fluctuation. One expects, therefore in a realistic, noisy environment that the growing surface of the aggregate will be very complicated, reflecting the history of fluctuations that it has experienced. Of course, a realistic environment will also provide stabilising processes which will limit the effect of the instability. At small scales surface tension, crystallinity effects and so on restrict the development of highly curved regions of the interface thus limiting the growth of "needle-like" fluctuations. At large scales the diffusion field cannot respond over lengths larger than the diffusion length and hence the growth of coherent,large "mountain-like" distortions will be limited by the diffusion process itself. However, between "needles" and "mountains" there are many length scales at which distortions can grow in an unconstrained manner. Thus the resulting aggregate will grow with structure on many length scales and will appear fractal to observations not able to probe the scales at which the stabilsing mechanisms dominate.

The DLA model usually employed is a modification of the one described above. An additional assumption is made that the size of the aggregate is less than diffusion length. The outer scale limit does not therefore arise. Moreover the diffusing field in the neighbourhood of the aggregate can be assumed to respond adiabatically to fluctuations so that the explicit time dependence in the diffusion can be dropped. Paradoxically, this results in a model of rather general form. One now has a field $\varphi(r)$ which satisfies Laplace's equation: $\nabla^2\varphi=0$, and is subject to the boundary conditions given above. The moving boundary is represented by: $\langle v\rangle\propto\nabla\varphi(r)$+noise, where the explicit noise is included to drive the instability. Formulated in this way the model clearly has strong connections with electrostatics. Indeed it has been used to describe the morphology of dielectric breakdown at point electrodes buried in insulating material (Niemeyer, Pietronero, Wiesmann 1984) . However, Laplace's equation can occur more generally, for example the model has recently been used to describe viscous fingering in interpenetrating fluids (Nittmann, Daccord, Stanley 1985). In this case $\varphi(r)$ is a pressure field and Laplace's equation arises from the application of d'Arcy's law to the mass continuity equation for the fluid.

Much of the interest in DLA has arisen from the ease with which it may be implemented as a computer model. Fig. 3 shows a DLA cluster of 100,000 sites grown numerically by Meakin (1985). The model consists of assuming a square lattice on which the aggregate can grow from an initial seed. The diffusion process is represented by a single random walker which is started a large distance from the aggregate. Depending upon the implementation the random walker may travel on or off the lattice. On arrival at a lattice site adjacent to the aggregate that site is incorporated into the aggregate and a new walker initiated. Clearly the inner scale here is given by the lattice spacing, while the single

Fig. 3
A DLA aggregate (off-lattice walks, on-lattice growth, 100,000 sites)

random walker constraint is equivalent to the slow growth/adiabatic assumption. The noise enters the problem as shot noise when the walker decides to which lattice it belongs as it attaches to the aggregate.

Apparently from Fig. 3 the aggregate resulting from DLA simulations is a fractal - this is in accord with the stability considerations above. The fractal dimension of these aggregates generally converges to a value which is independent of the particular realization of the random walk used, and of whether or not the walk was on or off lattice. The general method of calculation is to relate the radius of gyration to the number of sites on the aggregate - this is the mass radius scaling relation which has already been discussed. Generally it is found that $d_f=2.495\pm.06$ in 3-dimensions, and $d_f=1.67$ on a plane. These values are consistent with the fairly open structure manifest in Fig. 3. The openess results from the aggregate shielding its own inner structure from the diffusional flux.

Recent work on the morpholgy of DLA clusters provides an example of the use of measures defined on fractal set (Turkevich, Scher 1985, and Halsey, Meakin, Procaccia 1985). In this case the relevant measure is harmonic measure on the aggregate. This is the analogue of ∇u for the computer model and is calculated by studying the incidence and disposition of contacts of random walkers with an aggregate of fixed size. The harmonic measure is generally very singular with the singularities corresponding to growing tips. Halsey et al have analysed it in some detail using the Henschel-Procaccia hierachy, and developed a scaling theory to describe their results.

Several experimental systems for which DLA should be good model have been studied. These involve viscous fingering (Nittmann et al 1985), dielectric breakdown (Pietronero, Wiesmann 1984), and electrodeposition of zinc metal in a chemically engineered 2-dimensional environment (Matsushita, Sano, Hayakawa, Honjo, Sawada 1984). Brady and Ball (1984) looked at the electrodeposition of copper at a point electrode under circumstances carefully controlled to ensure that the process is diffusion limited. They note that the diffusion current onto a spherical shell is proportional to the radius of the shell, while the mass of copper deposited is proportional to the total charge passed by the cell. Thus by monitoring the current through the cell and its time integral, continuous measurement of mass of copper deposited and effective radius of the deposite could be made. In this way an experimental mass radius scaling exponent was obtained. The value found, $d_f \sim 2.43$, compares well with Meakin's computer result.

The general question of the origin of fractals in nature has yet to be fully answered, however unstable growth processes must surely be important. In particular, the lack of "detailed balance" between the instability and corresponding limiting processes in the DLA model results in a broad band of unstable modes. Recent molecular dynamics work on phase separation by the spinodal decomposition mechanism (Desai 1985) provides another good example of this. In the situation that the homogeneous phase is linearly unstable to fluctuations having a broad range of wavelengths it is to be expected that the separating inhomogeneous structure will appear to be fractal. This is indeed found to be the case in Desai's computer experiments. Finally, a fractal description of strong turbulence in fluids (Mandelbrot 1982) is suggested by a similar lack of "detailed balance". Here, instabilities occurring over a range of length scales feed energy to spacial modes with shorter and shorter wavelength until finally at small scales dissipation becomes dominant. The Kolmogorov power law spectrum is seen to support this viewpoint.

REFERENCES

Mandelbrot, B.B. 1982. The fractal geometry of nature. W.H.Freeman and Company.
Lovejoy, S. 1982. Science. 216, pp 185-187.

Farmer, J.D., Ott, E., Yorke, J.A. 1983. Physica 7D, p 153.
Family, F., Landau, D.P. 1984, (editors) Kinetics of aggregation and gelation, North-Holland.
Stanley, H.E., Ostrowsky, N. 1985, (editors) On growth and form. Martinus Nijhoff.
Hentschel, H.G.E., Procaccia, I. 1983. Physica 8D, p 435.
Farmer, J.D., 1982, Physica 4D, p 366.
Guckeheimer, J., Holmes, P. 1983. Nonlinear oscillations, dynamical systems and bifurcations of vector fields, Springer-Verlag.
Berry, M.V. 1977, J.Phys. A10, pp2061-2081
Berry, M.V. 1979, J.Phys. A12, pp781-797
Jakeman, E. 1982, J.Opt.Soc.Am. 72, pp1034-1041
Turcotte, D.L., Smalley, R.F., Sola, S.A. 1985, Nature 313, pp671-672.
Alexander, S. ,Orbach, R. 1982, J.Physique 43, L-625
Voss, R.F., Laibowitz, R.B., Allessandrini, E.I. 1982, Phys.Rev.Lett. 49, p1441.
Kapitulnik, A., Deutscher, G. 1982, Phys.Rev.Lett. 49, p1445.
Schaefer, D.W., Martin, J.E., Wiltzius, P., Cannell, D.S. 1984, Phys.Rev.Lett. 52, p2371.
Witten, T.A., Sander, L.M. 1981, Phys.Rev.Lett. 47, p1400
Ball, R.C. 1985, in Stanley, Ostrowsky (1985)
Niemeyer, L., Pietronero, L., Wiesmann, H.J. 1984, Phys.Rev.Lett. 52, p1033.
Nittmann, J., Daccord, G., Stanley, H.E. 1985, Nature 314, p141.
Meakin, P. 1985, in Stanley, Ostrowsky (1985)
Turkevich, L., Scher, H. 1985, to appear Phys.Rev.Lett.
Halsey, T.C., Meakin, P., Procaccia, I. 1985, to appear Phys.Rev.Lett.
Matsushita, M., Sano, M., Hayakawa, Y., Honjo, H., Sawada, Y. 1984, Phys.Rev.Lett. 53, p286
Brady, R.M., Ball, R.C. (1984), Nature 309, p225.
Desai, R. 1985, in Stanley, Ostrowsky (1985)

ACKNOWLEDGEMENTS

The author would like to acknowledge his indebtedness to S. Lovejoy, B.B. Mandelbrot and P. Meakin for permission to reproduce their figures in this paper.

5th International Symposium on Continuum Models of Discrete Systems / Nottingham / 14-20 July 1985

The balance equations of continuum mechanics from a corpuscular viewpoint

A.I.Murdoch
University of Strathclyde, Glasgow, UK

ABSTRACT: The balance relations of continuum mechanics are motivated on the basis of a small number of physically-plausible assumptions concerning particle interactions and thermal motions. Some implications for mixture theory are indicated.

1 INTRODUCTION

The basic concepts and balance relations of continuum mechanics can be motivated [1,2] by invoking a small number of physically-plausible assumptions concerning the thermal motions of, and interactions between, fundamental discrete entities (molecules, ions, or atoms). Such an approach provides some insight into the assumptions implicit in, and the approximative nature of, continuum modelling, and the physical interpretation of macroscopic fields.

2 PRELIMINARIES

Fundamental discrete entities are modelled as point masses P_i ($i=1,2,..$) whose masses and velocities are denoted by m_i and $\underset{\sim}{v}_i$. To discuss the spatial averaging necessary to establish macroscopic quantities it is useful to introduce the notion of an 'ε-cell centred at (a geometrical) point $\underset{\sim}{x}$.' This is essentially a simply-connected regular region with centroid at $\underset{\sim}{x}$ such that all boundary points $\underset{\sim}{y}$ satisfy $\varepsilon/2<|\underset{\sim}{x}-\underset{\sim}{y}|<3\varepsilon/2$.

Consider $\sum_i{}' m_i/V_\varepsilon$, where the sum is taken over all P_i within some cell of volume V_ε centred at $\underset{\sim}{x}$. (All cell sums are indicated by a superposed prime.) If this ratio is essentially insensitive to variations in ε appropriate to cell regions which are microscopically large yet macroscopically small (say $\varepsilon \sim 10^{-5}\text{m}=10^5\text{Å}$, for definiteness) and also insensitive to cell shape, then its value is denoted by the 'pseudo-limit' $\lim_\varepsilon\{\sum_i{}' m_i/V_\varepsilon\}$. This number is identified with $\rho(\underset{\sim}{x})$, the mass density at $\underset{\sim}{x}$ (at any given instant: we suppress time dependence for brevity). The total mass associated with particles within a region R of macroscopic proportions (which may thus be decomposed into very many ε-cells with $\varepsilon \sim 10^{-5}$m) is

$$\sum_{P_i \text{ in } R} m_i = \sum_{\text{cells}} (\textstyle\sum_i{}' m_i/V_\varepsilon)V_\varepsilon = \sum_{\text{cells}} \rho(\underset{\sim}{x})V_\varepsilon .$$

The last expression is a Riemann sum which is identified with the integral $\int_R \rho$.

In analogous manner, the macroscopic velocity field $\underset{\sim}{v}$ is defined by

$$\underset{\sim}{v}(\underset{\sim}{x}) := \lim_\varepsilon\{\textstyle\sum_i{}' m_i\underset{\sim}{v}_i / \sum_i{}' m_i\}.$$

The momentum to be associated with particles within R is accordingly identified with $\int_R \rho\underset{\sim}{v}$.

3 GENERAL BALANCE RELATIONS

A body is regarded to be a distinguished set of interacting point masses. Let R be a region of macroscopic proportion strictly within the region B in which ρ and $\underset{\sim}{v}$ are defined for this body. (We do not here consider boundary effects or conditions.) If $\underset{\sim}{f}_{i\ell}$ denotes the force exerted upon P_i by P_ℓ, and $\underset{\sim}{b}_i$ denotes the <u>net</u> force upon P_i due to all agencies other than the interactions with all other particles of the body, then the motion of P_i is governed by

$$\sum_\ell \underset{\sim}{f}_{i\ell} + \underset{\sim}{b}_i = \frac{d}{dt}(m_i\underset{\sim}{v}_i). \tag{1}$$

Here the sum is over <u>all</u> particles of the body, whether inside <u>or</u> outside R. It might be thought that summing equations (1) for all P_i in R, decomposing R into a honeycomb of ε-cells, and then identifying the resulting sums with Riemann integrals, should suffice to establish the linear momentum balance relation. However, with such an analysis it is impossible to take due account of particle diffusion across ∂R, the boundary of R, and it proves necessary to introduce time averaging in addition to spatial averaging.

The Δ-time average $\phi_\Delta(t)$ of a quantity ϕ at instant t is defined by

$$\phi_\Delta(t) := \frac{1}{\Delta}\int_{t-\Delta}^{t} \phi(\tau)d\tau.$$

Consider again relations (1). Let R_τ denote that region deforming with the body (as prescribed by $\underset{\sim}{v}$: cf. [1], p.167) which coincides with R at instant t. Sum equations (1) over all P_i instantaneously within R_τ and then take the Δ-time average at instant t of the resulting τ-dependent expression. Here Δ is taken to be the duration of a microscopically-long, yet macroscopically-short, time interval: for definiteness a notional value of 10^{-6}s may be kept in mind. The result, modulo the existence of local space-time averages, is the balance relation

$$\int_R (\underset{\sim}{f}+\underset{\sim}{b}) = \frac{d}{dt}\{\int_{R_t} \rho\underset{\sim}{v}\}+\int_{\partial R} \underset{\sim}{d}. \qquad (2)$$

Here the interaction body force $\underset{\sim}{f}$, and the external body force $\underset{\sim}{b}$, are local volume-time averages associated with sums of the forms $\sum_i{}'\sum_j \underset{\sim}{f}_{ij}$, and $\sum_i{}'\underset{\sim}{b}_i$, respectively. The quantity $\underset{\sim}{d}$ represents the net rate of loss of momentum associated with the deforming region R_t (instantaneously coincident with R), per unit surface area of ∂R, due to particles crossing the boundary ('diffusion'). On invoking mass conservation and assuming that $\underset{\sim}{d}$ depends continuously upon position and orientation, a standard argument implies the existence of an orientation-independent tensor $\underset{\sim}{D}$ (the diffusive stress tensor) such that

$$\underset{\sim}{d} = \underset{\sim}{D}\underset{\sim}{n}, \qquad (3)$$

where $\underset{\sim}{n}$ denotes the outward unit normal to R. Accordingly (2) may be written as

$$-\int_{\partial R} \underset{\sim}{D}\underset{\sim}{n}+\int_R (\underset{\sim}{f}+\underset{\sim}{b}) = \frac{d}{dt}\{\int_{R_t} \rho\underset{\sim}{v}\}. \qquad (4)$$

In order to arrive at balance relations for moment of momentum and for energy we make two assumptions related to the random nature of thermal motions. The thermal velocity $\tilde{\underset{\sim}{v}}_i$ of P_i is $\underset{\sim}{v}_i-\underset{\sim}{v}(\underset{\sim}{x}_i)$: that is, its velocity relative to the macroscopic motion at its instantaneous location. The thermal motion assumptions are:

T.M.1. $\sum_i{}' m_i\tilde{\underset{\sim}{v}}_i = \underset{\sim}{0}$,

and

T.M.2 $\sum_i{}'(\underset{\sim}{x}_i-\underset{\sim}{x}_\circ) \otimes m_i\tilde{\underset{\sim}{v}}_i = \underset{\sim}{0}$.

Here the sums are taken over any microscopically-large region (an ε-cell with $\varepsilon \sim 10^{-5}$m or union of such cells) and $\underset{\sim}{x}_\circ$ is an arbitrary fixed point. These assumptions we consider to be reasonable for solid, liquid, or gaseous phases, when point masses represent molecules.

The moment of momentum (energy) balance relation is obtained by tensorially (scalarly) multiplying (1) by $(\underset{\sim}{x}_i-\underset{\sim}{x}_\circ)$ [$\underset{\sim}{v}_i$], summing over all such relations for P_i in R_τ at instant τ, and then time averaging the result. These balances take the forms

$$-\int_{\partial R} \{\underset{\sim}{M}+(\underset{\sim}{x}-\underset{\sim}{x}_\circ)\otimes\underset{\sim}{D}\underset{\sim}{n}\}+\int_R\{\underset{\sim}{\ell}+\underset{\sim}{G}+\underset{\sim}{K}+\rho\underset{\sim}{v}\otimes\underset{\sim}{v}+(\underset{\sim}{x}-\underset{\sim}{x}_\circ)\otimes(\underset{\sim}{f}+\underset{\sim}{b})\} = \frac{d}{dt}\{\int_{R_t} (\underset{\sim}{x}-\underset{\sim}{x}_\circ)\otimes \rho\underset{\sim}{v}\} \qquad (5)$$

and

$$-\int_{\partial R} \{k+\underset{\sim}{D}\underset{\sim}{n}.\underset{\sim}{v}+\underset{\sim}{M}\cdot\underset{\sim}{L}^{\top}\}+\int_R \{Q+r+(\underset{\sim}{f}+\underset{\sim}{b}).\underset{\sim}{v}+(\underset{\sim}{\ell}+\underset{\sim}{G})\cdot\underset{\sim}{L}^{\top}\} = \frac{d}{dt}\{\int_{R_t} \rho(h+\underset{\sim}{v}^2/2)\}. \qquad (6)$$

Here $\underset{\sim}{M}$ and k denote diffusive quantities which are expressible, modulo continuous dependence upon position and orientation in the forms (cf. (3))

$$\underset{\sim}{M} = \underset{\sim}{M}\underset{\sim}{n} \quad \text{and} \quad k = \underset{\sim}{k}.\underset{\sim}{n}, \qquad (7)$$

where $\underset{\sim}{M}$ ($\underset{\sim}{k}$) denotes an orientation-independent rank three tensor (vector) field. We term $\underset{\sim}{M}$ the generalised diffusive couple-stress tensor and $\underset{\sim}{k}$ the diffusive heat-flux vector. The symbols $\underset{\sim}{\ell}$ and $\underset{\sim}{G}$ denote the generalised interaction and external body couples respectively, and $\underset{\sim}{K}$ denotes the thermal tensor. These three quantities are all local volume-time averages, the latter associated with cell sums of the form $\sum_i{}' m_i\tilde{\underset{\sim}{v}}_i \otimes \tilde{\underset{\sim}{v}}_i$. The terms Q and r are local volume-time averages, thermal in nature (their definitions involve thermal velocity), and represent interaction and external, heat supply, respectively. The heat content per unit volume, ρh, is merely the thermal kinetic energy density

($= \frac{1}{2}\operatorname{tr}\mathbf{K}$). Finally, $\mathbf{L}$ denotes the velocity gradient.

It should be noted that the foregoing balance relations are in a number of respects more general than those usually encountered. Most significantly, no assumptions whatsoever have been made concerning the nature of interactions. Further, moment of momentum balance is here a relation involving rank two tensor fields. The usual rotational momentum balance involves skew tensors, or equivalently, axial vectors, and is to be associated with the skew part of (5). This generality of (5) enables us, modulo smoothness assumptions, to obtain local forms of (4), (5) and (6) which enable the latter to be reduced to the form

$$-\operatorname{div}\mathbf{k}-\mathbf{M}\cdot\nabla(\mathbf{L}^T)+Q+r-\mathbf{K}\cdot\mathbf{L}^T=\rho\dot{h}. \tag{8}$$

Notice that $\mathbf{k}$ and $\mathbf{M}$ are quantities associated with diffusion (and hence thermal motions) and thus thermal in character, as also are Q, r, $\mathbf{K}$, and h. Accordingly, (8) may be regarded as a local balance of thermal energy.

4 REDUCED BALANCE RELATIONS

More familiar forms of the balance relations follow as a consequence of interaction assumptions. In respect of linear momentum balance two such assumptions seem to be necessary:

I.1 $\sum_i{}'\sum_k{}'\mathbf{f}_{ik}=\mathbf{0}$ for sums over all pairs of point masses within any microscopically-large region

and

I.2 $\sum_\ell \mathbf{f}_{i\ell}=\sum_n \mathbf{f}_{in}$, where $P_iP_n<\delta\ll\varepsilon\sim10^{-5}$m.

Since the double sum is expressible as $\frac{1}{2}\sum_{i\neq k}{}'\sum{}'(\mathbf{f}_{ik}+\mathbf{f}_{ki})$, I.1 is trivially true if interactions satisfy $\mathbf{f}_{ik}+\mathbf{f}_{ki}=\mathbf{0}$ for each pair of particles P_i, P_k in the region. However, I.1 is very much weaker than this, merely requiring that there be no net self-force associated with particles in any microscopically-large region. An interesting consequence of I.1 is that net interactions between particles within two distinct, microscopically-large, regions are balanced: that is, action and reaction are equal but opposite at the level of net interactions between microscopically-large regions within a body. Assumption I.2 implies that in determining the net force upon P_i only interactions with particles not too far away need be considered. This is consistent with molecular physics in which interactions with effective ranges of 1000 Å ($= 10^{-7}$m) are considered extremely long-range. Of course, I.2 does not imply that <u>individual</u> interactions become negligible at separations in excess of δ (for example, this would not be the case for ion-ion interactions) but rather reflects a co-operative effect. As a consequence of I.1 and I.2 it may be shown that there exists an interaction traction $\mathbf{t}$ in terms of which

$$\int_R \mathbf{f} = \int_{\partial R} \mathbf{t}. \tag{9}$$

Together, (4) and (9), upon assuming $\mathbf{t}$ depends continuously upon position and orientation, imply there exists an orientation-independent (interaction stress) tensor $\mathbf{T}^-$ such that

$$\mathbf{t} = \mathbf{T}^-\mathbf{n}.$$

Accordingly (4) may be written in the conventional form

$$\int_{\partial R}\mathbf{T}\mathbf{n}+\int_R \mathbf{b} = \frac{d}{dt}\{\int_{R_t}\rho\mathbf{v}\}, \tag{10}$$

where the Cauchy stress tensor

$$\mathbf{T} := \mathbf{T}^- - \mathbf{D} \quad . \tag{11}$$

Relation (11) indicates the separate contributions to stress which derive from interactions and diffusion.

A rather more general balance of rotational momentum than usual, involving an interaction couple-stress (rank three) tensor $\hat{\mathbf{C}}$, is obtainable from the skew part of (5) modulo the assumption

I.3 $\sum_i{}'\sum_k{}'(\mathbf{x}_i-\mathbf{x}_o)\wedge\mathbf{f}_{ik} = \mathbf{0}$ for any sum taken over all pairs of particles within any microscopically-large region.

We observe that I.3 trivially holds if $\mathbf{f}_{ik}$ is parallel to $\overline{P_iP_k}$, and may be phrased in terms of the vanishing of the self-couple associated with particles within microscopically-large regions. The balance relation in question may now be expressed in the form

$$\int_{\partial R}\{\mathbf{C}\mathbf{n}+(\mathbf{x}-\mathbf{x}_o)\wedge\mathbf{T}\mathbf{n}\}+\int_R\{\operatorname{sk}\mathbf{G}+(\mathbf{x}-\mathbf{x}_o)\wedge\mathbf{b}\}$$
$$= \frac{d}{dt}\{\int_{R_t}(\mathbf{x}-\mathbf{x}_o)\wedge\rho\mathbf{v}\}. \tag{12}$$

Here $\mathbf{C} := \hat{\mathbf{C}} - \hat{\mathbf{M}}$, where $\hat{\mathbf{M}}\mathbf{n} = \operatorname{sk}(\mathbf{M}\mathbf{n})$, and $\operatorname{sk}\mathbf{G}$ is to be interpreted as external body couple.

Upon making a further interaction

assumption I.4., which turns out to be equivalent to a balance of heat conduction exchange rates between adjoining microscopically-large regions, energy balance (6) simplifies to

$$\int_{\partial R} -\underset{\sim}{q}.\underset{\sim}{n}+\underset{\sim}{C}^{+}\underset{\sim}{n}\cdot\underset{\sim}{L}^{T}+\underset{\sim}{T}\underset{\sim}{n}.\underset{\sim}{v}\}+\int_{R}\{r+\underset{\sim}{b}.\underset{\sim}{v}+G\cdot L^{T}\} = \frac{d}{dt}\{\int_{R_t}\rho(\varepsilon+v^2/2)\}. \qquad (13)$$

Here the heat flux vector $\underset{\sim}{q}:=\underset{\sim}{q}^{-}+\underset{\sim}{k}$, where $\underset{\sim}{q}^{-}$ is the power exerted by interactions exerted across ∂R in the thermal motions of particles within R, and $\underset{\sim}{k}$ (cf.(7)) is the diffusive flux of thermal kinetic energy. The internal energy $\rho\varepsilon:=\rho(\beta+h)$ is the sum of an interaction energy density $\rho\beta$ with thermal kinetic energy density ρh. Further, $\underset{\sim}{C}^{+}:=\hat{\underset{\sim}{C}}-\underset{\sim}{M}$.

5 IMPLICATIONS FOR MIXTURE THEORY

A mixture, from the discrete viewpoint, is a set of interacting point masses which may be subdivided into subsets of 'like' particles. Such subsets form the 'constituents' or 'species' of the mixture. The motions of particles of a single species may be treated as in the foregoing, provided interactions with other particles are taken into account. (For simplicity we do not here consider reactions.) Thus the mass density ρ_α and macroscopic velocity field $\underset{\sim}{v}_\alpha$ for species α may be obtained as local spatial averages. Linear momentum balance for this species takes the form (cf. (4))

$$-\int_{\partial R}\underset{\sim}{D}_\alpha\underset{\sim}{n}+\int_{R}(\underset{\sim}{f}_\alpha+\underset{\sim}{b}_\alpha)=\left(\frac{d}{dt}\right)_\alpha\{\int_{R_t}\rho_\alpha\underset{\sim}{v}_\alpha\}. \qquad (14)$$

Here $\underset{\sim}{D}_\alpha$ denotes the α-species diffusive stress tensor, $\underset{\sim}{f}_\alpha$ the corresponding interaction body force (which includes the effect of all other species in addition to that of species α), and $\underset{\sim}{b}_\alpha$ the external body force upon species α. The time derivative is taken over a region which deforms with the motion of species α and instantaneously coincides with R.

As in the case of single continua, provided assumptions I.1 and I.2 hold for α-α interactions, it is possible to obtain a reduced linear momentum balance(cf.(10)):

$$\int_{\partial R}\underset{\sim}{T}_\alpha\underset{\sim}{n}+\int_{R}(\underset{\sim}{f}_\alpha^{-}+\underset{\sim}{b}_\alpha)=\left(\frac{d}{dt}\right)_\alpha\{\int_{R_t}\rho_\alpha\underset{\sim}{v}_\alpha\}. \qquad (15)$$

However, it is important to note that an individual species may well consist of ions of the same polarity. In such cases I.2 will most definitely not hold and (14) is the only form of linear momentum balance possible. Further, when (15) is valid the interpretations of $\underset{\sim}{T}_\alpha$ and $\underset{\sim}{f}_\alpha^{-}$ differ from those commonly accepted [3,4]. Here the quantity $\underset{\sim}{T}_\alpha\underset{\sim}{n}$ involves only species α diffusion and α-α interactions across ∂R, and $\underset{\sim}{f}_\alpha^{-}$ represents the net effect of all other species upon species α. The usual interpretation of $\underset{\sim}{T}_\alpha\underset{\sim}{n}$ is of the effect of the whole mixture outside R upon species α within R. This latter interpretation is not consistent with the corpuscular viewpoint even when interactions are pairwise balanced. Indeed, the conventional interpretation leads to a paradox [5,6]. It would seem that the above yields a clear resolution of this problem.

The contents of section 5 represent some results taken from joint work with Prof. A. Morro. Helpful comments by Profs. C. Cercignani and H. Cohen are gratefully acknowledged.

6 REFERENCES

[1] Murdoch, A.I. 1983. The motivation of continuum concepts and relations from discrete considerations. Q.J.Mech.Appl. Math. 36, 163-187.
[2] Murdoch, A.I. 1985. A corpuscular approach to continuum mechanics: basic considerations. Arch.Rational Mech. Anal. 88, 291-321.
[3] Truesdell, C. 1969. Rational thermodynamics. New York: McGraw-Hill.
[4] Bowen, R.M. 1976. Theory of mixtures. In continuum physics, Vol.3 (ed. A.C. Eringen). New York - San Francisco - London: Academic.
[5] Williams, W.O. 1973. On the theory of mixtures. Arch.Rational Mech. Anal. 51, 239-260.
[6] Gurtin, M.E., M.L. Oliver & W.O. Williams, 1973. On balance of forces for mixtures. Quart.Appl. Math. 30, 527-530.

CORRIGENDUM

I.3 trivially holds if $\underset{\sim}{f}_{ik}$ is parallel to $\overrightarrow{P_iP_k}$ and $\underset{\sim}{f}_{ik}+\underset{\sim}{f}_{ki}=\underset{\sim}{0}$.

5th International Symposium on Continuum Models of Discrete Systems / Nottingham / 14-20 July 1985

Existence of non-negative entropy production

W.Muschik
Institut für Theoretische Physik, Technische Universität Berlin, FR Germany

ABSTRACT: Starting out with an outline of non-equilibrium Thermodynamics of discrete systems in which non-equilibrium analogues of temperature and entropy are defined following items are derived: Non-equilibrium entropy S increases in isolated systems $C \rightarrow B(eq): S_B^{eq} - S_C \geq 0$, beyond that $\dot{S} \geq 0$ can be always achieved by special material dependent choice, and therefore material dependent dissipation inequalities in field formulation do always exist.

1 INTRODUCTION

Formulations of the Second Law can be classified according to the following scheme:

$$\left\{\begin{matrix}\text{Temperature}\\ \text{and}\\ \text{entropy}\end{matrix}\right\} \left\{\begin{matrix}\text{are}\\ \text{defined}\\ \text{as}\end{matrix}\right\}$$

$$\left\{\begin{matrix}\text{equilibrium}\\ \text{or}\\ \text{non-equilibrium}\end{matrix}\right\} \quad \text{quantities,}$$

and the

$$\text{formulation is} \quad \left\{\begin{matrix}\text{local}\\ \text{global}\end{matrix}\right\} \quad \text{in}$$

$$\left\{\begin{matrix}\text{time and/or}\\ \text{position}\end{matrix}\right\}.$$

So field formulations of Thermodynamics use often dissipation inequalities which are local in time and position and in which temperature and entropy are primitive concepts, i.e. the question wether they are defined as equilibrium or non-equilibrium quantities is not subject of discussion. On the contrary in Thermodynamics of discrete systems dissipation inequalities are used which are not local in time (e.g. Clausius' inequality) and in which no entropy appears in whatever way being defined.

Of course these different formulations of the Second Law are not equivalent because if time integrated quantities are definite, the definiteness of the quantity itself cannot be derived tritely. Therefore a non-negative entropy production which is local in time and position is either axiomatically presupposed or it is to derive from more general statements being global in time. The aim of this paper will be to show in two different ways the existence of non-negative entropy productions being local in time and position if Clausius' inequality is presupposed.

2 THERMODYNAMICS OF DISCRETE SYSTEMS

2.1 Preliminaries

We consider a compact generally time dependent domain $\mathfrak{G}(t)$ in R^3 which interacts through its surface $\partial\mathfrak{G}(t)$ with its vicinity $\bar{\mathfrak{G}}(t)$ by exchanging at time t heat $\dot{Q}(t)$, power $\dot{W}(t)$ and mass $\underline{\dot{n}}^e(t)$ (time rate of mole numbers by external exchange). Such a discrete system is denoted due to Schottky (1929). $\mathfrak{G}(t)$ will be called

isolated, if $\dot{Q} = 0$, $\dot{W} = 0$, $\underline{\dot{n}}^e = \underline{0}$,

adiabatic, if $\dot{Q} = 0$, $\underline{\dot{n}}^e = \underline{0}$,

closed, if $\underline{\dot{n}}^e = \underline{0}$

are satisfied for all times t.

We now consent

Thermostatics is presupposed to be known.

Therefore we know expressions such as

$\underline{a}$ work variables,

$\underline{A}^{eq}$ generalized forces,

U internal energy,

T thermostatic temperature,

$\underline{n}$ mole numbers,

$\underline{\mu}^{eq}$ chemical potentials,

S^{eq} equilibrium entropy,

and other equilibrium quantities. By use of Thermostatics of discrete systems we are able to formulate their Thermodynamics.

The concept of state is well known in Thermostatics due to the Zeroth Law. It states we need the mechanical variables $\underline{a}$, the mole numbers $\underline{n}$, and one additional thermal variable e.g. U, if the system is thermally homogeneous:

$$\underline{Z}^o = (\underline{a},\underline{n},U) \in \mathfrak{Z}^o \subseteq R^n.$$

In non-equilibrium the dimension of the state space $\mathfrak{Z}$ belonging to is much greater than that of $\mathfrak{Z}^o$:

$$\underline{Z} = (\underline{a},\underline{n},U,\ldots.) \in \mathfrak{Z} \subseteq R^m.$$

$$m > n$$

The additional non-equilibrium variables may be relaxation variables (Kestin 1979) or derivatives in time (or position) of the equilibrium variables $\underline{Z}^o$.

We have to distinguish two classes of state spaces called large and small (Muschik 1981). This division is caused by the dependence of the material properties represented by the material mapping $\mathbb{M}$ on the chosen state space. If $\mathbb{M}$ is local in time

$$\mathbb{M} : \underline{Z}(t) \longmapsto M(t),$$

we call $\mathfrak{Z}$ large, and if $\mathbb{M}$ depends on the history of a process

$$\underline{Z}^t(s) := \underline{Z}(t-s),\ t \text{ fixed},\ s \geq 0$$

i.e.

$$\mathbb{M} : \underline{Z}^t(\cdot) \longmapsto M(t),$$

we call $\mathfrak{Z}$, $\underline{Z}(t-s) \in \mathfrak{Z}$ for each fixed s, a small state space. Especially thermodynamic theories including after-effects are formulated in small state spaces.

We now decompose the system's surface $\partial\mathfrak{G}(t)$ into non-overlapping partial areas F^α

$$\mathbb{D} : \partial\mathfrak{G}(t) = \bigcup_\alpha F^\alpha(t)$$

$$F^\alpha \cap F^\beta = \phi,\ \alpha \neq \beta.$$

The exchange quantities between $\mathfrak{G}(t)$ and $\bar{\mathfrak{G}}(t)$ through $F^\alpha(t)$ are denoted by $\dot{Q}^\alpha(t)$, $\dot{W}^\alpha(t)$, and $\underline{\dot{n}}^{e\alpha}(t)$. Using them we define three kinds of contacts according to the definition of Schottky systems. We presuppose that $\mathfrak{G}(t)$ has a certain non-equilibrium state $\underline{Z}(t)$ whereas its vicinity $\bar{\mathfrak{G}}(t)$ is in equilibrium at state $\underline{Z}^o(t) \in \bar{\mathfrak{Z}}^o$.

The state dependent exchange quantities are

$\dot{Q}^\alpha(\underline{Z}(t),\ \bar{\underline{Z}}^o(t))$, $\dot{W}^\alpha(\underline{Z}(t),\ \bar{\underline{Z}}^o(t))$,

$\underline{\dot{n}}^{e\alpha}(\underline{Z}(t),\ \bar{\underline{Z}}^o(t))$,

or if all possible combinations of states in $\mathfrak{Z} \times \bar{\mathfrak{Z}}^o$ are considered:

$\dot{Q}^\alpha(\mathfrak{Z},\bar{\mathfrak{Z}}^o)$, $\dot{W}^\alpha(\mathfrak{Z},\bar{\mathfrak{Z}}^o)$, $\underline{\dot{n}}^{e\alpha}(\mathfrak{Z},\bar{\mathfrak{Z}}^o)$.

The set

$$\mathfrak{C}_t := \{\partial\mathfrak{G},\ \mathfrak{Z} \times \bar{\mathfrak{Z}}^o,\ \dot{Q}^\alpha(\mathfrak{Z},\bar{\mathfrak{Z}}^o),\ \text{for all } \mathbb{D}\}$$

$$\dot{W}^\alpha(\mathfrak{Z},\bar{\mathfrak{Z}}^o) = 0,\ \underline{\dot{n}}^{e\alpha}(\mathfrak{Z},\bar{\mathfrak{Z}}^o) = \underline{0},$$

is called a thermal contact (Muschik 1980). Analoguously we define: The set

$$\mathfrak{C}_m := \{\partial\mathfrak{G},\ \mathfrak{Z} \times \bar{\mathfrak{Z}}^o,\ \dot{W}^\alpha(\mathfrak{Z},\bar{\mathfrak{Z}}^o),\ \text{for all } \mathbb{D}\}$$

$$\mathfrak{G} \text{ adiabatic},$$

is called a mechanical contact, and the set

$$\mathfrak{C}_p := \{\partial\mathfrak{G},\ \mathfrak{Z} \times \bar{\mathfrak{Z}}^o,\ \underline{\dot{n}}^{e\alpha}(\mathfrak{Z},\bar{\mathfrak{Z}}^o),\ \text{for all } \mathbb{D}\}$$

$$\dot{Q}^\alpha(\mathfrak{Z},\bar{\mathfrak{Z}}^o) = 0,\ \dot{W}^\alpha(\mathfrak{Z},\bar{\mathfrak{Z}}^o) = 0.$$

is called a phase contact. These three kinds of contacts can be characterized by intensive non-equilibrium quantities called contact quantities.

2.2 Contact quantities

We now consider a certain non-equilibrium state $\underline{Z}$ of $\mathfrak{G}$ which is determined by a special preparation of the system.

Axiom: To each $\underline{Z}$ of $\mathfrak{G}$ there exists exactly one state $\underline{Z}^o$ of $\bar{\mathfrak{G}}$ so that

$\dot{Q}(\underline{Z},\bar{\underline{Z}}^o) = 0$, for the thermal contact,

$\dot{W}(\underline{Z},\bar{\underline{Z}}^o) = 0$, for the mechanical contact,

$\underline{\dot{n}}^e(\underline{Z},\bar{\underline{Z}}^o) = \underline{0}$, for the phase contact.

These unique zeros for the different contacts are correlated with a change of sign.

Because for a thermal contact no power and mass exchange takes place the relevant equilibrium variable of $\bar{\mathfrak{G}}$ is only its thermostatic temperature T^*. Because $\dot{Q}(\underline{Z},\bar{\underline{Z}}^o)$ changes its sign at its unique zero we can define a non-equilibrium contact quantity $1/\theta$ characterizing the non-equilibrium state $\underline{Z}$ of $\mathfrak{G}$ by

$$\dot{Q}(\underline{Z},\bar{\underline{Z}}^o)\left(\frac{1}{\theta} - \frac{1}{T^*}\right) \geq 0.$$

θ is called contact temperature of $\underline{Z}$ of $\mathfrak{G}$ belonging to the thermal contact $\mathfrak{C}_t$. Of course θ does not depend only on the states $\underline{Z}$ and $\bar{\underline{Z}}^o$ but in contrary to equilibrium also on the thermal properties of the contact.

In equilibrium the contact temperature becomes the thermostatic temperature because the definition for measuring Θ is identical to that for measuring T^* by the transitivity of thermal equilibrium. The contact temperature satisfies all properties of a classical measuring quantity (Muschik 1977).

Analoguously to the contact temperature we get the dynamic generalized forces $\underline{A}$ by

$$\dot{a}_j(A_j^{eq}-A_j) \geq 0, \text{ for all } j.$$

A trivial example is (j = 1)

$$\dot{a} = \dot{V},\ A^{eq} = -p^{eq},\ A = -p,$$

$$\dot{V}(p-p^{eq}) \geq 0.$$

The dynamic chemical potential μ_k is defined by

$$\dot{n}_k^e(\mu_k^{eq}-\mu_k) \geq 0, \text{ for all } k.$$

2.3 Processes and projections

A process of a discrete system is a mapping (Muschik 1979)

$$\underline{Z} : [i,f] \in R^1 \longrightarrow \mathfrak{Z}$$

$$\underline{Z}(t) = (\underline{a},\underline{n},U,\theta,\ldots;\underline{\varphi}^*)(t)$$

with

$$\underline{\varphi}^* = (T^*, \underline{A}^*, \underline{\mu}^*)$$

characterizing $\bar{\mathcal{G}}$ which is always assumed to be in equilibrium and therefore to have reservoir properties. We define projections of the process $\underline{Z}(t)$ onto the equilibrium subspace $\mathfrak{Z}^o$ (Fig. 1):

1. U-projection:

$$P_U\underline{Z}(t) = (\underline{a},\underline{n},U)(t),$$

2. Θ-projection:

$$P_\Theta\underline{Z}(t) = (\underline{a},\underline{n},U_\Theta)(t),$$

3. $*$-projection (Kestin 1985):

$$P_*\underline{Z}(t) = (\underline{a},\underline{n},U^*)(t).$$

Here U_Θ and U^* are given by the equilibrium caloric equation of state

$$T = T(\underline{a},\underline{n},U),$$

$$\Theta = T(\underline{a},\underline{n},U_\Theta),\ T^* = T(\underline{a},\underline{n},U^*).$$

According to their definition we get in equilibrium

$$P_U\underline{Z}^o = P_\Theta\underline{Z}^o = P_*\underline{Z}^o = \underline{Z}^o.$$

The definitions of the temperature Θ and T^* are quite symmetric: given $\mathcal{G}$ the contact temperature Θ is that thermostatic temperature of $\bar{\mathcal{G}}$ which causes the heat exchange to be vanished, whereas for given $\bar{\mathcal{G}}$ the heat exchange will be zero, if the contact temperature Θ of $\mathcal{G}$ is T^*.

Due to Keller (1971) projections of $\underline{Z}(t)$ are called accompanying process to $\underline{Z}(t)$. They are represented by trajectories in equilibrium subspace. These accompanying processes are those which Thermostatics deals with (Fig. 1).

2.4 First Law and entropies

We consider the discrete system $\mathcal{G}(t)$ at time t and its vicinity $\bar{\mathcal{G}}(t)$. Because

$$\Theta = T^* \quad \text{and} \quad \underline{A} = \underline{A}^*$$

by choice of the state $\underline{\bar{Z}}^o(t)$ of $\bar{\mathcal{G}}(t)$ we have

$$\dot{Q} = 0 \quad \text{and } \underline{\dot{a}} = \underline{0}.$$

Further we assume that no chemical reactions take place in $\mathcal{G}(t)$

$$\underline{\dot{n}}^i = \underline{0}$$

but external mass exchange between $\mathcal{G}(t)$ and $\bar{\mathcal{G}}(t)$ is possible. Therefore the internal energy of $\mathcal{G}(t)$ will change by an amount which depends on $\underline{\dot{n}}^e$

$$\dot{U} = \underline{h}\cdot\underline{\dot{n}}^e.$$

So the dynamical molar enthalpy $\underline{h}$ is defined by

$$\underline{h} := \frac{\partial U}{\partial \underline{n}},\ \dot{Q} = 0,\ \underline{\dot{a}} = \underline{0},\ \underline{\dot{n}}^i = \underline{0}.$$

Hence the First Law for a non-equilibrium process writes

$$\dot{U} = \dot{Q} + \underline{A}\cdot\underline{\dot{a}} + \underline{h}\cdot\underline{\dot{n}}^e$$

and e.g. for the $*$-projection

$$\dot{U}^* = \dot{Q}^* + \underline{A}^*\cdot\underline{\dot{a}} + \underline{h}^*\cdot\underline{\dot{n}}^e.$$

Here in equilibrium

$$\underline{h}^* = \frac{\partial}{\partial \underline{n}}(U-\underline{A}^*\cdot\underline{a})(\underline{A}^*,\underline{n},T^*)$$

holds as usual.

The well known "time rate" of the equilibrium entropy

$$S^{eq}(t) = S^{eq}(\underline{a},\underline{n},U)(t)$$

$$T\dot{S}^{eq} = \dot{U} - \underline{A}_U\cdot\underline{\dot{a}} - \underline{\mu}_U\cdot\underline{\dot{n}}$$

can be interpreted as generated by an U-projection of a non-equilibrium process. Both the other projections yield

$$\Theta\dot{S}^{eq} = \dot{U}_\Theta - \underline{A}_\Theta\cdot\underline{\dot{a}} - \underline{\mu}_\Theta\cdot\underline{\dot{n}},$$

$$T^*\dot{S}^{eq} = \dot{U}^* - \underline{A}^*\cdot\underline{\dot{a}} - \underline{\mu}^*\cdot\underline{\dot{n}}.$$

Using the contact quantities Θ, $\underline{A}$, and $\underline{\mu}$

the time rate of a non-equilibrium entropy should have the following shape

$$\Theta\dot{S} := \dot{U} - \underline{A}\cdot\underline{\dot{a}} - \underline{\mu}\cdot\underline{\dot{n}} + \Theta\Sigma$$

where Σ is the entropy production whose definition fixes $\dot{S}$. In general S is a functional defined on histories of processes in

$$S(t) = \mathfrak{S}(\underline{Z}^t(\cdot))$$

so that $\mathfrak{Z}$ is a small state space. Analoguously we have

$$\underline{A}(t) = \mathfrak{A}(\underline{Z}^t(\cdot)),\ \underline{\mu}(t) = \mathfrak{N}(\underline{Z}^t(\cdot)).$$

$\mathfrak{S}$, $\underline{\mathfrak{A}}$, and $\mathfrak{N}$ are special material mappings. Introducting the First Law into the time rate of entropy we get

$$\Theta\dot{S} = \dot{Q} + \Theta\underline{s}\cdot\underline{\dot{n}}^e - \underline{\mu}\cdot\underline{\dot{n}}^i + \Theta\Sigma$$

where we used

$$\underline{\dot{n}} = \underline{\dot{n}}^e + \underline{\dot{n}}^i\ ,$$

and

$$\Theta\underline{s} := \underline{h} - \underline{\mu}\ .$$

Because in equilibrium no entropy production and no chemical reactions take place the entropy rate for e.g. the $*$-projection becomes

$$T^*\dot{S}^{eq} = \dot{Q}^* + T^*\underline{s}^*\cdot\underline{\dot{n}}^e$$

Another non-equilibrium entropy is

$$T^*\dot{S}' := \dot{Q} + T^*\underline{s}^*\cdot\underline{\dot{n}}^e - \underline{\mu}_\Theta\cdot\underline{\dot{n}}^i + \Theta\Sigma'$$

In this definition the exchange quantities are referred to $\tilde{\mathcal{G}}$ whereas the entropy producting quantities belong to $\mathcal{G}$ where the entropy is in fact produced.

2.5 Clausius' inequality and embedding axiom

The usual formulation of the Second Law for non-closed (open) systems by Clausius' inequality refers to a non-equilibrium trajectory (Haase 1963) having the actual heat exchange $\dot{Q}$ and to its $*$-projection with T^* and s^*

$$\oint\left(\frac{\dot{Q}}{T^*} + \underline{s}^*\cdot\underline{\dot{n}}^e\right)dt \le 0.$$

An extended formulation of the Second Law using contact quantities is possible (Muschik 1983)

$$\oint\left(\frac{\dot{Q}}{\Theta} + \underline{s}\cdot\underline{\dot{n}}^e\right)dt \le 0.$$

The essential difference between both formulations is that T^* and $\underline{s}^*$ are related to $\tilde{\mathcal{G}}$ whereas Θ and $\underline{s}$ refer to $\mathcal{G}$ itself. But in both the cases $\dot{Q}$ is the actual heat exchange which is of course different from $\dot{Q}^*$ which refers to the $*$-projection:

$$\oint\dot{S}^{eq}dt = 0 = \oint\left(\frac{\dot{Q}^*}{T^*} + \underline{s}^*\cdot\underline{\dot{n}}^e\right)dt.$$

Whatever the definition of non-equilibrium entropy S may be it must not contradict the presupposed equilibrium entropy S^{eq}. Hence we have to demand the embedding axiom as in equilibrium (Muschik 1979)

$$\overset{+}{\oint}\dot{S}\,dt = 0 = \overset{+}{\oint}\left(\frac{\dot{Q}}{\Theta} + \underline{s}\cdot\underline{\dot{n}}^e\right)dt + \overset{+}{\oint}\left(\Sigma - \frac{1}{\Theta}\underline{\mu}\cdot\underline{\dot{n}}^i\right)dt$$

The $+$ indicates that the closed path in state space $\mathfrak{Z}$ includes at least one equilibrium state. From the extended Clausius' inequality follows that the second term is not negative.

$$\overset{+}{\oint}\left(\Sigma - \frac{1}{\Theta}\underline{\mu}\cdot\underline{\dot{n}}^i\right)dt \ge 0$$

Applying the embedding axiom to $\dot{S}'$ we get

$$0 = \overset{+}{\oint}\left(\frac{\dot{Q}}{T^*} + \underline{s}^*\cdot\underline{\dot{n}}^e\right)dt + \overset{+}{\oint}\left(\frac{\Theta}{T^*}\Sigma' - \frac{1}{T^*}\underline{\mu}_\Theta\cdot\underline{\dot{n}}^i\right)dt$$

with

$$\overset{+}{\oint}\frac{1}{T^*}\left(\Theta\Sigma' - \underline{\mu}_\Theta\cdot\underline{\dot{n}}^i\right)dt \ge 0.$$

3 NON-NEGATIVE ENTROPY PRODUCTION

3.1 Entropy production in isolated systems

We consider an arbitrary process (Fig. 2) $\mathcal{P}$: A(eq)$\to$B starting out from equilibrium. If the system is in state B, it will be isolated and will go into the equilibrium state C(eq)

$$\mathcal{B} : B \to C(eq)\ :\ \dot{Q} = 0,\ \underline{\dot{a}} = \underline{0},\ \underline{\dot{n}}^e = 0.$$

We now close the process with an equilibrium trajectory

$$\mathcal{C} : C(eq) \to A(eq)\ :\ \Sigma = 0,\ \underline{\dot{n}}^i = \underline{0}.$$

The dissipation inequality along this process $A \to B \to C \to A$ is

$$\overset{+}{\oint}\left(\Sigma - \frac{1}{\Theta}\underline{\mu}\cdot\underline{\dot{n}}^i\right)dt = \mathcal{P}\!\int_A^B\left(\Sigma - \frac{1}{\Theta}\underline{\mu}\cdot\underline{\dot{n}}^i\right)dt +$$

$$+ \mathcal{B}\!\int_B^C \dot{S}\,dt + 0 \ge 0,$$

from which follows

$$\mathcal{P}\!\int_A^B\left(\Sigma - \frac{1}{\Theta}\underline{\mu}\cdot\underline{\dot{n}}^i\right)dt \ge S_B - S_C^{eq}\ .$$

Analoguously we get for $\dot{S}'$

$$\mathcal{P}\!\int_A^B\frac{1}{T^*}\left(\Theta\Sigma' - \underline{\mu}_\Theta\cdot\underline{\dot{n}}^i\right)dt \ge S'_B - S_C^{eq}\ .$$

Both the inequalities demonstrate that the change of entropy $S_B - S_C^{eq}$ cannot be estimated because we do not know the sign of the left hand integrals. Therefore Clausius' inequality does not determine the sign of entropy production in isolated systems because it only gives a statement on cyclic processes. Of course we get immediately

$$S_C^{eq} \geq S_A^{eq}$$

which only relates equilibrium states but does not give any information on the entropy production for the partial process $\mathfrak{J}$ of the isolated system from non-equilibrium to equilibrium. Therefore we need additional hints how to select entropy production Σ and Σ' so that entropy increases in isolated systems.

3.2 Non-negative entropy production rate

Along $\mathfrak{J}$ we have

$$\mathfrak{J}\int_B^C (\Sigma - \frac{1}{\Theta}\, \underline{\mu}\cdot\underline{\dot{n}}^i)dt = S_C^{eq} - S_B ,$$

$$\mathfrak{J}\int_B^C \frac{T}{T^*}(\Theta\Sigma' - \underline{\mu}_\Theta\cdot\underline{\dot{n}}^i)dt = S_C^{eq} - S_B'$$

which gives the connection between the non-equilibrium entropies S_B and S_B' and their defining quantities Σ and Σ'. Because B is a non-equilibrium state

$$\underline{Z}(t_B) = (\underline{a}^B, \underline{n}^B, U^B, \Theta^B \ldots\ldots; \underline{\varphi}_B^*)$$

the integrals on the left hand side are functionals of the process history until t_B

$$\mathfrak{J}\int_B^C (\Sigma - \frac{1}{\Theta}\, \underline{\mu}\cdot\underline{\dot{n}}^i)dt =$$

$$= -\mathbb{F}(U_\Theta^B - U^C, \underline{n}^B - \underline{n}^C, \ldots\ldots\ldots\ldots; \underline{Z}^{t_B}(\cdot)),$$

and because the system is isolated along $\mathfrak{J}$ the functional $\mathbb{F}$ depends only on the difference of the state variables between the initial state B and the final equilibrium state C. Because of the isolation along $\mathfrak{J}$ the history influences the value of the integral only until the time t_B from which the isolation begins (Fig. 2).

We abbreviate

$$\xi(t) := U_\Theta(t) - U^C , \quad \underline{\eta}(t) := \underline{n}(t) - \underline{n}^C .$$

Along $\mathfrak{J}$ the total differential of $\mathbb{F}$ is

$$D\,\mathbb{F} = \frac{\partial\mathbb{F}}{\partial\xi}(\xi,\underline{\eta},\ldots;\underline{Z}^{t_B}(\cdot))\dot{\xi} +$$

$$+ \frac{\partial\mathbb{F}}{\partial\underline{\eta}}(\xi,\underline{\eta},\ldots;\underline{Z}^{t_B}(\cdot))\cdot\dot{\underline{\eta}} + \ldots .$$

because the history $\underline{Z}^{t_B}(\cdot)$ is only parameter along $\mathfrak{J}$. If we now choose on $\mathfrak{J}$

$$(\Sigma - \frac{1}{\Theta}\, \underline{\mu}\cdot\underline{\dot{n}}^i)(t) = D\,\mathbb{F}(\xi(t), \underline{\eta}(t), \ldots; \underline{Z}^{t_B}(\cdot))$$

which integrated along $\mathfrak{J}$ from B to C will

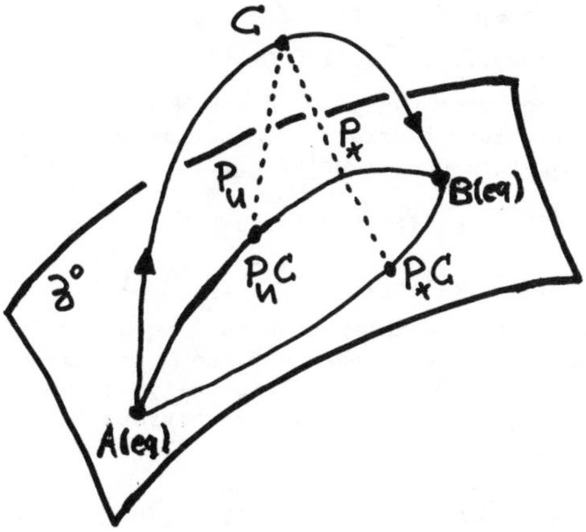

Fig. 1 A process $\underline{Z}$ starting out from an equilibrium state A(eq) goes through the non-equilibrium state C and ends in B(eq). $P_U C$ and $P_* C$ are the U-projection or the $*$-projection of C. The trajectory in $\mathfrak{J}^\circ$ through A(eq) - $P_U C$ - B(eq) is the accompanying U-process belonging to $\underline{Z}$. The accompanying $*$-process from A(eq) through $P_* C$ to B(eq). The path parameter is transferred by projection from $\underline{Z}$ to its accompanying process.

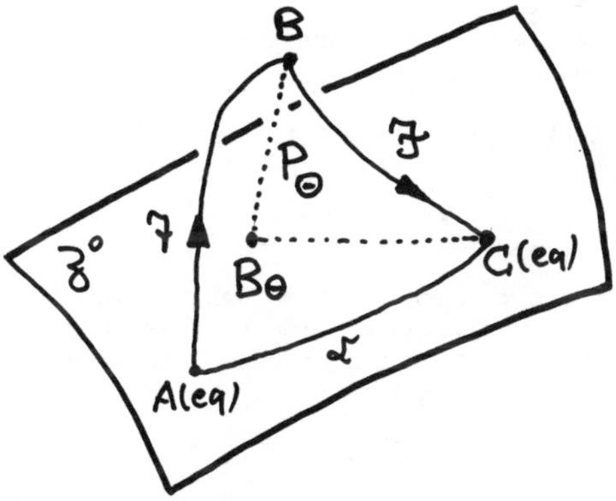

Fig. 2 Along $\mathfrak{J}$ the process is an arbitrary one. Whereas along the system is isolated. $\mathcal{L}$ is an arbitrary trajectory in $\mathfrak{J}^\circ$. By Θ-projection we get B_Θ from B. Because $U = U^C$ along $\mathfrak{J}$ by isolation of the system $-\mathbb{F}(U_\Theta^B - U^C, \underline{n}^B - \underline{n}^C, \ldots . Z^{t_B}(\cdot))$ is the entropy production along $\mathfrak{J}$ between B and C(eq).

yield $\mathbb{F}$, if we put

$$\mathbb{F}(0,\underline{0},\ldots;\underline{Z}^{t_B}(\cdot)) = 0$$

The "process speed" along $\mathfrak{F}$, $\dot{\xi}(t)$ and $\dot{\underline{\eta}}(t)$, is determined by material properties. Therefore constitutive equations exist

$$\dot{\xi}(t) = \mathbb{X}(\xi(t),\ \underline{\eta}(t),\ldots;\underline{Z}^{t_B}(\cdot))$$

$$\dot{\underline{\eta}}(t) = \mathbb{E}(\xi(t),\ \underline{\eta}(t),\ldots;\underline{Z}^{t_B}(\cdot))$$

which may be relaxation rate equations in special cases. Because $\mathbb{X}$ and $\mathbb{E}$ are given and $\mathbb{F}$ can be chosen arbitrarily we choose for getting a non-negative entropy production (Fig. 3)

$$\text{sgn}\ \frac{\partial\mathbb{F}}{\partial\xi} = \text{sgn}\ \dot{\xi}\ ,\quad \text{sgn}\ \frac{\partial\mathbb{F}}{\partial\underline{\eta}} = \text{sgn}\ \dot{\underline{\eta}}$$

This choice is always possible because the functionals $\mathbb{F}$, $\mathbb{X}$, and $\mathbb{E}$ have the same domain. The entropy production and therefore the entropy itself are not uniquely determined in non-equilibrium because we only have to choose signs which do not fix $\mathbb{F}$. We get

$$(\Sigma - \frac{1}{\Theta}\,\underline{\mu}\cdot\underline{n}^i)(t) \geq 0, \text{ along } \mathfrak{F}$$

from which

$$S_C^{eq} - S_B \geq 0,\quad S_C^{eq} - S_B' \geq 0,$$

and

$$\dot{S} \geq 0,\ \dot{S}' \geq 0, \text{ along}$$

follows.

Because Σ and $\underline{\mu}$ are functionals of the process history

$$\Sigma(t) = \mathbb{\Sigma}(Z^t(\cdot)),\quad \underline{\mu}(t) = \boldsymbol{\mu}(Z^t(\cdot))$$

the expression

$$\Sigma - \frac{1}{\Theta}\,\underline{\mu}\cdot\dot{\underline{n}}^i$$

does not depend on time derivatives of the state variables at time t. Therefore it is independent of the process direction in state space $\mathfrak{z}$ and

$$\Sigma - \frac{1}{\Theta}\,\underline{\mu}\cdot\dot{\underline{n}}^i \geq 0$$

holds generally and not only along $\mathfrak{F}$. So we get

$$\dot{S} \geq \frac{\dot{Q}}{\Theta} + \underline{s}\cdot\dot{\underline{n}}^e,\quad \dot{S}' \geq \frac{\dot{Q}}{T^*} + \underline{s}^*\cdot\dot{\underline{n}}^e\ ,$$

two dissipation inequalities for discrete systems with different non-equilibrium entropies and different external entropy exchange by heat and mass.

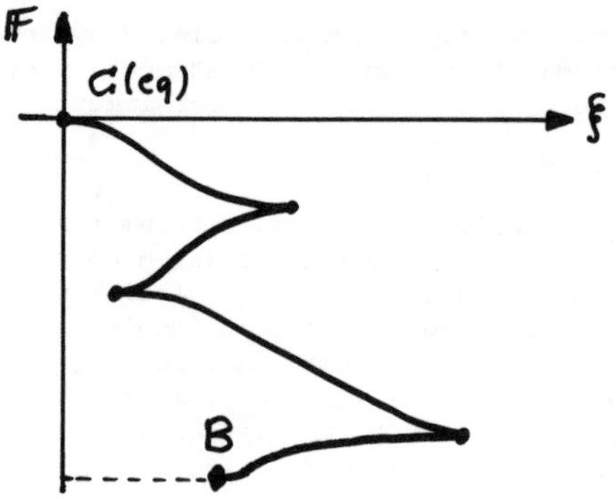

Fig. 3 Example for $\mathbb{F}$ for fixed history and one variable: Along $\mathfrak{F}$ $\dot{\xi}$ changes three times the sign which induces a change of sign of $\partial F/\partial\xi$, too. Because in equilibrium $\mathbb{F} = 0$ is put F^B is negative. The functional $\mathbb{F}$ is not unique in ξ and consists in general of several branches.

3.3 Dissipation inequalities are not unique

We had considered two equations for the time rate of different entropies

$$\Theta\dot{S} = \dot{Q} + \Theta\underline{s}\cdot\dot{\underline{n}}^e - \underline{\mu}\cdot\dot{\underline{n}}^i + \Theta\Sigma,$$

$$T^*\dot{S}' = \dot{Q} + T^*\underline{s}\cdot\dot{\underline{n}}^e - \underline{\mu}_\Theta\cdot\dot{\underline{n}}^i + \Theta\Sigma'.$$

Both the expressions have non-negative entropy production by suitable choice. These entropy productions define the time rate of an entropy. But other entropies can be defined by other entropy production by adding terms which satisfy the embedding axiom:

$$\Theta\dot{S}'' = \dot{Q} + \Theta\underline{s}\cdot\dot{\underline{n}}^e - \underline{\mu}\cdot\dot{\underline{n}}^i + \Theta\Sigma + \Theta R,$$

$$\overset{+}{\oint} R(t)dt \geq 0.$$

If we choose $R = \dot{S}^{eq} - \dot{S}$,

it will satisfy the embedding axiom, and we get $\dot{S}'' = \dot{S}^{eq}$

$$\Theta\dot{S}^{eq} = \dot{Q} + \Theta\underline{s}\cdot\dot{\underline{n}}^e - \underline{\mu}\cdot\dot{\underline{n}}^i + \Theta\tilde{\Sigma}.$$

But $\tilde{\Sigma}$ is not a functional of the process history (as Σ is) because it depends on the time derivatives of the state variables. Therefore its definiteness cannot be concluded and an inequality analoguous to those for $\dot{S}$ and $\dot{S}'$ are not derivable.

However, dissipation inequalities are not unique because different definitions of entropy production and consequently different entropy rates are possible.

3.4 Dissipation inequalities in field formulation

In field formulation the discrete quantities become

$$S := \int_{\mathcal{G}(t)} (\sum_\alpha \rho_\alpha \hat{s}_\alpha) dV,$$

ρ_α = partial density
$\hat{s}_\alpha$ = specific entropy

$$\frac{\dot{Q}}{\Theta} := \oint_{\partial\mathcal{G}(t)} \underline{\Phi}\cdot d\underline{f}, \quad \frac{\dot{Q}}{T^*} := - \oint_{\partial\mathcal{G}(t)} \underline{\Phi}^*\cdot d\underline{f}$$

$\underline{\Phi}$ = entropy flux density

$$\underline{s}\cdot\underline{\dot{n}}^e := - \oint_{\partial\mathcal{G}(t)} (\sum_\alpha \hat{s}_\alpha \underline{m}_\alpha)\cdot d\underline{f},$$

$\underline{m}_\alpha$ = mass flux vector

With the usual technique we get (Muschik/Müller 1983)

$$\frac{\partial}{\partial t} \sum_\alpha \rho_\alpha \hat{s}_\alpha + \nabla\cdot(\underline{\Phi} + \sum_\alpha \rho_\alpha \hat{s}_\alpha \underline{v}_\alpha) \geq 0,$$

and

$$\frac{\partial}{\partial t} \sum_\alpha \rho_\alpha \hat{s}'_\alpha + \nabla\cdot(\underline{\Phi}^* + \sum_\alpha \rho_\alpha \hat{s}'_\alpha \underline{v}_\alpha) \geq 0,$$

where $\underline{v}_\alpha$ is the velocity of the component α. Because entropy flux densities are different in these inequalities also in the field formulation of Thermodynamics no unique dissipation inequality exist. Therefore the old controversy how the entropy flux density looks like is without object because Θ and T^* cannot be both the temperature "measured by a thermometer". The only purpose for introducing dissipation inequalities is to take into account the Second Law with regard to the constitutive equations which must not contradict it.

4. CLOSURE

Starting out with Clausius' inequality in its classical form or in its extended formulation the sign of entropy production in the course of an irreversible process is not determined. It is always possible, but not in a unique manner, to make a material dependent definition of entropy production so that it will be definite. Therefore dissipation inequalities can be formulated which are local in time (and position). But in non-equilibrium no special definition of an entropy is distinguished.

REFERENCES

Haase, R. 1963. Thermodynamik der irreversiblen Prozesse §1.10. Darmstadt: Steinkopff.

Keller, J.U. 1971. Ein Beitrag zur Thermodynamik fluider Systeme. Physica 53:602.

Kestin, J. 1979. A Course in Thermodynamics Vol. II 24.20.7. New York: McGraw Hill.

Kestin, J. 1985. Private communication. Seminar für Theoretische Physik, TU Berlin.

Muschik, W. 1977. Empirical foundation and axiomatic treatment of non-equilibrium temperature. Arch. Rat. Mech. Anal. 66: 379-401.

Muschik, W. 1979. Fundamentals of dissipation inequalities. J. Non-Equilib. Thermodyn. 4:277-294.

Muschik, W. 1980. Entropies of heat conducting discrete or multi-temperature systems with use of non-equilibrium temperatures. Int. J. Engng. Sci. 18: 1399-1410.

Muschik, W. 1981. Thermodynamische Theorien, Überblick und Vergleich. ZAMM 61:T213-219.

Muschik, W. 1983. Extended formulation of the Second Law for open discrete systems. J. Non-Equilib. Thermodyn. 8:219-228.

Muschik, W. & W.H. Müller 1983. Bilanzgleichungen offener mehrkomponentiger Systeme. J. Non-Equilib. Thermodyn. 8: 47-66.

Schottky, W. 1929. Thermodynamik. Erster Teil. §1. Berlin: J. Springer.

5th International Symposium on Continuum Models of Discrete Systems / Nottingham / 14-20 July 1985

On the continuum mechanics of crystal slip

K.S.Havner
North Carolina State University, Raleigh, USA

ABSTRACT: Aspects of continuum modeling of metal plasticity at both the local, individual crystal and macroscopic, polycrystalline levels are addressed. Regarding the first level, fundamental kinematics, stress analysis and structure of constitutive laws are concisely reviewed from the dual perspectives of gross material and underlying lattice. Mathematical considerations and empirical evidence (from single crystal experiments) serve as guides for focus upon one family of slip-system hardening parameters among several alternative sets, leading to a class of hardening theories that have proved useful. Regarding the second level, basic connections between the individual crystal and polycrystalline aggregate at finite strain are traced, corresponding to the model of an array of identical unit cubes in the reference state. These connections include stress and deformation measures, energy, elastic moduli, generalized normality, and the plastic potential.

1 INTRODUCTION

This presentation is intended as an expository, albeit concise, review of theoretical work over the past fifteen years by Hill, Rice and myself on basic mechanics of crystals and polycrystalline metals at finite strain, including reference to relevant experimental studies. One objective is the illustration, in a particular case, of how the interrelation of general mathematical structure with established experimental evidence can lead to useful theory. We shall begin with fundamental kinematics and stress analysis of local crystal behavior and proceed through an ordered succession of topics, including axial load experiments on single crystals, to the establishment of plastic potentials and generalized normality for polycrystalline aggregates.

2 FUNDAMENTAL KINEMATICS AND STRESS ANALYSIS OF CRYSTALS

In Fig. 1 are depicted the basic transformations, or mappings, of crystal and lattice that are our starting point, following Hill and Havner (1982). Here an asterisk (*) refers to quantities related to the underlying lattice, but on a gross continuum scale such that 1* represents, say, 10^1-10^2 atomic spaces. In contrast, "1" corresponds to the differential neighborhood of a crystal material point and is of the order 10^3 atomic spaces, or greater. In the figure, which is limited to coplanar kinematics for simplicity, deformation gradient $\tilde{A}$ maps the reference crystal neighborhood into the "intermediate" configuration shown, with no change in area (volume) and no distortion or rotation of the lattice relative to the reference frame. Deformation gradient A^* maps this configuration into the actual current configuration, with crystal and lattice deforming and rotating as one (carefully constructed to scale in Fig. 1). Note from the included formulas that $\underset{\sim}{h}^*_b$ is an embedded basis vector for the lattice (on the gross continuum scale defined) which equals unit vector $\underset{\sim}{b}_o$ in the reference state, whereas $\underset{\sim}{h}^*_{(n)}$ is a reciprocal basis vector which initially equals unit vector $\underset{\sim}{n}_o$ (i.e., $\underset{\sim}{h}^*_b$, $\underset{\sim}{h}^*_{(n)}$ are respectively covariant and contravariant vectors). Also, ε is the Eulerian strain-rate of the lattice, and ω is lattice spin relative to the reference frame. Γ^s obviously represents the contributions of slip

$$h^*_{\underset{\sim}{(n)}} = \underset{\sim}{n}_o A^{*-1} = \underset{\sim}{n} / \lambda^*_{(n)}$$

$$h^*_{\underset{\sim}{b}} = A^* \underset{\sim}{b}_o = \lambda^*_b \underset{\sim}{b}$$

$$A = A^* \tilde{A} = \frac{\partial \underset{\sim}{x}}{\partial \underset{\sim}{x}_o}$$

$$\Gamma = \Gamma^* + \Gamma^s = \frac{\partial \dot{\underset{\sim}{x}}}{\partial \underset{\sim}{x}}$$

$$\Gamma^* = \varepsilon + \omega$$

$$\Gamma^s d\theta = \sum (\underset{\sim}{b} \otimes \underset{\sim}{n} \, d\gamma)_j \quad , \qquad dA^* = (\Gamma^* d\theta) A^*$$

$$d\tilde{A} = \sum (\underset{\sim}{b}_o \otimes \underset{\sim}{n}_o \, d\tilde{\gamma})_j \, \tilde{A} \quad , \qquad (\lambda^*_b d\tilde{\gamma})_j = (\lambda^*_{(n)} d\gamma)_j$$

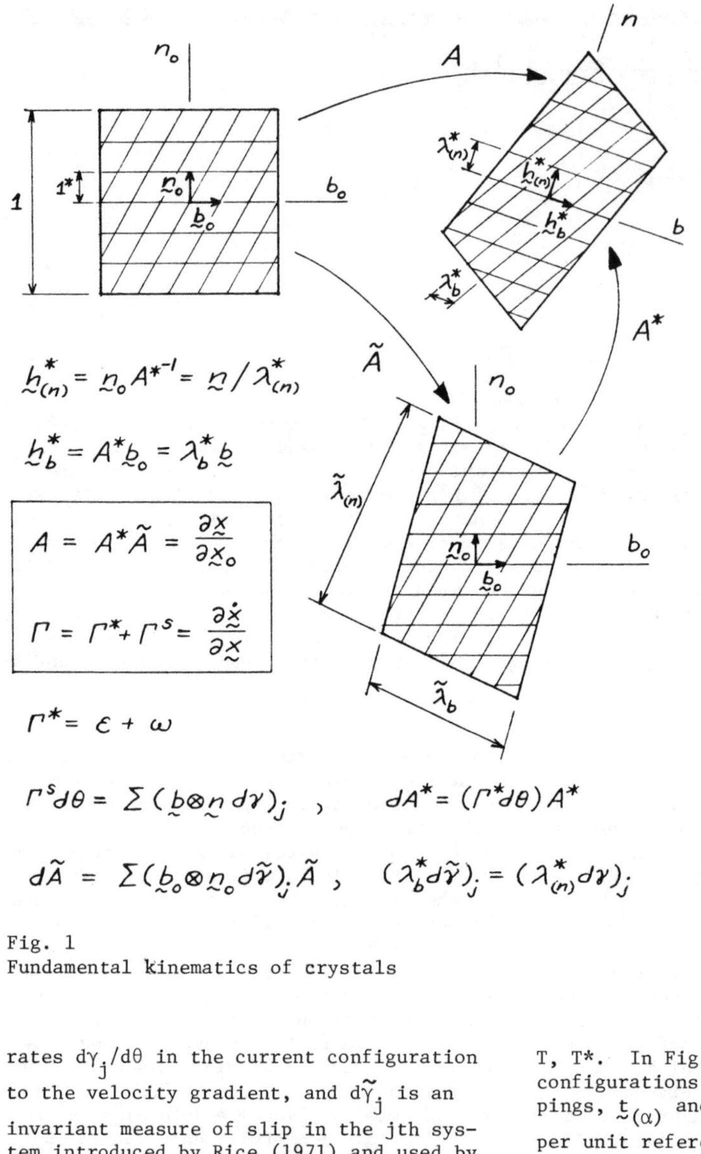

Fig. 1
Fundamental kinematics of crystals

rates $d\gamma_j / d\theta$ in the current configuration to the velocity gradient, and $d\tilde{\gamma}_j$ is an invariant measure of slip in the jth system introduced by Rice (1971) and used by Hill and Rice (1972), Havner (1974), and Hill and Havner (1982). ($\int d\gamma_j$ depends upon lattice strain history but $\int d\tilde{\gamma}_j$ does not.)

We now turn to differential kinematics and basic stress analysis, adopting for the present the Green strain measure for both material and lattice:

$$E = \frac{1}{2}(A^T A - I) \ , \ E^* = \frac{1}{2}(A^{*T}A^* - I) \ . \qquad (1)$$

The associated work-conjugate stress is, of course, contravariant Kirchhoff stress T, T*. In Fig. 2, which encompasses five configurations and three differential mappings, $\underset{\sim}{t}_{(\alpha)}$ and $\underset{\sim}{t}^*_{(\alpha)}$ represent tractions per unit reference area (defined by reference direction α) acting across the respective material and lattice planes in the current configuration. Correspondingly, T, T* comprise the resolutions of the nominal tractions $\underset{\sim}{t}_{(\alpha)}$, $\underset{\sim}{t}^*_{(\alpha)}$ ($\alpha = 1,2,3$) on the embedded basis vectors $\underset{\sim}{h}_\beta$, $\underset{\sim}{h}^*_\beta$ ($\beta = 1,2,3$) of material and lattice, as displayed for coplanar deformation in the figure.

Regarding the differential mappings (Fig. 2), distinctions among ordinary differentials of unstarred quantities, differentials of starred (*) quantities, and starred differentials of unstarred

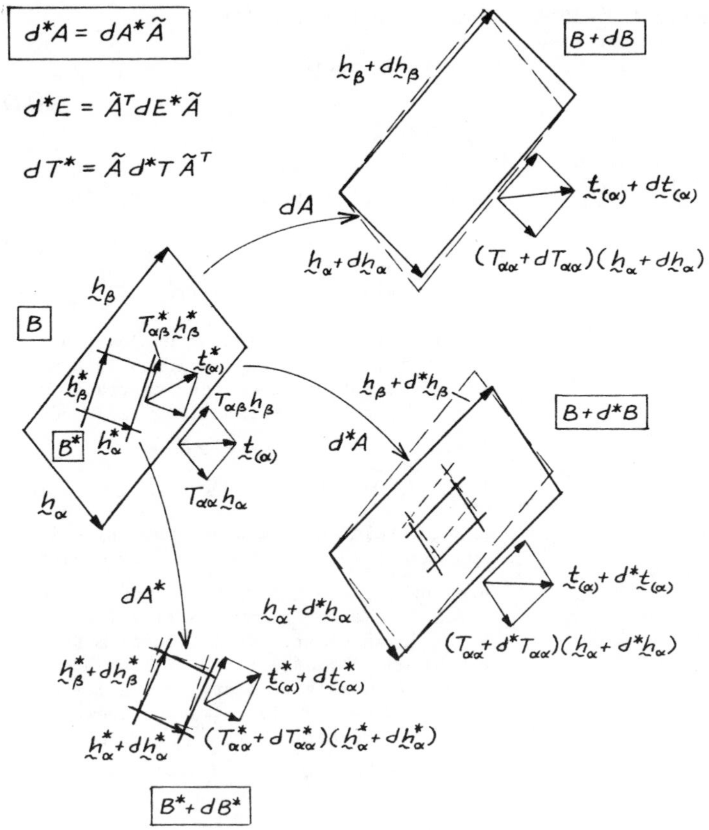

Fig. 2
Differential kinematics and stress analysis

quantities should be carefully noted and comprehended: dA is the ordinary material differential of deformation gradient A (referred to the initial reference state B_o) from current state B of the crystal element; dA* is the differential of deformation gradient A* of the underlying lattice from its current state B*; and d*A is the differential change of an element of lattice volume chosen to have the same configuration as the gross material element in current state B (as if the lattice element had deformed with the material from the reference state). Thus dA takes material state B into B + dB; dA* takes lattice state B* into B* + dB*; and d*A takes a lattice element having the configuration of material state B into configuration B + d*B. Equivalently, B + d*B would be the new state if material and lattice deformed incrementally as one from the current state (ie. the differential change with all $d\tilde{\gamma}_j$ held fixed). Both the stress changes dT, dT*, d*T indicated in Fig. 2 and the strain changes dE, dE*, d*E may be similarly understood. (For simplicity, dA* and d*A are constructed to scale as pure rotations in the figure.) Starred differentials of A, E and T are connected to the differentials dA*, dE* and dT* through the slip transformation $\tilde{A}$ as shown (Hill & Havner, 1982).

2.1 Kinematic and stress approximations

For cubic crystals, which respond isotropically to hydrostatic pressure, we may write

$$A^* = R^*(\lambda^*(p)I + \xi^*) \ , \qquad (2)$$

where R* is a proper orthogonal transformation, $\lambda^*(p) \approx 1$ at moderate pressures, and strain tensor ξ^* is of order 10^{-4}-10^{-3}. Consequently, we have the useful approximations $A^* \approx R^*$, $dA^* \approx (\omega d\theta)R^*$, and

$$A \approx R^*\tilde{A} \tag{3}$$

(at other than very high pressures), ω being the spin of the lattice frame relative to the observer frame. Moreover,

$$\underset{\sim}{h}^*_b \approx R^*\underset{\sim}{b}_o \approx \underset{\sim}{b} \ , \ \underset{\sim}{h}^*_{(n)} \approx \underset{\sim}{n}_o R^{*T} \approx \underset{\sim}{n} \ ,$$

$$\Gamma^s d\theta \approx R^*(d\tilde{A}\tilde{A}^{-1})R^{*T} \ , \ d\tilde{\gamma}_j \approx d\gamma_j \ (\text{all } j) \ , \tag{4}$$

$$\text{and } dA \approx (\Sigma(\underset{\sim}{b}\otimes\underset{\sim}{n}d\gamma)_j + \omega d\theta)A \ . \tag{5}$$

For Green strain E and its conjugate stress T:

$$E \approx \frac{1}{2}(\tilde{A}^T\tilde{A} - I) \ , \ E^* \approx \xi^* \approx 0 \ ,$$

$$T^* = \tilde{A}T\tilde{A}^T \approx R^{*T}\sigma R^* \ , \tag{6}$$

with σ denoting Cauchy stress. Thus

$$dT^* \approx R^{*T}\mathcal{D}^*\sigma R^* \tag{7}$$

$$\text{where } \mathcal{D}^*\sigma \equiv d\sigma + (\sigma\omega + \omega\sigma)d\theta \ , \tag{8}$$

the lattice co-rotational increment in Cauchy stress (whence $dT^* \approx d\sigma$ in a lattice co-rotational reference frame).

3 GENERAL RELATIONSHIPS FOR ARBITRARY MEASURES: PLASTIC POTENTIALS

Consider now arbitrary, symmetric, conjugate stress-strain measures t, e and t*, e* for material and lattice (Hill, 1968; also see Hill, 1978, and Havner, 1982), of which Green strain and contravariant Kirchhoff stress are a particular case. It is assumed that there exists a Green-elastic potential energy function $\phi^*(e^*)$ for the lattice; hence we can write

$$td^*e = t^*de^* = d\phi^* \tag{9}$$

$$\text{with } de - d^*e = \mathcal{K}(dA - d^*A) \tag{10}$$

and $\mathcal{K} = \partial e/\partial A^T$ (following Hill, 1984, who introduced this kinematic transformation tensor for the Green measure). From the equations for A, $\tilde{A}$ and d^*A in Figs. 1 and 2,

$$dA - d^*A = A\Sigma(Bd\tilde{\gamma})_j \ ,$$

$$B_j = \tilde{A}^{-1}(\underset{\sim}{b}_o\otimes\underset{\sim}{n}_o)_j\tilde{A}. \tag{11}$$

Thus, from (10) and (11),

$$de - d^*e = \Sigma(\nu d\tilde{\gamma})_j \ ,$$

$$\nu_j = \mathcal{K}(AB_j) \ . \tag{12}$$

We also define (Hill and Havner, 1982)

$$dt - d^*t = \Sigma(\alpha d\tilde{\gamma})_j \ , \tag{13}$$

$$dw_p = tde - d\phi^* = \Sigma(\tilde{\tau}d\tilde{\gamma})_j \ , \tag{14}$$

the latter of which (differential plastic work per unit reference volume) is a scalar invariant under change in strain measure. There follows from (9), (12) and (14):

$$\tilde{\tau}_j = t\nu_j = (\det A^*)\underset{\sim}{h}^*_{(n)}\sigma\underset{\sim}{h}^*_b \ , \tag{15}$$

the generalized Schmid stress, which is approximately equal to $(\sigma_{nb})_j$ at ordinary pressures (i.e. the resolved Cauchy shear stress). Moreover, by equating invariant works in an elastic variation from states B + dB and B + d*B of material and lattice (Fig. 2), namely $(t + dt)\delta(e + de) = (t + d^*t)\delta(e + d^*e)$, it may be shown that (Hill and Havner, 1982)

$$\alpha_j = -t\frac{\partial\nu_j}{\partial e} \tag{16}$$

in which the gradient is taken at fixed slips (i.e. constant $\tilde{A}$).

The plastic potential equations for local crystal behavior now can be established, following Hill and Havner (1982). We define

$$d^pt = \mathcal{L}de - dt \tag{17}$$

the plastic decrement in stress after a (virtual) strain cycle, and

$$d^pe = de - \mathcal{M}dt \ , \tag{18}$$

the plastic increment in strain after a (virtual) stress cycle, where

$$\mathcal{L} = \frac{\partial^2\phi}{\partial e\partial e} \ , \ \phi(e,\tilde{A}) = \phi^*(e^*) \tag{19}$$

and

$$\mathcal{M} = \mathcal{L}^{-1} = \frac{\partial e}{\partial t} \tag{20}$$

at fixed slips. Then, from (12), (13), (17), (18) and $d^*t = \mathcal{L}d^*e$ (which follows from (9) and (19)) we have

$$d^P t = \Sigma(\Lambda d\tilde{\gamma})_j \ , \ d^P e = \Sigma(M d\tilde{\gamma})_j \ , \tag{21}$$

with

$$\Lambda_j = \mathcal{L}\nu_j - \alpha_j \ , \ M_j = \nu_j - \mathcal{M}\alpha_j \ . \tag{22}$$

From $\tilde{\tau}_j = t\nu_j$ and (16), (20) and (22) we find

$$\frac{\partial\tilde{\tau}_j}{\partial e} = \Lambda_j \ , \ \frac{\partial\tilde{\tau}_j}{\partial t} = M_j \ . \tag{23}$$

Thus

$$d^P t = \frac{\partial}{\partial e}\Sigma(\tilde{\tau}d\tilde{\gamma})_j \ , \ d^P e = \frac{\partial}{\partial t}\Sigma(\tilde{\tau}d\tilde{\gamma})_j \ , \tag{24}$$

the crystal plastic potential equations (Hill and Havner, 1982).

4 CRITICAL STRENGTHS AND HARDENING LAWS: NORMALITY AND CONSTITUTIVE INEQUALITIES

Let

$$\tilde{\tau}_j(T,\tilde{A}) \equiv \tilde{\tau}_j(E^*) = \tilde{\tau}_j^c \ , \tag{25}$$

$j = 1,\cdots,n$, define the n critical slip systems in which incremental slips $d\tilde{\gamma}_j > 0$ are possible from current state B, B* of material and lattice, with the material parameter $\tilde{\tau}_j^c$ called the critical strength of the jth system. (Note that $\tilde{\tau}_j^c \approx (\sigma_{nb})_j$, the classical Schmid stress of the critical system.) We now make the basic assumption that the critical strengths $\tilde{\tau}_k^c$, $k = 1,\cdots,N$ (the total number of slip systems), are independent of lattice strain, hence of current stress T*. Thus (Hill and Havner, 1982),

$$d\tilde{\tau}_k^c = \sum_j H_{kj}d\tilde{\gamma}_j \ , \tag{26}$$

$j = 1,\cdots,n$; $k = 1,\cdots,N$; the general hardening law for the N crystallographic slip systems. The critical system inequalities then may be written

$$d\tilde{\tau}_k \leq \sum_j H_{kj}d\tilde{\gamma}_j \ , \ j,k = 1,\cdots,n \ , \tag{27}$$

with the equality holding in each active system ($d\tilde{\gamma}_k > 0$). The hardening moduli H_{kj} are in general functionals of the slip-history of all systems (not just those momentarily critical) and functions of the current stress state relative to the lattice axes.

Consider now alternative forms of the critical slip-system inequalities. As $\tilde{\tau}_k$ is a scalar invariant, we may write

$$d\tilde{\tau}_k = \frac{\partial\tilde{\tau}_k}{\partial e}d^*e = \frac{\partial\tilde{\tau}_k}{\partial t}d^*t \ , \tag{28}$$

the gradients again taken at fixed slips. Consequently, using (12), (13), (15), (22) and (28), we can express (27) in the equivalent forms (Hill and Havner, 1982; also see Hill & Rice, 1972):

$$d(\nu_k t) \leq \sum_j H_{kj}d\tilde{\gamma}_j \ , \ d(\nu_k t) \approx (\underset{\sim}{n}\mathcal{D}^*\sigma\underset{\sim}{b})_k \ , \tag{29}$$

$$\Lambda_k de \leq \sum_j g_{kj}d\tilde{\gamma}_j \ , \ g_{kj} = H_{kj} + \Lambda_k\nu_j \ , \tag{30}$$

$$M_k dt \leq \sum_j h_{kj}d\tilde{\gamma}_j \ , \ h_{kj} = H_{kj} + M_k\alpha_j \ , \tag{31}$$

$j,k = 1,\cdots,n$.

For increments δe, δt producing purely elastic response (i.e. $\delta t = \mathcal{L}\delta e$), we have from (23), (30) and (31)

$$\frac{\partial\tilde{\tau}_k}{\partial e}\delta e \leq 0 \ , \ \frac{\partial\tilde{\tau}_k}{\partial t}\delta t \leq 0 \ , \tag{32}$$

$k = 1,\cdots,n$; whence, upon multiplying each of (32) by the respective nonnegative $d\tilde{\gamma}_k$ and summing, there follows from (24)

$$d^P t\delta e \leq 0 \ , \ d^P e\delta t \leq 0 \ . \tag{33}$$

These are the normality laws in e- and t-space respectively, which are invariant under change in strain measure (Hill, 1972; Hill and Rice, 1972; also see Havner, 1982). They are a consequence of the definition (25) of critical slip systems and the assumption (26) that critical strengths $\tilde{\tau}_k^c$ are independent of lattice strain. Moreover, they are equivalent inequalities from (20)-(22) and the definition of δe, δt (i.e. $d^P t\delta e = d^P e\delta t$).

In contrast to (33), for de, dt producing one or more $d\tilde{\gamma}_k > 0$, the two scalar products are not equal and only the first is invariant under change in strain measure. That is,

$$d^P t de = \Sigma\Sigma g_{kj}d\tilde{\gamma}_k d\tilde{\gamma}_j \quad \text{(invariant)},$$
$$dt d^P e = \Sigma\Sigma h_{kj}d\tilde{\gamma}_k d\tilde{\gamma}_j \quad \text{(not invariant)}. \tag{34}$$

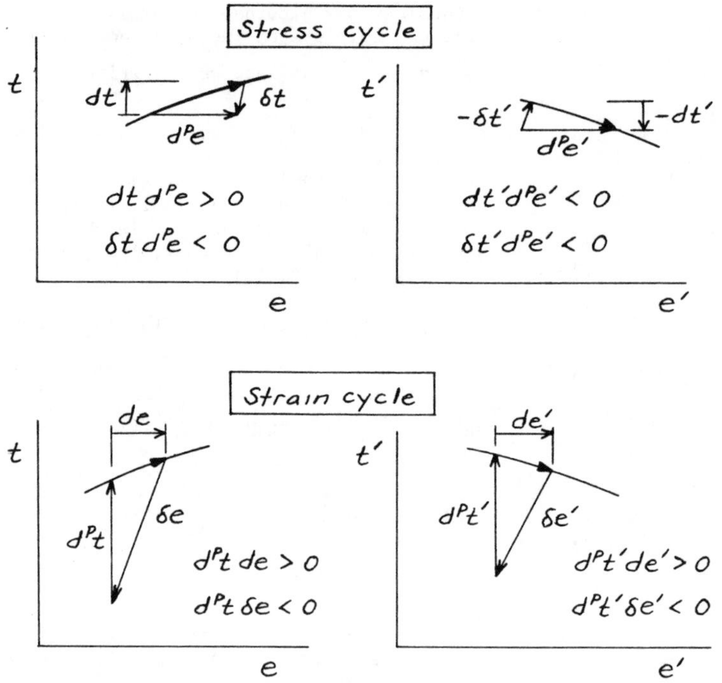

Fig. 3
Differential cycles in uniaxial loading

Parameters g_{kj} (as moduli H_{kj}) are therefore invariant, but parameters h_{kj} are not (Hill and Rice, 1972; Hill and Havner, 1982). Nevertheless, these latter parameters (for a particular measure) will prove a useful focus of emphasis, as we shall see.

The respective invariant and non-invariant quality of the two scalar products (34) may be more readily understood by reference to Fig. 3, which illustrates differential cycles in uniaxial loading for two different strain measures, one having a rising and the other a falling stress-strain curve. We assume there always will be a rising curve for any metal crystal corresponding to the logarithmic measure (i.e. the "true" stress-strain curve). Accordingly, the cycles in the figure suggest that it may be reasonable to postulate the following general constitutive inequalities for local crystal behavior during incremental slip:

$$d^ptde > 0 \quad \text{(invariant)},$$
$$dt_o d^pe_o > 0 \quad \text{(logarithmic measure)}. \tag{35}$$

From (34), the first of these inequalities will be satisfied if the matrix of parameters g_{kj} is positive-definite over critical systems. This also is a sufficient condition to ensure that slip-system inequalities (30) yield a unique solution for the $d\tilde{\gamma}_j$ for a prescribed increment in strain (as proved by Hill and Rice, 1972, with the current state as reference). Again from (34), the second of inequalities (35) will be satisfied if the matrix of parameters h^o_{kj} (logarithmic measure) is positive-definite, or if the quadratic form $(34)_2$ is positive-definite in that measure for all non-negative $d\tilde{\gamma}_k$ (which is less restrictive). From (31), positive definite h_{kj} ensure a unique set of incremental slips for a given dt (Hill & Rice, 1972). Physically meaningful progress beyond these mathematical points requires the consideration of experimental information from axial load tests of single crystals, to which we now turn.

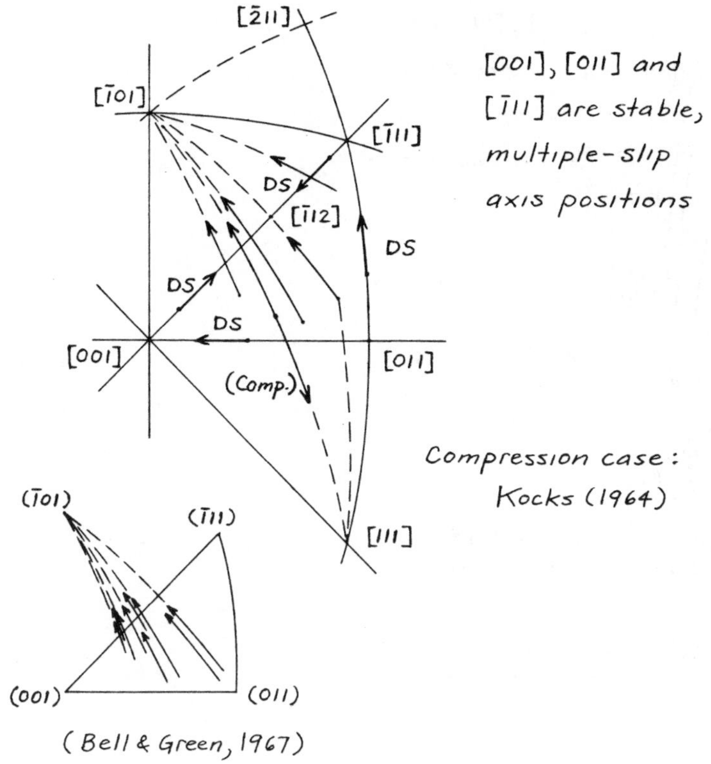

Fig. 4
Axis rotations in f.c.c. crystals in tension

5 AXIAL LOAD EXPERIMENTS ON F.C.C. AND B.C.C. CRYSTALS

In Fig. 4 is shown a partial [001] stereographic projection for f.c.c. crystals. The stereographic triangle within which the initial loading axis positions lie corresponds to slip system [$\bar{1}$01](111) (in Miller index notation) as the most highly stressed, hence active one. (The exceptional positions are the double-slip orientations on the boundaries, for each of which both this system and the respective adjacent one are equally stressed.) All indicated axis rotations correspond to single slip in primary system [$\bar{1}$01](111), save for those designated "DS" for double-slip. Except for the compression case, taken from a specific experiment on aluminum crystals by Kocks (1964), the rotations shown in the main projection are merely representations of what has commonly been found by experimentalists: namely, the tensile axis "overshoots" the crystallographic symmetry line, and single slip is perpetuated during finite uniform deformation. In the lower diagram, however, the rotation curves are smooth means through successive axis positions of aluminum crystals determined experimentally by Bell and Green (1967). Those authors also analyzed 152 other f.c.c. single crystal experiments (reported in the literature over a 40-year span) in which rotations reaching the symmetry line were measured. They found that in almost all cases "the main features of the resolved deformation are given by the initial primary plane behaviour" (Bell and Green, 1967). It also is noteworthy that, in an earlier study of this phenomenon, Piercy, Cahn and Cottrell (1955) concluded that "overshooting should be regarded as a normal characteristic of the plastic deformation of crystals, and that the intriguing feature is that it ever should be absent."

Implicit in the phenomenon of overshooting, as represented in Fig. 4, is the concept that the latent slip-system into whose

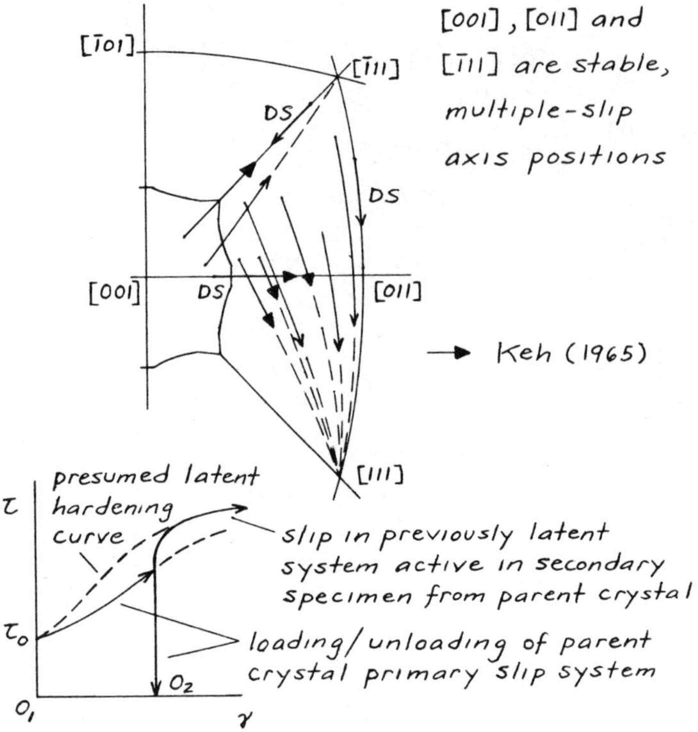

Fig. 5
Axis rotations in b.c.c. crystals in tension

stereographic triangle the axis rotates, as single slip proceeds, must harden substantially more than the active system. If this were not the case, one of the following (which are in fact quite rare) would likely be observed. (1) Equal double-slip would begin when the axis reached the symmetry line, and the axis rotation would abruptly turn toward the vector sum of the separate limiting directions of the two systems (e.g. [$\bar{1}$12] in a tensile test), as originally hypothesized by Taylor in his 1923 Bakerian Lecture to the Royal Society (Taylor and Elam, 1923). (2) The axis would "undershoot" the symmetry line and begin rotating toward some positive linear combination of these limiting positions, as would be predicted by theories that adopt positive-definite matrices of hardening moduli H_{kj} (see the discussion in Havner, 1981).

A similar situation regarding overshooting and implications for latent hardening obtains in the case of b.c.c. crystals, for which a partial [001] stereographic projection is shown in Fig. 5. Here the projected region is subdivided on the basis of slip in one of the <111> family of directions on any of the {110}, {112} or {123} families of planes. (For a further subdivision of the region that distinguishes among the planes, see Havner and Baker, 1979, or Havner, 1982.) The small region in the upper right quadrant of the projection corresponds to initial axis positions with [$\bar{1}$11] as slip direction. The partial stereographic triangle encompassing the balance of axis positions corresponds to slip direction [111]. The arrows with solid heads depict actual rotations determined from tensile experiments on iron crystals by Keh (1965). The other indicated rotations are merely representative of the overshooting that has been observed in various axial load tests of b.c.c. crystals, beginning with the classic experiments of Fahrenhorst and Schmid (1932). Note, in particular, that Keh's measured rotation along the [001]-[$\bar{1}$11] symmetry line is in the opposite direction from that corresponding to double slip (DS) in the (subsequently) most highly stressed systems (the adjacent, partial stereographic triangles, for which the resultant limiting direction is [001]). Implicit in Fig. 5, as in Fig. 4, is the necessarily greater latent hardening of those slip systems that become more highly stressed than the (original) active system as the axis rotates in single slip.

The apparent hardening of latent systems during single slip has been determined experimentally by a number of investigators (notably: Kocks, 1964; Ramaswami, Kocks and Chalmers, 1965; Jackson and Basinski, 1967; Franciosi, Berveiller and Zaoui, 1980). Typically, as illustrated schematically in the lower diagram in Fig. 5, a crystal is unloaded from a plastically deformed state (that resulted from single slip) without further slipping. Then, a differently oriented specimen taken from the deformed "parent" crystal is reloaded until, desirably, measureable slip is initiated in a system other than that one originally active. The unknown value of the critical strength developed in this previously latent system, as a result of the original deformation of the parent crystal, is taken to equal the respective resolved (Cauchy) shear stress on this system at the onset of significant slip in the reloaded specimen. In the overwhelming majority of such experimental studies, latent hardening determinations are consistent with the overshooting phenomenon. Moreover, most latent systems that have been investigated harden more than the initially active system in finite single-slip experiments (cf. Jackson and Basinski, 1967).

In contrast to single slip, initial loading axis positions in 4, 6 or 8-fold symmetry (Figs. 4 and 5) are generally stable, corresponding to equal slip in all critical systems (see, for example, Keh, 1965, and Vorbrugg, Goetting and Schwink, 1971). Moreover, the hardening of all slip systems for crystals in such high symmetry, multiple-slip modes may be essentially isotropic, based upon experiments on aluminum crystals conducted by Y. Nakada and described by Wonsiewicz and Chin (1970).

The qualitative features of single crystal, axial load experiments may be summarized as follows.

Single-slip orientations:

(a) overshooting phenomenon,

(b) greater hardening of latent systems (in general) than active system.

High-symmetry orientations:

(c) axis stability and equal multiple slip,

(d) approximately equal hardening of all systems (based upon limited data, however).

Accordingly, it seems reasonable to draw the following conclusions:

(1) hardening moduli H_{kj} depend explicitly upon stress state relative to lattice axes;

(2) anisotropy of hardening depends upon rotation of material and/or stress state relative to the lattice.

(In single-slip, material and axial stress state rotate separately with respect to the crystal lattice. In 6- and 8-fold multiple-slip, for which isotropic hardening has been found, neither rotates.)

Of the various recognized hardening theories that had been introduced in the literature through 1975: all permit (c) above and adopt $H_{kj} = H_{jk}$; only Taylor's classical isotropic rule gives (d); only a 2-parameter empirical rule gives either (a) or (b) (with several theories predicting the opposite); none incorporate either (1) or (2) above.

6 FORMULATION OF A HARDENING RULE

We now shall see how the empirical knowledge reviewed in Section 5 can be merged successfully with the mathematical structure of Section 4 to obtain a useful hardening theory that encompasses both qualitative features of experiment and postulated inequalities (35). To this end, we first express the hardening law (26) in alternative forms involving parameters g_{kj}, h_{kj}, and plastic strain-rate and spin. These latter are given by

$$D^p = \Gamma_{sym} - \varepsilon = \Sigma(N \frac{d\gamma}{d\theta})_j \; , \; N_j = sym(\underset{\sim}{b} \otimes \underset{\sim}{n})_j \; ,$$

$$\Omega^p = \Gamma_{skw} - \omega = \Sigma(\Omega \frac{d\gamma}{d\theta})_j \; , \; \Omega_j = skw(\underset{\sim}{b} \otimes \underset{\sim}{n})_j \; . \quad (36)$$

With the current state as reference, and to first order in infinitesimal lattice strain, it may be shown from the equations of Sections 2, 3 and 4 that:

$$d\tau_k^c = \sum_j H_{kj} d\gamma_j \; , \; j=1,\cdots,n \; ; \; k=1,\cdots,N \; ,$$

$$d\tau_k^c = \sum_j g_{kj} d\gamma_j - (N_k \mathcal{L}_o - 2\sigma\Omega_k) D^p d\theta \; , \quad (37)$$

$$d\tau_k^c = \Sigma h_{kj}^o d\gamma_j - 2N_k \sigma \Omega^p d\theta \; ,$$

with τ_k^c now denoting critical strength and suffix o designating the logarithmic measure as before.

Consider now the important task of defining the various sets of parameters in (37) in a manner consistent with the summarized experimental features (a) through (d) and the conclusions (1) and (2) of Section 5. Taylor's isotropic hardening rule,

$$H_{kj} = H \ , \ \text{all } k,j \ ; \ d\tau_k^c = H\Sigma d\gamma_j \ , \qquad (38)$$

is consistent only with (c) and (d), as already noted. It does not appear likely that any other direct specification of parameters H_{kj} according to a relatively simple rule can capture all the experimental features.

Turning to the g-moduli, we observe from $(37)_2$ that these are necessarily very large, of the same order of magnitude as elastic moduli, in order to offset the term $N_k \mathcal{L}_o D^p d\theta$ and give reasonable-size values for $d\tau_k^c$. There is no explicit dependence on relative rotation of material and lattice in $(37)_2$, and it is not apparent how any positive-definite set of g_{kj} could be directly defined by simple formulas that would encompass the qualitative empirical evidence.

Lastly, for the parameters h_{kj}^o, which I have named the "effective slip-system hardening moduli" (Havner, 1977), we see that the plastic spin directly appears in $(37)_3$ and is multiplied (tensor product) by stress state σ to give the explicit dependencies concluded in Section 5. Thus, this form of the general hardening law, expressed for the logarithmic measure, appears to be the most promising form within which theories capturing the essence of single and multiple-slip experiments could be posed. Moreover, mathematical considerations regarding potential functions and incremental boundary value problems have led to the proposal (Havner, 1977; Havner and Shalaby, 1977) that these effective hardening moduli should be symmetric (i.e. $h_{kj}^o = h_{jk}^o$). The simplest possible such theory is then (Havner and Shalaby, 1977):

$$h_{kj}^o = h \ , \ \text{all } k,j \ ; \ H_{kj} = h - 2N_k\sigma\Omega_j . \qquad (39)$$

Consequently,

$$d\tau_k^c = h\Sigma d\gamma_j - 2N_k\sigma\Omega^p d\theta \ , \qquad (40)$$

which I have called the "simple theory" of rotation dependent anisotropic hardening.

The "simple theory" (40) encompasses conclusions (1) and (2) (Section 5) and also features (c) and (d) from high-symmetry (4, 6 or 8-fold) multiple slip, corresponding to which it effectively reduces to isotropic hardening (as $\Omega^p = 0$). The theory also satisfies postulated inequality $(35)_2$ and, as one may reasonably expect $\mathcal{L}_o$ to be positive-definite, it satisfies $(35)_1$ as well. Whether (40) also satisfies the major features (a) and (b) of overshooting and greater latent than active hardening in single slip, for both f.c.c. and b.c.c. crystals, is not obvious, however, and requires analysis. Just such extensive analyses have been carried out by Havner and Shalaby (1977, 1978), Havner, Baker and Vause (1979), Havner and Baker (1979), and Vause and Havner (1979), who have established that the "simple theory" is indeed a universal theory of overshooting in f.c.c. and b.c.c. crystals. Thus, the theory encompasses all the experimental features and conclusions of Section 5 and satisfies the postulated constitutive inequalities of Section 4.

7 MACROSCOPICALLY UNIFORM DEFORMATION OF CRYSTALLINE AGGREGATES: THE AVERAGING THEOREM

We now turn to the macroscopic level and investigate analytical connections between single crystal and polycrystalline metal behavior. Consider the aggregate model of an extended array of identical, polycrystal "unit cubes" (say, 1 mm scale), as introduced in Havner (1971). Under conditions of quasi-static, macroscopically uniform finite deformation we have (Havner, 1974):

$$\underset{\sim}{x}(a_i^+) - \underset{\sim}{x}(a_i^-) = \overset{i}{\underset{\sim}{c}} \ , \ \underset{\sim N}{t}(a_i^+) = -\underset{\sim N}{t}(a_i^-) \ , \qquad (41)$$

where a_i^+, a_i^- are the pair of element faces that were normal to axis x_i in the reference state (refer to Fig. 6), and $\underset{\sim}{x}(a_i^+)$, $\underset{\sim N}{t}(a_i^+)$ indicate point-dependence over the respective face (with $\underset{\sim N}{t}$ the nominal traction vector). In Fig. 6, the macroscopic element dimension $\bar{l}$ is of order 1 mm, consistent with typical grain sizes of order 10^{-2}-10^{-1} mm in polycrystalline metals. The distortion and rotation of an individual crystal grain also are depicted in the figure, together with the rotation of its underlying lattice. (The intersecting families of lines in the deformed crystal represent slip lines.)

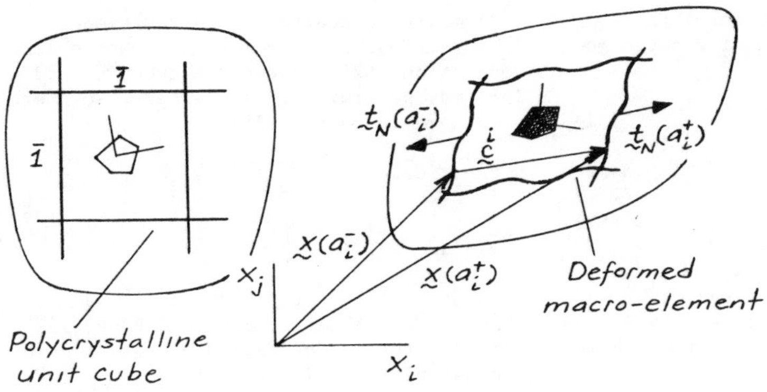

Fig. 6
Macroscopically uniform crystalline aggregate model

Let us define

$$\langle\cdots\rangle = \int_{B_o} (\cdots)dV_o \;,\; \int dV_o = 1 \tag{42}$$

(on scale 1 mm^3). From (41)-(42), the equations of local equilibrium, and the Green-Gauss transformation we obtain (Havner, 1974)

$$\langle A_{ik}\rangle = \overset{i}{c}_k \;,\; \langle N_{kj}\rangle = \int_{a_o} t_j^N da_k^+ \;, \tag{43}$$

in which $N = TA^T$, the nominal stress (i.e. $d\underset{\sim}{a}_o N = \underset{\sim}{t}_N da_o$). Then, from (41)-(43) and another application of the Green-Gauss transformation, there follows (Havner, 1974, 1982)

$$\langle AN\rangle = \langle A\rangle \langle N\rangle \tag{44}$$

for both scalar and tensor products. This is the averaging theorem at finite strain, first given by Hill (1972) and proved for the alternative boundary conditions of a macro-element embedded in either a uniformly strained or a uniformly stressed medium. (The most general conditions under which (44) holds may be found in Hill, 1984).

It should be clear from both (43) and (44) (with reference to Fig. 6), that appropriate macroscopic measures $\bar{A}$, $\bar{N}$ of deformation gradient and nominal stress are simply the volume averages of their local fields:

$$\bar{A} = \langle A\rangle \;,\; \bar{N} = \langle N\rangle \;. \tag{45}$$

Moreover, the averaging theorem may be extended to a work·equality and to the scalar product of arbitrary differential changes dN, dA (Hill, 1972; also see Havner, 1982):

$$\langle NdA\rangle = \langle N\rangle \langle dA\rangle = \bar{N}d\bar{A} \;, \tag{46}$$

$$\langle dNdA\rangle = \langle dN\rangle \langle dA\rangle = d\bar{N}d\bar{A} \tag{47}$$

(both scalar products). We also can write, for arbitrary, work-conjugate stress and strain measures t, e (Hill, 1972; Havner, 1982):

$$\langle tde\rangle = \langle NdA\rangle = \bar{t}d\bar{e} \;. \tag{48}$$

However, $\bar{t}$, $\bar{e}$ will not be the unweighted volume averages of t, e.

8 MACROSCOPIC PLASTIC POTENTIALS AND NORMALITY

From $de = \mathcal{K}\, dA$ and $tde = NdA$ (scalar product) we obtain

$$N = \mathcal{K}^T t \;,\; dN = \mathcal{K}^T dt + \mathcal{J} dA \;,$$
$$\mathcal{J} = \frac{t\partial^2 e}{\partial A^T \partial A^T} = \mathcal{J}^T \;. \tag{49}$$

Then, from (24) and (49), there follow the local plastic potential equations

$$d^P N = \mathcal{C}\, dA - dN = \frac{\partial}{\partial A^T} \Sigma(\tilde{\tau} d\tilde{\gamma})_j \;,$$
$$d^P A = dA - \mathcal{C}^{-1} dN = \frac{\partial}{\partial N^T} \Sigma(\tilde{\tau} d\tilde{\gamma})_j \tag{50}$$

in the respective nine-dimensional A- and N-spaces, with each gradient taken at fixed slips and $\mathcal{C}$ (assumed invertible) the fourth-order tensor of nominal elastic moduli of the crystal:

$$\mathcal{C} = \partial N/\partial A^T = \mathcal{K}^T \mathcal{L} \mathcal{K} + \mathcal{J} . \tag{51}$$

We also have

$$d^pN = \mathcal{K}^T d^pt , \tag{52}$$

hence d^pN, d^pt transform as N, t.

Consider now the influence tensor $\mathcal{A} = \partial A/\partial \bar{A}^T$ of elastic heterogeneity in the deformed "unit cube," in terms of which we can write (Hill, 1984)

$$\delta A = \mathcal{A} d\langle A\rangle , \quad \langle \delta A\rangle = d\bar{A} , \tag{53}$$

where δA represents (imagined) elastic response to $d\bar{A}$, with corresponding nominal stress change $\mathcal{C}\,\delta A$. From (45) and (53), the resulting macroscopic nominal stress change is determined as (Hill, 1984)

$$\delta \bar{N} = \bar{\mathcal{C}}\, \delta \bar{A} , \quad \bar{\mathcal{C}} = \langle \mathcal{C} \mathcal{A} \rangle . \tag{54}$$

Thus, the macroscopic tensor $\bar{\mathcal{C}}$ of nominal elastic moduli is the weighted volume average (by the influence tensor $\mathcal{A}$) of the local tensor $\mathcal{C}$.

To establish macroscopic plastic potential and normality equations, we first define plastic decrement in $\bar{N}$ and plastic increment in $\bar{A}$ in an analogous way to (50) (assuming the invertibility of $\bar{\mathcal{C}}$):

$$d^p\bar{N} = \bar{\mathcal{C}}\, d\bar{A} - d\bar{N}, \quad d^p\bar{A} = d\bar{A} - \bar{\mathcal{C}}^{-1} d\bar{N} . \tag{55}$$

Making use of (45), (47), (53) and (54), one obtains (Hill, 1984)

$$d^p\bar{N} = \langle \mathcal{A}^T d^pN \rangle , \tag{56}$$

and we see that the plastic decrement in macroscopic nominal stress is the elastically-weighted average of the local plastic decrement. Then, from the plastic potential equation (50) for d^pN and the chain rule of partial differentiation, we have (Havner, 1985)

$$d^p\bar{N} = \frac{\partial}{\partial \bar{A}^T} \langle \Sigma(\tilde{\tau} d\tilde{\gamma})_j \rangle . \tag{57}$$

Following similar lines, it can be concluded that

$$d^p\bar{A} = \frac{\partial}{\partial \bar{N}^T} \langle \Sigma(\tilde{\tau} d\tilde{\gamma})_j \rangle . \tag{58}$$

Thus, $\langle \Sigma(\tilde{\tau} d\tilde{\gamma})_j \rangle$, the volume average of the local increment in energy dissipation, is a macroscopic plastic potential for $d^p\bar{N}$ in $\bar{A}$-space and $d^p\bar{A}$ in $\bar{N}$-space. Furthermore, (57) may be converted to an equation for arbitrary macroscopic measures through the connection (52) between d^pN and d^pt, and (58) may be similarly converted. We therefore have (Havner, 1985)

$$d^p\bar{t} = \bar{\mathcal{L}} d\bar{e} - d\bar{t} = \frac{\partial}{\partial \bar{e}} \langle \Sigma(\tilde{\tau} d\tilde{\gamma})_j \rangle ,$$

$$d^p\bar{e} = d\bar{e} - \bar{\mathcal{M}} d\bar{t} = \frac{\partial}{\partial \bar{t}} \langle \Sigma(\tilde{\tau} d\tilde{\gamma})_j \rangle , \tag{59}$$

with $\bar{\mathcal{L}} = \partial \bar{t}/\partial \bar{e}$ (at fixed slips) and $\bar{\mathcal{M}} = \bar{\mathcal{L}}^{-1}$. (Also see Hill and Rice, 1973, for a different derivation of equations of macroscopic plastic potentials.)

The generalized normality conditions in macroscopic variables can now be given. From the invariance of bilinear form (33) (Hill, 1972) we may write

$$d^pN\delta A \le 0 , \quad d^pA\delta N \le 0 , \tag{60}$$

with $\delta A = \mathcal{A}\, \delta \bar{A}$ for an incremental change $\delta \bar{A}$ that produces overall elastic response of the polycrystalline element from its current deformed state. Then, from (47), (53), (55), (56) and the above, the dual normality conditions in $\bar{A}$ and $\bar{N}$-space readily follow:

$$d^p\bar{N}\delta\bar{A} \le 0 , \quad d^p\bar{A}\delta\bar{N} \le 0 . \tag{61}$$

(Note that the two left-hand sides are equal.) These may be converted to equations in arbitrary macroscopic measures $\bar{t}$, $\bar{e}$ from the invariance of the bilinear form (Hill, 1972; Havner, 1982).

ACKNOWLEDGEMENT

The work upon which this lecture is based was supported by the U.S. National Science Foundation, Solid Mechanics Program. The excellent typing was done by Mrs. Annette Maynard.

REFERENCES

Bell, J.F. & R.E. Green, Jr. 1967. An experimental study of the double slip deformation hypothesis for face-centred cubic single crystals. Phil.Mag.Ser.8. 15:469-476.

Fahrenhorst, W. & E. Schmid 1932. Über die plastiche Dehnung von α-Eisenkristallen. Z.Phys. 78:383-394.

Franciosi, P., M. Berveiller & A. Zaoui 1980. Latent hardening in copper and aluminium single crystals. Acta Metall. 28:273-283.

Havner, K.S. 1971. A discrete model for the prediction of subsequent yield surfaces in polycrystalline plasticity. Int.J.Solids Struct. 7:719-730.
Havner, K.S. 1974. Aspects of theoretical plasticity at finite deformation and large pressure. Z.angew.Math.Phys. 25:765-781.
Havner, K.S. 1977. On uniqueness criteria and minimum principles for crystalline solids at finite strain. Acta Mech. 28:139-151.
Havner, K.S. 1981. A theoretical analysis of finitely deforming f.c.c. crystals in the sixfold symmetry position. Proc.R. Soc.Lond.A 378:329-349.
Havner, K.S. 1982. The theory of finite plastic deformation of crystalline solids. In H.G. Hopkins & M.J. Sewell (eds.), Mechanics of solids, the Rodney Hill 60th anniversary volume, p. 265-302. Oxford: Pergamon.
Havner, K.S. 1985. Fundamental considerations in micromechanical modeling of polycrystalline metals at finite strain. In P. Germain & J. Zarka (eds.), Proc., Int. symp. on physical basis and modelling of finite deformation of aggregates. In press.
Havner, K.S. & G.S Baker 1979. Theoretical latent hardening in crystals - II. Bcc crystals in tension and compression. J.Mech.Phys.Solids 27:285-314.
Havner, K.S., G.S. Baker & R.F. Vause 1979. Theoretical latent hardening in crystals - I. General equations for tension and compression with application to f.c.c. crystals in tension. J.Mech.Phys. Solids 27:33-50.
Havner, K.S. & A.H. Shalaby 1977. A simple mathematical theory of finite distortional latent hardening in single crystals. Proc.R.Soc.Lond.A 358:47-70.
Havner, K.S. & A.H. Shalaby 1978. Further investigation of a new hardening law in crystal plasticity. J.appl.Mech. 45:500-506.
Hill, R. 1968. On constitutive inequalities for simple materials - I.,II.J.Mech.Phys. Solids 16:229-242, 315-322.
Hill, R. 1972. On constitutive macro-variables for heterogeneous solids at finite strain. Proc.R.Soy.Lond.A 326:131-147.
Hill, R. 1978. Aspects of invariance in solid mechanics. In C.-S. Yih (ed.), Advances in applied mechanics, p. 1-75. New York: Academic Press.
Hill, R. 1984. On macroscopic effects of heterogeneity in elastoplastic media at finite strain. Math.Proc.Camb.Phil.Soc. 95:481-494.
Hill, R. & K.S. Havner 1982. Perspectives in the mechanics of elastoplastic crystals. J.Mech.Phys.Solids 30:5-22.
Hill, R. & J.R. Rice 1972. Constitutive analysis of elastic-plastic crystals at arbitrary strain. J.Mech.Phys.Solids 20:401-413.
Hill, R. & J.R. Rice 1973. Elastic potentials and the structure of inelastic constitutive laws. SIAM J.appl.Math. 25:448-461.
Jackson, P.J. & Z.S. Basinski 1967. Latent hardening and the flow stress in copper single crystals. Can.J.Phys. 45:707-735.
Keh, A.S. 1965. Work hardening and deformation sub-structure in iron single crystals deformed in tension at 298°K. Phil. Mag.Ser.8. 12:9-30.
Kocks, U.F. 1964. Latent hardening and secondary slip in aluminum and silver. Trans.Met.Soc.AIME 230:1160-1167.
Piercy, G.R., R.W. Cahn & A.H. Cottrell 1955. A study of primary and conjugate slip in crystals of alpha-brass. Acta Metall. 3:331-338.
Ramaswami, B., U.F. Kocks & B. Chalmers 1965. Latent hardening in silver and an Ag-Au alloy. Trans.Met.Soc.AIME 233: 927-931.
Rice, J.R. 1971. Inelastic constitutive relations for solids: an internal-variable theory and its application to metal plasticity. J.Mech.Phys.Solids 19:433-455.
Taylor, G.I. & C.F. Elam 1923. The distortion of an aluminium crystal during a tensile test. Proc.R.Soc.Lond.A 102: 643-667.
Vause, R.F. & K.S. Havner 1979. Theoretical latent hardening in crystals - III. F.c.c. crystals in compression. J.Mech. Phys.Solids 27:393-414.
Vorbrugg, W., H.Ch. Goetting & Ch. Schwink 1971. Work-hardening and surface investigations on copper single crystals oriented for multiple glide. Phys.Stat.Sol. B 46:257-264.
Wonsiewicz, B.C. & G.Y. Chin 1970. Plane strain compression of copper, Cu 6Wt Pct Aℓ, and Ag 4Wt Pct Sn Crystals. Metall. Trans. 1:2715-2722.

5th International Symposium on Continuum Models of Discrete Systems / Nottingham / 14-20 July 1985

Coexistent austenitic and martensitic phases in elastic crystals

G.P.Parry
School of Mathematics, University of Bath, UK

ABSTRACT: Piecewise homogeneous configurations of a thermoelastic crystal which provide strong minima of the free energy functional satisfy the Maxwell relation, so that in the absence of loading the energy density is constant throughout the body. This rules out the possibility that phase equilibria of austenite/martensite type correspond to such strong minima, for these phases generally have different energy densities. However, weak minima can provide models of such phase mixtures, and we illustrate with an explicit construction.

1 INTRODUCTION

Variational models of phase equilibria have recently received considerable attention. Most of the models that have been analysed incorporate simplifying features which produce problems in fewer than three dimensions, and when the authors treat stability they generally do so in a pragmatic fashion, stipulating that equilibria are stable if they provide minima of a potential in some topology which is selected arbitrarily, without reference to dynamics. Furthermore, it is characteristic of these studies that the equilibria are realised as the extremals of some autonomous functional, so that when 'translations' of the equilibrium configuration provide admissible variations they render that functional stationary. It follows, in this case, that any equilibrium composed of homogeneous phases in some (unknown) arrangement must consist of phases with the same energy density, and the most likely situation is that the different phases are symmetry-related, Maddocks and Parry (7). Thus, such theories cannot model the situation where austenite and martensite coexist over a range of temperature, for it is recognised that these two phases have generally different strain energies.

Here we construct a theory which allows a particular three dimensional equilibrium configuration of phases with distinct energy densities to be reckoned as stable, in some cases. Fosdick and James (3) point out, in a different context, that one need only require that translations of the equilibria do not provide admissible variations. When the equilibria are piecewise homogeneous, it is sufficient to suppose that the stable equilibria are those which provide weak minima of the energy functional, so that Gibbs might call them metastable. The kinematics of the particular equilibria that we construct are given by James (6) in his work on the kinematics of coherent phase transformations, and this work amounts in part, then, to a gathering and placing in context of the studies of Fosdick and James.

The paper begins with a few well known results regarding properties of equilibria which correspond to strong and weak minima of the energy. Bearing in mind our aim of modelling coexistent austenite/martensite, we then agree to focus on just weak minima and simplify matters by considering only piecewise homogeneous configurations of the crystal. In the case where the crystal is modelled by a purely elastic (nonlinear) constitutive law, a coherent partition is seen to be stable in the sense described above if each individual phase provides a local minimum of the energy density, which is a function just of the deformation gradients. Thus the calculation of equilibria of this type from a given elastic strain energy function corresponding to (say) cubic symmetry of an austenitic phase involves two distinct steps. Firstly one calculates the local minima, which correspond to configurations which are themselves cubic or to sets of configurations which are 'cubically related' to each other. Secondly, choosing a coherent partition, one determines if these minima may be arranged so as to compose the distinct phases of that partition. We perform this calculation explicitly for one particular choice of smooth energy density, and demonstrate that there is a corresponding coherent equilibrium which may perhaps be described as cubic-hexagonal, consisting of symmetrically oriented monoclinic phases embedded in a cubic matrix.

2 STABILITY

In the context of three dimensional elasticity, one imagines that the equilibria of an unloaded material correspond to extremals of a functional

$$\Sigma(\mathbf{x}) \overset{\Delta}{=} \int_{B_O} \chi(F)\,d\mathbf{X} \quad , \tag{1}$$

where χ is the strain energy density, and

$$F = (F_{ij}) = \left(\frac{\partial x_i}{\partial X_j}\right) \quad , \tag{2}$$

is the matrix of deformation gradients, with $\mathbf{x} = (x_i)$ and $\mathbf{X} = (X_i)$ being vectors representing the current and reference positions of a material point. In (1), the integration extends over a reference configuration B_O . Further, stable equilibria are defined to correspond to the extremals of (1) which are also minima. To be precise, if

$$\Sigma(\mathbf{x}) \geq \Sigma(\mathbf{x}_O) \quad , \tag{3}$$

for all $\mathbf{x} - \mathbf{x}_O \in \zeta$, where ζ is some prescribed class of functions, then the configuration corresponding to the field $\mathbf{x}_O$ is called stable. It is clear that the definition of stability depends upon the choice of the class ζ.

James (6) and Gurtin (5) analyse, in particular, the case where

$$\zeta = \zeta_\varepsilon \overset{\Delta}{=} \left\{ \mathbf{f} \ : \ \sup_{\mathbf{X}\in B_O} |\mathbf{f}(\mathbf{X})| < \varepsilon \right\} \tag{4}$$

with ε sufficiently small, so that stable configurations provide strong minima, in the jargon of the calculus of variations. (In fact, Gurtin's ζ is a little smaller than ζ_ε.) In this case there is no restriction on the size of variations of F, and there are some well-known consequences of this choice of ζ which we choose to present with reference only to piecewise homogeneous configurations of the body. Suppose that the different phases which compose the body correspond to deformation gradients which we label F_i, then

α) χ is rank one convex at each F_i , so that

$$\chi(A) - \chi(F_i) \geq S(F_i).(A - F_i) \quad , \tag{5}$$

where $S = \frac{\partial \chi}{\partial F}$, provided $A - F_i$ is a tensor product.

β) χ obeys the Maxwell relation

$$\chi(F^+) - \chi(F^-) = S(F^\pm).(F^+ - F^-) \tag{6}$$

where F^+ and F^- correspond to adjacent phases in the body.

γ) If each phase borders every other phase, then S is constant throughout the body.

Since the external boundaries of the body are unloaded, then from (γ), $S = 0$ everywhere if each phase borders every other, and so from (β), $\chi(F_i) = \chi(F_j)$ for all i and j, in that case. Bearing in mind that the austenite and martensite phases have generally different energy densities, it follows that some modification of this structure is called for if we are to model the situation where these two phases coexist in an unloaded crystal.

So let us choose

$$\zeta = \zeta'_\varepsilon \overset{\Delta}{=} \left\{ \mathbf{f} : \sup_{\mathbf{X}\in B_O} |\mathbf{f}(\mathbf{X})| + \sup_{\mathbf{X}\in B_O} \|\nabla \mathbf{f}(\mathbf{X})\| < \varepsilon \right\} \tag{7}$$

for ε sufficiently small, where stable configurations provide weak minima of ξ. There are no results comparable to (α), (β) (γ) in this case, but the field $\mathbf{f}$ is certainly equilibrated, and traction is continuous across all boundaries. Also it is clear that since

$$\Sigma(\mathbf{x}_O) = \sum_i V_{B_i} \ \psi(F_i) \quad , \tag{8}$$

where V_{B_i} is the volume of that part of the body which is in the phase F_i, it follows that if each F_i provides a local minimum of χ, then $\mathbf{x}_O$ corresponds to a stable configuration (because admissible variations have the same surfaces of true discontinuity as $\mathbf{x}_O$).

We shall construct a stable ($\zeta = \zeta'_\varepsilon$) equilibrium configuration which violates (6) with $S = 0$ by arranging local minima of χ in such a fashion that the resulting deformation is continuous.

3 KINEMATICS

James (6) has catalogued what he calls the "simplest coherent local partitions", giving a systematic listing of continuous deformations with the property that the corresponding deformation gradients are continuous except on a collection of smooth surfaces. Central to his analysis is the proposition that if F^+ and F^- are the limiting values of deformation gradients on either side of a surface S with unit normal $\mathbf{n}$, then there exists a vector $\mathbf{a}$, the 'amplitude', such that

$$F^+ - F^- = \mathbf{a} \otimes \mathbf{n} \quad , \tag{9}$$

cf. Trusdell and Toupin (10).

We shall focus attention on just one of James's partitions, that one which he labels (6,4,4), illustrates in fig. 4, and calls a 'tetrad'. In this partition there are four regions with distinct values of F, arranged as in the following diagram.

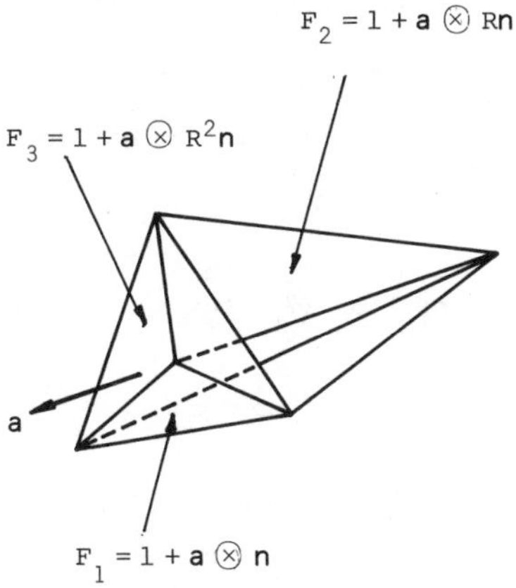

$$F_0 = 1 \ , \quad R^3 = 1 \ , \quad RR^T = 1 \ .$$

The 'end' face of the tetrahedron is imagined to form part of the boundary of B_o, and it is easy to check that the condition of coherence (9) is satisfied at each of the six surfaces of discontinuity. Explicitly, three of the required relations are immediate, whereas it is clear that there exist vectors $\mathbf{a}'$ and $\mathbf{n}'$ such that

$$F_2 - F_1 = \mathbf{a} \otimes R\mathbf{n} - \mathbf{a} \otimes \mathbf{n} = \mathbf{a}' \otimes \mathbf{n}' \quad , \tag{10}$$

for example.

Now suppose that χ is invariant under some material symmetry group $\mathcal{G}$, and suppose also that it is objective. Then

$$\chi(F) = \chi(Q\,F\,G) \quad , \quad G \in \mathcal{G} \quad , \ Q \text{ orthogonal} \tag{11}$$

so that it is clear that if $\overline{F}$ provides a local minimum of χ, so does $Q\overline{F}G$, for all choices of R and G. Of course this last statement is trivial if $\overline{F} = Q\overline{F}G$, in which case

$$\overline{F}^T\overline{F} = G^T(\overline{F}^T\overline{F})G \tag{12}$$

The group

$$\mathcal{P} \overset{\Delta}{=} \left\{ G \ : \ (12)\text{holds, for a given } \overline{F}^T\overline{F} \right\} \tag{13}$$

is the point group corresponding to the Cauchy Green tensor $\overline{F}^T\overline{F}$. It is sensible to ask if the F_i defined above can fall into this trivial category. If they do, then there exist orthogonal Q, orthogonal $G \in \mathcal{P}$ such that

$$1 + \mathbf{a} \otimes R\mathbf{n} = Q(1 + \mathbf{a} \otimes \mathbf{n})G = QG + Q\mathbf{a} \otimes G^T\mathbf{n} \ . \tag{13}$$

Clearly this has solution of the form $Q^T = G = R^T$ provided that $Q\mathbf{a} = R\mathbf{a} = \mathbf{a}$. Recall, from the figure, that $R^3 = 1$. Thus the F_i provide minima of this type if $R \in \mathcal{P}$ and if $\mathbf{a}$ lies along the axis of R. Thus if $\mathcal{P}$ is the cubic point group say, it is necessary to choose $\mathbf{a}$ along the 'body diagonal' of the usual unit cell.

4 MINIMA OF POTENTIALS WITH CUBIC SYMMETRY

It remains to demonstrate, by explicit construction, that there are indeed potentials with (at least) cubic symmetry having local minima at $F = 1$ and $F = 1 + \mathbf{a} \otimes \mathbf{n}$, with $\mathbf{a}$ parellel to (1,1,1), relative to the cubic axes. It is clear that such a construction is possible. For suppose that a potential χ is defined arbitrarily in some fundamental domain $\mathcal{D}$ (Parry (8)) and generated throughout the space of the Cauchy Green tensor via the material symmetry. With $\mathcal{D}$ corresponding to cubic symmetry, one can check that the points corresponding to $1 + \mathbf{a} \otimes \mathbf{n}$ generally lie in the interior of $\mathcal{D}$. Construct a ball $\mathcal{B}$ surrounding this point, strictly in $\mathcal{D}$, and suppose that χ increases from its value at $1 + \mathbf{a} \otimes \mathbf{n}$ to a constant value on $\partial\mathcal{B}$, and also that χ assumes that constant value elsewhere in $\mathcal{D}$. Then χ has local minima at 1 and $1 + \mathbf{a} \otimes \mathbf{n}$, though the first does not provide a strict local minimum. Thus, this rather pathological potential, which can be arbitrarily smooth, certainly allows stable piecewise homogeneous equilibria of the type indicated. However, what is not evident is the construction of coherent equilibria with minima lying on $\partial\mathcal{D}$, as Ericksen's calculations in (2) effectively demonstrate by showing that there are no coherent cubic-tetragonal equilibria.

It is fairly straightforward to provide a

more explicit example. Select a configuration where the material is cubic and write the nine invariants of this group as $\overline{I}_1, \ldots, \overline{I}_9$ (take the list of Green and Adkins (4), in order). It is clear that only six of these nine invariants can be functionally independent. Indeed, Smith (9) shows that any polynomial χ can be written in the form

$$\chi = \overline{W}(\overline{I}_1, \ldots, \overline{I}_6) + \alpha_i \overline{I}_i + \alpha_{ij}\overline{I}_i\overline{I}_j \qquad (i,j = 7,8,9) \tag{14}$$

where the coefficients α_i, α_{ij} depend only on the six functionally independent invariants $\overline{I}_1, \ldots, \overline{I}_6$. For ease of calculation we restrict χ so that

$$\chi = W(I_1, \ldots, I_6) \tag{15}$$

where the I_k, $k = 1,..6$ are given in terms of the Green strain e_{ij}, $i,j = 1,2,3$ via

$$\begin{aligned}
I_1 &= \overline{I}_1 = e_{11} + e_{22} + e_{33}\\
I_2 &= \overline{I}_2 = e_{11}e_{22} + e_{22}e_{33} + e_{33}e_{11}\\
I_3 &= \overline{I}_3 = e_{11}e_{22}e_{33}\\
I_4 &= \overline{I}_4 = e_{12}^2 + e_{23}^2 + e_{31}^2\\
I_5 &= \overline{I}_5 = e_{12}^2e_{23}^2 + e_{23}^2e_{31}^2 + e_{31}^2e_{12}^2\\
I_6 &= \overline{I}_6^2 = e_{12}^2e_{23}^2e_{31}^2
\end{aligned} \tag{16}$$

Notice that the two sets of invariants $(I_1, I_2, I_3), (I_4, I_5, I_6)$ are independent of each other, and each set satisfies an equation of the form

$$\lambda^3 - \lambda^2\, \mathrm{I} + \lambda^2\, \mathrm{II} - \mathrm{III} = 0 \quad , \tag{17}$$

in the notation of Ericksen (1). We shall construct energies w having local minima at $e_{ij} = 0$ and $e_{ij} = e_{ij}^0$, with corresponding sets of invariants $I_k = 0$, $I_k = I_k^0$ respectively. To simplify matters yet further, assume that exactly two of $(e_{11}^0, e_{22}^0, e_{33}^0)$ are distinct, and likewise that exactly two of $(e_{12}^{0\,2}, e_{23}^{0\,2}, e_{31}^{0\,2})$ are distinct. It follows that both corresponding sets of invariants are such that the discriminant deriving from (17) vanishes. Thus, for example,

$$\left(\frac{1}{3}\, I_2^0 - \frac{1}{9} I_1^{0\,2}\right)^3 + \left\{\frac{1}{6}\left(3I_3^0 - I_1^0 I_2^0\right) + \frac{1}{27} I_1^{0\,3}\right\}^2 = 0 \quad . \tag{18}$$

With these assumptions, one has the freedom to prescribe, say, I_2^0, I_1^0, I_4^0, I_3^0 compatible with $3I_2^0 - I_1^{0\,2} < 0$, $3I_5^0 - I_4^{0\,2} < 0$, and it is elementary to construct a class of functions w with minima at $e_{ij} = 0$, $e_{ij} = e_{ij}^0$ where e_{ij}^0 corresponds to invariants prescribed as above, consistently with equations like (18). (For example, let $f(x,\alpha)$ be such that $\frac{\partial f}{\partial x} = x(x - \theta\alpha)(x - \alpha)$, with $0 < \theta < 1$, and then define $w = \sum_{t=1}^{6} f(I_k, I_k^0)$.) It is not so easy to construct energy densities which are 'realistic' or 'generic', but here we just demonstrate the nature of the problem.

It remains to check that such a non-trivial local minimum of w can correspond to a deformation gradient of the form $1 + \mathbf{a} \otimes \mathbf{n}$, with $\mathbf{a}$ parallel to (1,1,1) and $|\mathbf{n}| = 1$. Let $\mathbf{a} = \nu(1,1,1)$, so that

$$\beta = (\beta_{ij}) \overset{\Delta}{=} 1 + \mathbf{a} \otimes \mathbf{n} = 1 + (1,1,1) \otimes \nu\mathbf{n}, \tag{19}$$

and put $\underset{\sim}{\mu} \overset{\Delta}{=} \nu\mathbf{n}$. We ask if the system

$$\tfrac{1}{2}(\beta^T \beta - 1) = (e_{ij}^0) \tag{20}$$

has a solution for $\underset{\sim}{\mu}$, with e_{ij}^0 as prescribed. Equation (20) can be rearranged as

$$\begin{aligned}
&\mu_k + \frac{3}{2}\mu_k^2 = e_{kk}^0 \quad , \quad \text{no sum},\\
&\frac{1}{2}(\mu_i + \mu_j) + \frac{3}{2}\mu_i\mu_j = e_{ij}^0 \quad , \quad i \neq j,
\end{aligned} \tag{21}$$

where exactly two of $(e_{11}^0, e_{22}^0, e_{33}^0)$ are distinct, and exactly two of $(|e_{12}^0|, |e_{23}^0|, |e_{31}^0|)$ are distinct. Clearly, the first set of three equations in (21) determine the μ_k, and the remaining equations provide three constraints. Suppose that $e_{22}^0 = e_{33}^0$. Then it is easy to show that (21) has a solution with $\mu_2 = \mu_3$ provided that

$$\begin{aligned}
&\left(\frac{2}{3}e_{11}^0 + \frac{1}{9}\right)\left(\frac{2}{3}e_{22}^0 + \frac{1}{9}\right) = \left(\frac{2}{3}e_{12}^0 + \frac{1}{9}\right)^2 \quad ,\\
&\left(\frac{2}{3}e_{22}^0 + \frac{1}{9}\right)\left(\frac{2}{3}e_{33}^0 + \frac{1}{9}\right) = \left(\frac{2}{3}e_{23}^0 + \frac{1}{9}\right)^2 \quad ,\\
&\frac{2}{3}e_{kk}^0 + \frac{1}{9} > 0 \quad ,\\
&e_{12}^0 = e_{13}^0 \quad .
\end{aligned} \tag{22}$$

Thus, given $(e_{11}^0, e_{22}^0, e_{22}^0)$ consistent with $(22)_3$, this triple determines $(e_{12}^0, e_{23}^0, e_{31}^0)$ through the remaining equations of (23). Finally, then, given I_1^0, I_2^0 consistent with $3I_2^0 - I_1^{0\,2} < 0$, the coherence condition effectively determines all the remaining invariants. So it is clear that there is at least a two-parameter family of strain energies where equilibria having the form of tetrads exist stably. Perhaps it is sensible to suppose that strain energies which are in some sense close to such w will admit stable equilibria which are in some related sense close to that one which we have described, though it is likely that they will not be piecewise homogeneous.

One last point is of interest. We have constructed an equilibrium configuration where, in one of the subregions of the tetrad, the nontrivial Green strain has the form

$$E_0 \triangleq \begin{bmatrix} e_{11}^0 & e_{12}^0 & e_{12}^0 \\ e_{12}^0 & e_{22}^0 & e_{23}^0 \\ e_{12}^0 & e_{23}^0 & e_{22}^0 \end{bmatrix} . \qquad (23)$$

Introduce the notation $R_{\mathbf{k}}^{\phi}$ for a rotation of ϕ about the axis $\mathbf{k}$. Then in symmetry related subregions, the Green strain is $R_{\mathbf{a}}^{2\pi/3} E_0 \left(R_{\mathbf{a}}^{2\pi/3}\right)^T$. Notice that the point group corresponding to E_0 includes $R_{\mathbf{f}}^{\pi}$, where $\mathbf{f} = (0,1,-1)$, so that each separate subregion is at least monoclinic. Also $\mathbf{a}.\mathbf{f} = 0$. In some loose sense then, the symmetric arrangement of congruent subregions can be regarded as hexagonal, and the whole system might be described as a coherent cubic-hexagonal phase mixture.

ACKNOWLEDGEMENTS

I am grateful to Roger Fosdick and John Maddocks for helpful conversation, and to the Institute for Mathematics and its Applications, University of Minnesota for hospitality.

REFERENCES

1 Ericksen, J.L. 1960. Tensor fields. Handbuch der Physik III/I, ed. S. Flügge, Springer-Verlag, Berlin-Göttingen-Heidelberg.

2 Ericksen, J.L. Constitutive theory for some constrained elastic crystals, I.M.A. Preprint series #123.

3 Fosdick, R.L. and R.D. James 1981. The elastica and the problem of pure bending for a non-convex stored energy function. J. Elasticity 11 pp.165-186.

4 Green, A.E. and J. Adkins 1970. Large elastic deformations. 2nd edition, Clarendon Press, Oxford.

5 Gurtin, M.E. 1983. Two phase deformations of elastic solids. Arch. Rational Mech. Anal., 84, pp.1-29.

6 James, R.D. 1981. Finite deformation by mechanical twinning. Arch. Rational Mech. Anal, 77, pp.143-176.

7 Maddocks, J.H. and G.P. Parry. A model for twinning. I.M.A. Preprint series #125, to appear in J. Elasticity.

8 Parry, G.P. 1976. On the elasticity of monatomic crystals. Math. Proc. Cambridge Phil. Soc., 80, pp.189-211.

9 Smith, G.F. 1962. Further results on the strain energy function for anistropic elastic materials. Arch. Rational Mech. Anal., 10, pp.108-118.

10 Truesdell, C. and R. Toupin 1960. The classical field theories, Handbuch der Physik III/I, ed. S. Flügge, Springer-Verlag, Berlin-Göttingen-Heidelberg.

5th International Symposium on Continuum Models of Discrete Systems / Nottingham / 14-20 July 1985

The stability of plastic flow in crystals*

M.Boček

Kernforschungszentrum Karlsruhe, Institut für Material- und Festkörperforschung II, FR Germany

ABSTRACT: In general, the deforming crystals are very complex, non-linear dynamical systems. The latter are suggested as networks of particular elements. Their inputs are feeded by the loading conditions ($\dot{\varepsilon}/\sigma$) through which the generator (supply of mobile dislocations) as well as different structurizers are activated. The generator, operating as a closed loop, is connected to a particular structurizer. Analytically the generator is usually described by means of one dimensional, non-linear first order differential equations. The only solution therefrom are dislocation densities converging monotoneously to a stationary value. In order to reveal more about the behaviour of plastic flow in crystalline solids the analysis of the generator is performed by means the time-discrete difference calculus. Two dislocation models originally formulated in the continuous approach (differential calculus) were transposed into the time discrete concept (difference calculus). Depending upon the loading conditions and upon particular structural parameters the time discrete approach leads to a variety of behaviour. Besides the case that dislocation density becomes stationary in time - the only one case predicted by the time continuous treatment - the time discrete approach predicts also diverging or oscillating densities. Hence the latter approach gives the more general results. In view of these results it seems reasonable to reconsider the explanation of particular dynamical phenomena in crystal plasticity.

1 INTRODUCTION

Plastically deforming crystals reveal characteristics of non-linear dynamical systems. Their analytical description occurs by means of a proper differential strain rate equation which besides the stress, temperature and their respective time derivatives contains so-called structure parameters. Because the latter are in general time dependent, hence each of the structure parameter is described by means of a differential equation. In order to obtain the so-called constitutive equation such a set of simultaneous differential equations must be combined with the strain rate equation. The solution of the constitutive equation finally gives the desired strain/time relationship. Although when isotropy and homogeneity is assumed - i.e. the constitutive equations do not contain the coordinates - their solution is possible only under very simplified conditions /1/. Accordingly these solutions cover only certain aspects of plastic behaviour. In order to reveal more about the complex nature of plasticity it seems reasonable to seek for new appropriate descriptions which are easier accessible to an analytical treatment.

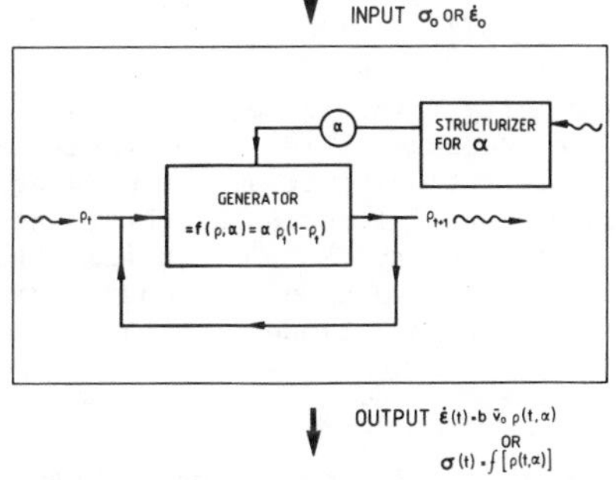

Fig. 1
Network system. Two elements are shown. The connections to other elements are shown by arrows

* In memory of a grand man, the Czech scientist Professor Zdenek Frankenberger

A plastically deforming crystal - as shown in Fig. 1 - can be replaced by a system of elements forming a network (see e.g. /2/). One of these elements the generator provides the system with mobile dislocations which are prerequisite for plastic flow. There are other elements, so-called structurizers. In this information about deformation relevant material properties (e.g. dislocation substructure) and their change during plastic deformation is stored. Macroscopically plastic flow appears as a continuous phenomenon. More thorough inspection on a finer scale, however, reveals that, dissimilar to viscous flow of fluids, plastic deformation of solid crystals occurs stepwise i.e. discontinuously in time (see e.g. /3/). This feature reflects the kinematic properties of dislocations moving through a crystalline material. In situ transmission-electron microscopic (TEM) investigations leave no doubt that plastic glide is a time discontinuous phenomenon which basically depends on the generation and annihilation of mobile dislocations. Experiments indicate that the generator itself is a nonlinear dynamical closed loop sub-system (see Fig. 1) the behaviour of which we shall analyse in the present investigation.

The change in density $d\rho$ of mobile dislocations is determined by the production rate $\dot{\rho}^+ \equiv d\rho^+/dt$ and exhaustion rate $\dot{\rho}^- \equiv d\rho^-/dt$ respectively. The decrease in mobile dislocation density can be due to several reasons. Some of the dislocations will escape from the crystal, other (opposite charged) can disappear (annihilate) by recombination and some will be immobilized by glide obstacles. The relative importance of the individual contributions is mainly dependent upon the loading conditions i.e. stress σ and temperature T and is naturally influenced also by substructural properties like total dislocation density, grain size, alloying elements etc. Because dislocations themselves act as dislocation sources, the production rate $\dot{\rho}^+$ will be a function of the total dislocation density (mobile and immobile) i.e. $\dot{\rho}^+$ will depend upon the loading conditions. The time dependence $\rho(t)$ follows from

$$\dot{\rho} = \dot{\rho}^+ + \dot{\rho}^- \qquad (1)$$

were $\dot{\rho} = \dot{\rho}$ (loading conditions, substructure). According to Eq. (1) steady state i.e. $\rho = \rho_s = \text{const.} \neq f(t)$ will establish for $\dot{\rho}^+ = \dot{\rho}^-$. In general ρ depends on coordinates as well, in this paper, however, we restrict to the description of its evolution in time.

Johnston and Gilman /4/ for the first time investigated the dynamical behaviour of dislocations experimentally. By means of a stress-pulsing method the mobility of individual dislocations and their dependency upon the loading conditions (stress and temperature) could be examined. These measurements for the first time verified the Orowan equation

$$\dot{\varepsilon} = b\rho\bar{v} \qquad (2)$$

on a microscopic scale. $\dot{\varepsilon}$ is the macroscopical strain rate, b is the modul of the Burgers vector, ρ is the density of mobile dislocations and $\bar{v}$ is the mean dislocation velocity. The time dependency of ρ turned out to be of the form

$$d\rho/dt \equiv \dot{\rho} = \alpha_a\rho - \beta\rho^2. \qquad (3)$$

The solution of this non-linear first order differential equation gives

$$\rho(t)/\rho_o = \frac{\exp(\alpha t)}{1-(\rho_o\beta/\alpha_a)[\exp(\alpha t)-1]} \qquad (4)$$

where α_a, β are constants which, as shown by experiments /4/, depend upon the applied stress σ and temperature T; ρ_o is the initial mobile dislocation density.

The comparison of Eqs. (1) and (3) shows, that the production and exhaustion rates, depend upon the mobile dislocation density as well as upon the loading conditions. Whatever the loading conditions are, ρ limits monotoneously towards a steady state value $\rho_s = \alpha_a/\beta$. The time in which ρ_s is established depends on α_a only. We conclude therefrom that plastic flow evolves into a stable equilibrium. This in many cases is actually confirmed by experiments. However, dynamical phenomena in crystal plasticity like strain ageing /5/, dynamic recrystallization /6/ as well as effects associated with local instabilities (manifesting e.g. by propagation of Lüders-band /7/) or giving rise to localized plastic flow /8/ are very frequent and common phenomena which hardly can keep with the idea of stationary deformation behaviour.

2 THE TIME-DISCRETE APPROACH

2.1 General remarks

Deforming crystals are physical systems whose time dependence is deterministic in the sense that it is possible to predict their future behaviour from the past. For instance the change in mobile dislocation density usually described by dif-

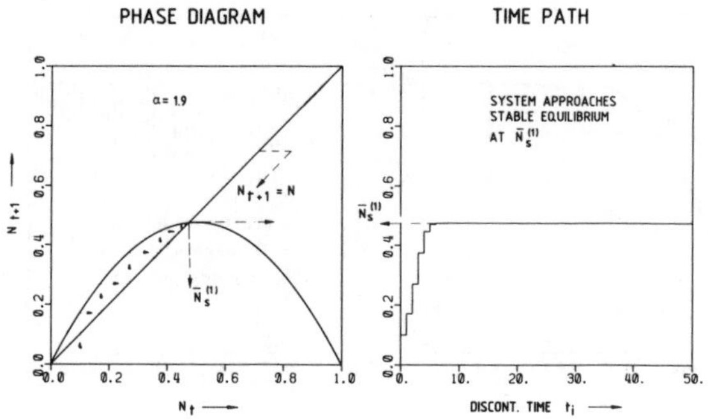

Fig. 2
Phase diagram for $\alpha = 1.9$. The arrows indicate the iteration steps of $F(N_t)$ and the corresponding time path.

ferential equations is considered as rather regular because due to the picture generated by the differential calculus successive states evolve continuously from each other.

As recently stressed by May /9/ there are many situations, in many disciplines, which can be described, at least approximately, by a simple first order difference equation. Studies of the dynamical properties of such models usually consist of finding equilibrium solutions and to determine their stability with respect to small perturbations. Substituting in Eq. (3) $\rho = N\alpha_a/\beta$ this is rewritten in the form

$$dN/dt \equiv \dot{N} = \alpha\, N(1-N). \qquad (5)$$

The transposition of Eq. (5) into a difference equation gives /10/

$$N_{t+1} = \alpha N_t(1-N_t) = F(N_t) \qquad (6)$$

where N_{t+1} is the change of N per time interval $\Delta t = 1$ and the frequency factor $\alpha = 1+\alpha_a$, $[\alpha] = 1/s$, is the only one system parameter. In the time discrete approach we interpret the values of t as refering to periods - rather than points - of time, with t = 1 denoting period 1 and t = 2 denoting period 2, and so forth.

This apparently simple equation, which appears in many contexts, shows the tendency for the variable N to increase from the time period t to the next period t+1 when N is small, and to decrease when N is large. The non-linearity of $F(N_t)$ is tuned by α, the parameter which determines the steepness of the hump in the $F(N_t)$ curve. In order to study the dynamical behaviour of the system described by Eq. (6) we must determine the stable (stationary), equilibrium values of N in that equation. As shown in Fig. 2 these can be found graphically as the points ($\bar{N}$), where the curve F(N) that maps N_t into N_{t+1} intersects the line $N_{t+1} = N_t$. The stability of the equilibrium can be determined from the time path of the system. This procedure is nothing else but the iteration of the function $F(N_t)$, namely: the result of the calculation i.e. N_{t+1} is feeded back as $N_t(=N_{t+1})$. When the N_t-values limit towards the stationary value $\bar{N}_s$ the corresponding point in the phase diagram is called a fixed point. A rigorous analysis of the stability condition and an account of the material, covered so far, the reader will find in ref. /11/.

Increasing the system parameter α the curve $F(N_t)$ becomes more and more steeply humped and for a certain α-value the equilibrium point $\bar{N}$ is no longer stable. At exactly the stage when this occurs there are born two new fixed points of period 2 (see Fig. 3a), between which the system alternates in a stable cycle of period 2. As with increasing α the hump in $F(N_t)$ continues to steepen, again for a certain α-value both the fixed points become unstable - as shown in Fig. 3b - each of them splits up (bifurcates) into two new fixed points to give a stable cycle of period 4. This in turn gives way to a cycle of period 8, and hence to a hierarchy of bifurcating stable cycles of periods $16,32,..,2^n$. Although this process produces an infinite sequence of cycles with periods 2^n, the range of the α-value wherein any one cycle is stable progressively diminishes, so that the entire process is a convergent one, being bounded above by some limiting value $\alpha^{\mathbf{x}}$. For Eq. (8) it is $\alpha^{\mathbf{x}} = 3.5700$. Beyond this point ($\alpha > \alpha^{\mathbf{x}}$) there is an infinite number of fixed points with different periodicties and an infinite number of

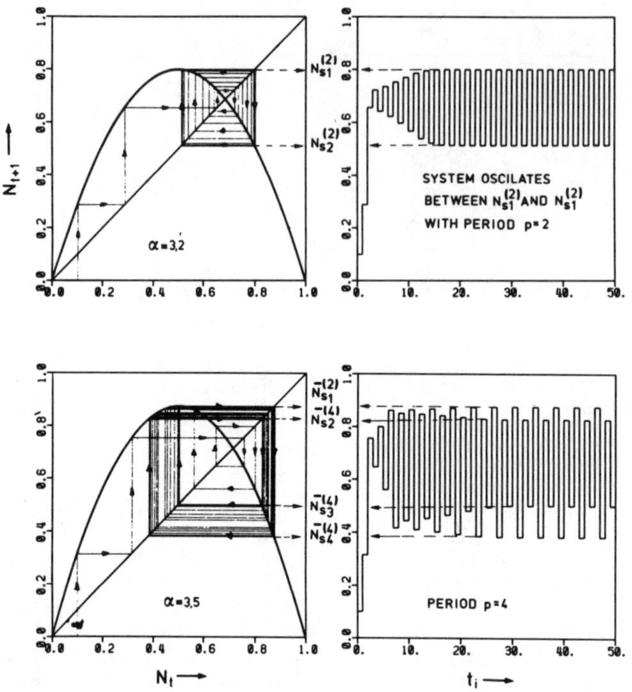

Fig. 3
(a) Phase diagram and time path for $\alpha = 3.2$
(b) the same for $\alpha = 3.5$

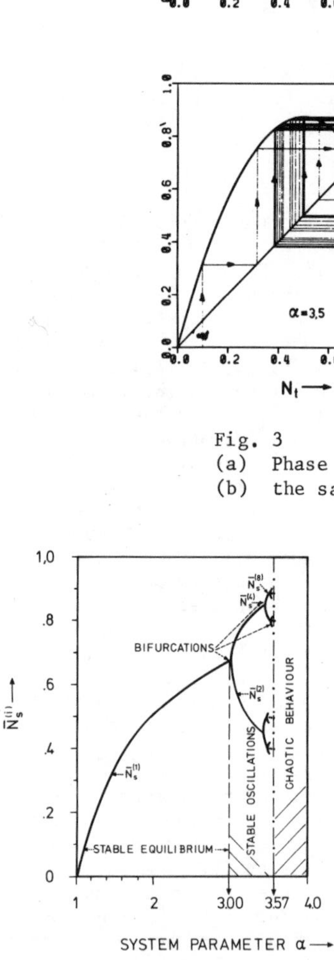

Fig. 4
The "bifurcation tree"

different periodic cycles. Such a situation where an infinite number of orbits in the phase diagram can occur has been coined as "chaotic" behaviour /12/. The bifurcation phenomenon for Eq. (8) is depicted in Fig. 4. It should be emphasized that this behaviour is generic to most functions $F(N_t)$ with a hump the steepness of which is tunable by only one system parameter α. From the above and by inspection of Fig. 4 several conclusions can be drawn: i) The dynamical behaviour of the system is completely determined by the value of a single parameter α. For $1 \leq \alpha \leq 3$ the system variable limits to a (dynamically) stable equilibrium value. For $\alpha \geq 3$ the behaviour of the system is fundamentally different from that predicted by the continuous approach. ii) Close to distinct values α^1 small perturbations in α cause a sudden change in the behaviour of the system. iii) Besides the case of a stable equilibrium (the only one behaviour predicted by the continuous approach), the discrete treatment predicts also oscillating equilibriums and "chaotic" behaviour resembling that of random processes. Obviously the latter approach contains more information.

2.2 Connection with crystal plasticity

In the following we shall paradigmatically examine the behaviour of two time continuous dislocation models in the time discrete approach. The models considered are concerned with the evolution of mobile and immobile dislocation density respectively.

2.2.1 The model of Essmann and Mughrabi

In order to interpret experimental results the above authors /13/ suggested a model for the strain dependency of mobile dislocation density which is analytically concordant with that of Johnston and Gilman (see Eq. (3)). According to that, the density change dρ of edge dislocations caused by a strain increase dε is expressed as

$$d\rho = (2/bL_e)d\varepsilon-(2y_ec_e\rho/b)d\varepsilon \qquad (7)$$

Changing the variable by means of Eq. (2) we obtain Eq. (3) where now

$$\alpha_a = (2/bL_e)\dot{\varepsilon} \qquad (8a)$$

and

$$\beta = (2y_ec_e/b)\dot{\varepsilon}. \qquad (8b)$$

L_e is the distance travelled by the edge component of the dislocation loop with the mean velocity $\bar{v}$, y_e is a critical distance between glide planes such that dislocations of opposite sign which meet at a distance $y \leq y_e$ will annihilate and c_e denotes the fraction of the strain ε carried by edge dislocations. L_e, y_e, c_e are substructural quantities and hence can change in the course of deformation. The above parameters α_a and β contain also the loading condition $\dot{\varepsilon} = \dot{\varepsilon}(\sigma)$.

Performing the deformation at a constant strain rate $\dot{\varepsilon}_o$ the loading condition by means of Eq. (2) is then expressed as

$$\dot{\varepsilon}_o = b\rho_o\bar{v}_o \qquad (9)$$

$\bar{v}_o$ is the mean velocity of edge dislocation segments in the following assumed to be constant throughout the test and ρ_o is the initial mobile dislocation density. Combining Eqs. (8a,b) with Eq. (11) we finally have

$$\alpha_a = 2\bar{v}_o/L_e \qquad (10)$$

and

$$\beta = 2c_ey_e\bar{v}_o. \qquad (11)$$

Therefrom one obtains for the steady state density in the EMM $\rho_s = \alpha_a/\beta = (L_ey_ec_e)^{-1}$. Hence, ρ_s is explicitely independent of the loading conditions and is solely determined by substructural parameters L_e, c_e and y_e. In transition to the time discrete picture we transpose Eq. (3) into the corresponding difference equation and we end up with Eq. (6) where now it is

$$\alpha = 1+\alpha_a = 1+2\bar{v}_o/L_e = 1+(2\dot{\varepsilon}_o/\rho_ob^2)(b/L_e) \qquad (12)$$

and

$$N_t = (\beta/\alpha_a)\rho_t = c_ey_eL_e\rho_t \qquad (13)$$

The system parameter α is determined by the loading condition ($\dot{\varepsilon}_o$) and by the substructure parameters ρ_o and L_e. The latter can change in the course of deformation (e.g. due to work hardening L_e may decrease). $\bar{v}_o$ is the microscopic control parameter which is connected by Eq. (9) with the macroscopic strain rate $\dot{\varepsilon}_o$. The density $\rho_o \neq f(t)$ defines the initial condition of the generator. For the loading procedure considered the corresponding output quantity of the network system (Fig. 1) is the stress σ. The time dependency of σ is determined by that of the generator and α-structurizer resp. i.e. $\sigma(t) = f[\rho(t),\alpha(t)]$. Alternatively, choosing σ(= constant) as the input quantity (creep experiment) the corresponding output quantity is the strain rate, the behaviour of which is determined by the Orowan equation $\dot{\varepsilon}(t) = b\bar{v}_o\rho(t,\alpha)$.

According to the results of the analysis depicted in Fig. 4 for $1 \leq \alpha \leq 3$ in concordance with the time continuous approach the density ρ converges to a stable equilibrium. Principally differently behaves the system for $3 < \alpha < 3.57$, the density ρ periodically oscillates between certain fixed values and for $\alpha \geq 3.57$ the oscillations reveal "chaotic" behaviour. Suggesting the values $\alpha^1 = 3.0$ and $\alpha^* = 3.57$ as "critical" values, from Eq. (12) one obtains the relationship between the corresponding values of the structure parameter L_e and loading condition $\dot{\varepsilon}_o$, namely

$$L_e^1/b = X_o/2b^2 \qquad (14a)$$

and

$$L^*/b = X_o/2.57b^2 \qquad (14b)$$

where

$$X_o = 2\dot{\varepsilon}_o/b^2\rho_o. \qquad (14c)$$

In Fig. 5 for different $\dot{\varepsilon}_o$-values graphs of the function $\alpha(L_e)_{\dot{\varepsilon}_o}$ are shown (for $\rho_o = 1.10^{10}m^{-2}$ and $b = 3.10^{-10}m$). The figure is horizontally divided into 3 regions I,II,III. Points below the line $\alpha^1 = 3$ (region I) belong to systems in which ρ approaches a stable equilibrium value ρ_s. For points between the lines $\alpha^1 = 3$ and $\alpha^* = 3.56$ (region II) ρ oscil-

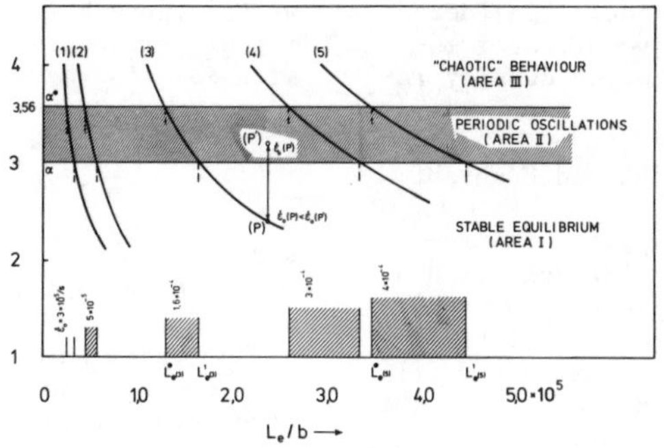

Fig. 5
The $\alpha(L_e)$-dependency for different $\dot{\varepsilon}_o$-values. ($b = 3.10^{-10}$m; $\rho_o = 1.10^{10}m^{-2}$)

lates periodically and for points above the line $\alpha > 3.56$ (region III) the oscillations become "chaotic". Whereas the transitions I $\rightleftarrows$ II and I $\rightleftarrows$ III are expected to manifest experimentally, the transitions within region II (period doubling) and transitions II $\rightleftarrows$ III probably elude observation in deformation tests. As compared to steady state behaviour (region I) the periodic density oscillations (region II) occur within a narrow band of L_e-values. For the width of these bands we obtain by means of Eqs. (14)

$$\overline{\Delta L}_e/b \equiv (L_e^1 - L_e^*/b) = 0.11(X_o/b^2) = 0.22\,(L_e^1/b) \sim \dot{\varepsilon}_o \quad (15)$$

Thus at otherwise unchanged conditions the width of the range of L_e-values within which ρ periodically oscillates increases linearly with $\dot{\varepsilon}_o$. Particularly with regard to experimental investigations of "flow instabilities" the sensitivity of α against a change of the substructure parameter L_e is of interest. This, defined as $d\alpha/dL_e \equiv \tilde{\alpha}$, gives together with Eq. (12)

$$\tilde{\alpha} = -(X_o/b^3)(b/L_e)^2 \quad (16a)$$

At the transition I $\rightleftarrows$ II i.e. for $\alpha = \alpha^1$ it is

$$\tilde{\alpha}^1 \equiv (d\alpha/dL_e)\alpha^1 = -(4b/X_o) \sim 1/\dot{\varepsilon}_o . \quad (16b)$$

Hence, for given $\dot{\varepsilon}_o$, $\tilde{\alpha}$ increases with $1/L_e^2$ and for given L_e, $\tilde{\alpha}$ is proportional to $1/\dot{\varepsilon}_o$.

In conclusion the following should be emphasized: i) In difference to the time continuous approach, in the discrete approach close to distinct values of the system parameter α even small variations of the structure parameter L_e lead to a fundamental change in $\rho(t)$. ii) The criticality of $\rho(t)$ i.e. the transition from steady state to oscillating behaviour (or vice versa; above denoted by I $\rightleftarrows$ II) can be achieved for distinct L_e- or $\dot{\varepsilon}_o$-values.

With regard to experimental investigations it is important to realize that as indicated in Fig. 5, criticality can be simply encompassed by a change of the loading conditions. Similarily, if the respective structure parameter (L_e) varies during deformation, the system can develop to criticality. The transition to criticality can cause a change in L_e which in turn will effect the α-value. The above mechanism can give rise to a behaviour in which critical α-values alternate with non critical ones i.e. ... $\alpha(L_e) \rightarrow \alpha(L_e^1) \rightarrow \alpha(L_e) \rightarrow \alpha(L_e^1) \rightarrow$ ....

Obviously this mechanism can manifest as "irregularities" in the output quantity (here σ) of the system considered. There is abundant experimental evidence /6,14,15/ for "irregular" behaviour in plastic flow, e.g. undulations in stress/strain- or creep curves respectively which usually are associated with repeated changes in grain- or dislocation structure. Work-hardening, recovery, grain growth, recrystallization, mechanisms leading to variations in phase- and chemical composition, or irradiation induced effects, these all are characterized by the change of a particular substructural parameter which may influence the corresponding system parameter α.

For $\alpha < 3$ the mobile dislocation density attains a steady state value $\rho_s = (\alpha_a/\beta)N_s = (c_e y_e L_e)^{-1} N_s$. From Eq. (6) it is for $\alpha = 3$ (criticality) $\dot{N} = 0.67$.

An estimate of ρ_s for $c_e = 0.5; {}^s y_e/b \approx 5$ and $L_e/b \approx 3.10^5 \div 1.10^4$ gives $\rho_s = (10^{13} \div 4.10^{15})\ m^{-2}$, i.e. a range of values which is observed experimentally (see e.g. /13/).

2.2.2 The model of Prinz and Argon /16/

During the course of deformation a certain fraction of mobile dislocations are immobilized on glide obstacles. With proceeding deformation (and especially at higher temperatures) the dislocations form "dislocation walls". Surrounded by the matrix with low density these walls usually compose a cellular substructure. The model under consideration (in the following denoted by PAM) investigates the dislocation flux balances between the cell interiors and the cell walls at steady state deformation. This is done by calculating dislocation storage and reduction rates in both these regions respectively. The net change of stored dislocation density $\hat{\rho}$ in the cell interior is given by the difference of the storage and annihilation term respectively, namely

$$d\hat{\rho}/d\varepsilon = (2\pi\omega/bK^2)\hat{\rho}^{1/2} - (L_a/bK^2)\hat{\rho} \quad (17a)$$

Using $\hat{\rho} = \hat{\rho}\ [\varepsilon(t)]$ we can rewrite the above nonlinear first order differential equation in the form

$$\dot{\hat{\rho}} \equiv d\hat{\rho}/dt = \dot{\varepsilon} d\hat{\rho}/d\varepsilon = P\hat{\rho}^{1/2} - R\hat{\rho} \quad (17b)$$

The corresponding rate factors are $P = (2\pi\omega/bK^2)\dot{\varepsilon}_o$ and $R = (L_a/bK^2)\dot{\varepsilon}_o$, where ω, K are numerical factors. The only one structure parameter L_a is the line length per dislocation loop eliminated by annihilation. Considering again $\dot{\varepsilon} = \dot{\varepsilon}_o$ as the constant input parameter, the solution of Eq. (17b) gives

$$\hat{\rho}(t)/\hat{\rho}_o = \{\exp(-Rt/2) - (P/R\hat{\rho}_o^{1/2})\,[1-\exp(-Rt/2)]\}^2 \quad (18)$$

where $\hat{\rho}_o$ is the initial density. For $t >> 2/R$ we obtain the steady state value $\hat{\rho}_s = (P/R)^2 = (2\pi\omega/L_a)^2$ which, independent of the loading conditions, is governed by the structure parameter L_a. Similar to the mobile dislocation density ρ, $\hat{\rho}(t)$ first increases with t. Then the increase decelerates and finally $\hat{\rho}$ becomes stationary.

With the substitution $Z = \hat{\rho}^{1/2}$ Eq. (17b) can be linearized and we have

$$dZ/dt \equiv \dot{Z} = P/2 - RZ/2 \quad (19)$$

The transposition of this differential equation into the corresponding difference equation gives /10/

$$Z_{t+1} = p - rZ_t \quad (20)$$

where $p = P/2 = (\pi\omega/bK^2)\dot{\varepsilon}_o$, in $(ms)^{-1}$ and $r = (R/s-1) = (L_a/2b^2K^2)\dot{\varepsilon}_o - 1$, in $(1/s)$.

Two cases are considered:

i. for $r = -1$ Eq. (20) has the solution

$$Z_t = Z_o + pt \quad \text{and} \quad (21)$$

ii. for $r \neq -1$, the solution of Eq.(20) is

$$Z_t = [Z_o - p/(1+r)](-r)^t + p/(1+r) \quad (22)$$

Obviously, because the condition $r=-1$ requires $\dot{\varepsilon}_o = 0$, the solutions given by Eq. (21) have no physical significance. Hence, the only one equation we have to deal with is Eq. (22). Because $\dot{\varepsilon}_o > 0$ the restriction holds that $r > -1$. Resubstituting into Eq. (22) we have

$$\hat{\rho}_t/\hat{\rho}_o = \{[1-u_o/1+r)](-r)^t + u_o/(1+r)\}^2 \quad (23)$$

where $u_o = P/\hat{\rho}_o^{1/2} = (\pi\omega/bK^2\hat{\rho}_o^{1/2})\dot{\varepsilon}_o (1/s)$.

Whether a dynamically stable stationary dislocation density $\hat{\rho}_s/\hat{\rho}_o = u_o^2/(1+r)^2$ will establish depends on the term $[1-u_o/(1+r)](1+r)](-r)^t$ as t is increased indefinitely. Obviously the value of r - which contains both, the structure parameter (L_a) and the loading conditions ($\dot{\varepsilon}_o$) - is of the utmost importance in this regard.

For the present purpose 6 dinstinct regions of r-values are considered. These are marked off in Fig. 6 on a vertical scale with r = +1,0 and -1 as the demarcation points, which in themselves constitute the regions A, C and B respectively. The time paths of $\hat{\rho}_t/\hat{\rho}_o$ for several r-values are reproduced as schematical drawings in Fig. 6.

The principal feature of $\hat{\rho}(t)$ in the time continuous approach is that irrespective of r (=R/2 - 1), the density $\hat{\rho}$ converges monotonically to a stationary value $\hat{\rho}_s$. In contrast to that in the time discrete approach, depending of the sign and value of the system parameter r, the features of $\hat{\rho}_t$ are basically different. In the regions A and B both the approaches give the same results. However, for $r \geq 0$ (regions C to F) the time discrete approach predicts a variety of behaviour. Because for $-1 < r \leq 0$ (region C) $\hat{\rho}_t$ approaches monotoneously a steady state value, r = o can be suggested as a critical value above which the descriptions in the respective pictures foundamentally differ. For r > o and with increasing r the density $\hat{\rho}(t)$ successively exhibits damped (D) stable (E) and diverging oscillations (F). The r-values defining the particular regions depend on the structure parameter

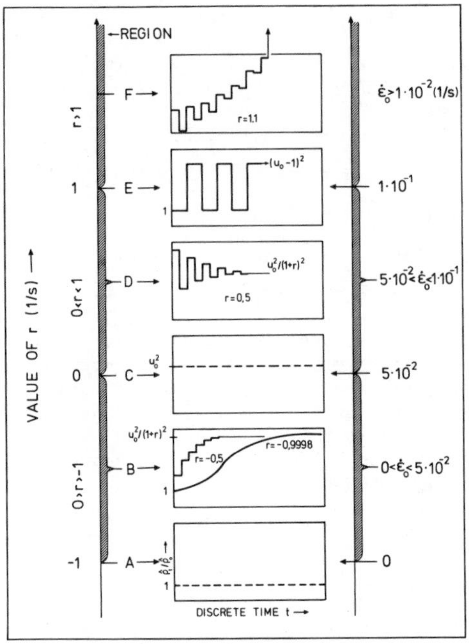

Fig. 6
Solutions of the PAM in the time discrete approach (schematically)

(L_a) as well as upon the loading condition ($\dot{\varepsilon}_o$).

In conclusion: In the above model (as in the case of EMM), the time discrete approach contains a system parameter (r). Distinct r-values characterize distinct behaviour of the system considered. Although the model equations ((6) and (30)) analyzed are quadratic first order difference equations, the results therefrom are analytically different. Whereas Eq. (6) has solutions which, with increasing value of the system parameter α, are represented by the so-called Feigenbaum's "bifurcation tree"/9,11,17/ (see Fig. 4). With increasing parameter r, first stable solutions exist of Eq. (30), which exhibits only one bifurcation (for r = 1) of period two (see Fig. 6). There are no stable solutions for r > 1.

Common to both the models is: i) that below a certain (critical) value of the system parameter, the behaviour of the system is identical with that predicted by the time continuous approach, i.e. evolution into steady state; ii) that above this critical value, the time discrete approach predicts fluctuating or diverging behaviour respectively; iii) that the respective system parameter contains the loading conditions as well as substructural parameters.

Obviously these points deserve the attention of the experimentalists.

3 SUMMARY AND CONCLUSIONS

The paper examines the stability of plastic flow in crystals. In general, the deforming specimen is a very complex, nonlinear dynamical system. The latter is suggested as a network of elements. Its input is feeded by the loading conditions ($\dot{\varepsilon}/\sigma$) through which the generator (supply of mobile dislocations) as well as different structurizers are activated. The generator, operating as a closed loop, is connected to a particular structurizer. Analytically the generator is usually described by means of nonlinear first order differential equations. The only solution therefrom are dislocation densities converging monotoneously to a stationary (steady state) value.

In order to reveal more about the behaviour of plastic flow in crystalline solids the analysis of the generator is performed by means the time-discrete difference calculus. Two dislocation models known from literature were paradigmatically transposed into the time discrete picture and the solutions of the respective non-linear first order difference equations were discussed and compared with solutions resulting from the time-continuous approach. The main characteristics of the time discrete approach are: i) The dynamical behaviour of the system is completely determined by a single system parameter m, containing the loading conditions as well as substructural quantities. ii) Close to distinct values m^i small perturbations in m lead to foundamental changes in system's behaviour. iii) Below a critical m-value the behaviour of the system is the same as that predicted by the time continuous approach, i.e. the dislocation density limits monotoneously to a steady state value. For m-values larger than the critical one the time discrete approach predicts - depending of the model considered - fluctuating or diverging dislocation densities.

Depending upon the loading conditions and upon a particular structural parameter the time discrete approach (difference calculus) leads to a variety of behaviour. Besides the case that dislocation density becomes stationary in time - the only one behaviour predicted by the time continuous treatment (differential calculus) - the time discrete approach predicts also diverging or oscillating densities. Hence, the latter approach gives the more general results. These probably are manifested by macroscopic phenomena.

In view of these considerations it seems reasonable to reconsider the explanation of particular dynamical phenomena in crystal plasticity like strain ageing, recovery and recrystallization etc. An understanding of these macroscopic effects, however, requires in addition to the above considerations about the time evolution of dislocation assemblies necessarily also the description of their spacial interaction.

The idea to this paper originated in discussions with Dr.Michael Michelitsch from IBM Deutschland in Böblingen. In particular the author is indebted to Dr.Boris Spruŝil from the Department of Metal Physics, Charles University Prague, for supporting criticism. Sincere thanks are expressed to Mr.Kleiser for revising the manuscript and to Mr.D.Piel (both from the above institute) for help in computational and graphical work.

4 REFERENCES

/1/ Kocks, U.F., Argon, A.S., Ashby, N.F.
Thermodynamics and kinetics of slip; Progress in materials sciences.
Pergamon Press, Vol. 13, 1975, p.1

/2/ Ilschner, B.
Hochtemperatur-Plastizität.
Springer, Berlin-Heidelberg-New York 1973, p. 281

/3/ Nabarro, F.R.N.
Theory of dislocations.
Claredon Press, Oxford, 1967, p. 458, 694.

/4/ Johnston, W.G., Gilmann, J.J.
J. Appl. Phys. 30 (1959) 129

/5/ Mayer, M.
Die dynamische Reckalterung von CuZn-, CuNi- und CuNiZu-Legierungen.
PhD-Thesis, Faculty of Mechanical Engineering, University of Karlsruhe, Karlsruhe, 1978.

/6/ Sakai, T., Jonas, J.J.
Acta Metall. 32 (1984) 189

/7/ Schlipf, J.
Z. Metallk. 75 (1984) 517

/8/ Argon, A.S.
Stability of plastic flow, in the inhomogeneity of plastic deformation.
ASM, Metals Park, Ohio, 1973, p. 161

/9/ May, R.M.
Nature 261 (1976) 459

/10/Goldberg, S.
Introduction to difference equations.
J.Wiley and Sons, New York, 1958

/11/Schuster, H.G.
Deterministic chaos.
Physic Verlag, Weinheim (Federal Republic of Germany), 1984, p. 31

/12/Li, T.Y., Yorke, A.J.
Am. Math. Monthly 82 (1975) 985

/13/Essmann, U., Mughrabi, H.
Phil. Mg. 40A (1979) 731

/14/McQueen, H.J., Jonas, J.J.
Recovery and recrystallization during high temperature deformation in treaties on materials science and technology. (ed. J.Arsenault).
Academic Press, New York, 1975, Vol. 6, p. 393-493

/15/Jonas, J.J., Sakai, T.
Deformation, processing and structure. (ed. A.Krauss).
ASM, Metals Pare, Ohio, 1984, p. 185-230

/16/Prinz, F.B., Argon, A.S.
Acta Metall. 32 (1984) 1021

/17/Feigenbaum, M.J.
J. Stat. Phys. 19 (1978) 25.

5th International Symposium on Continuum Models of Discrete Systems / Nottingham / 14-20 July 1985

Crystals as ordered media

L.Mistura
Dipartimento di Energetica, Università 'La Sapienza', Roma, Italy

ABSTRACT: The stability of the characteristic triple periodic structure of crystals is due to the existence of long range correlations that appear, at the macroscopic level, as a sensitivity to boundary conditions. As a consequence, the applicability of the methods of classical thermoelasticity (e.g. the local equilibrium assumption) appear to be inadequate in a number of circumstances where one feels the principle of thermodynamics should be applicable. For example: structural phase transitions, including melting, inhomogeneous deformations and structure of defects. With this in mind I present in this paper some preliminary considerations on the possibility of describing a crystal by means of a local order parameter, in analogy with other ordered media, like magnets. The main result of the analysis is an expression of the elastic stress tensor in terms of the thermodynamic variables.

1 INTRODUCTION

The formalism of classical thermodynamics is usually developed only for homogeneous systems, like a liquid or a gas in the absence of any external field, whose equilibrium states are specified by a small number of global parameters that, like volume, number of particles and energy, make no reference to the structure of system. A basic equation of this formalism, which is indeed a direct consequence of the homogeneity assumption, is the one expressing the pressure p in terms of the other thermodynamic quantities, namely

$$(1) \quad p = \mu\rho - \varepsilon + T\sigma = \mu\rho - f$$

where μ is the chemical potential, ρ the density, ε the internal energy density, T the temperature, σ the entropy density and f the free energy density.
This formalism cannot be directly applied to discuss the thermal properties of crystalline solids because crystals, even perfect crystals, are intrinsically inhomogeneous systems, in the sense that the one-particle density is not uniform in equilibrium, even in the absence of external fields. The question then arises of the status of the homogeneity assumption and Eq. (1) in the framework of crystals energetics.

The most natural way to extend the applicability of the principles of thermodynamics to non-uniform states is to introduce the so called local equilibrium assumption. Namely, if the inhomogeneities are very small over molecular distances, as those due to gravity or other external forces usually are, they can be treated by considering the system as made up of relatively small homogeneous parts in local equilibrium. As a consequence one can, for example, define a local free energy density, that will depend on the local values of temperature and mass density, exactly as it does for a macroscopic homogeneous system.
It is clear, however, that on this basis one can only describe those non-uniform configurations which are continuously mapped onto the homogeneous one by turning off the external field. On the other hand, for intrinsically non-uniform states, like the triple periodic crystal structure, which is characterized by the existence of long range correlations, appearing at the macroscopic level, as a sensitivity to boundary conditions, a local free energy density, in the sense of the local equilibrium postulate, cannot exist.
These complications, attendent upon applying the methods of thermodynamics to crystals, have not received the attention I believe they require. This may be because

there is a natural way to generalize the thermodynamics of homogeneous systems to crystals, which is usually referred to as thermoelasticity.
However, the range of applicability of classical thermoelasticity is restricted to homogeneous deformations, not necessarily small, but in any case within the range of stability of each crystalline phase. Thus a number of typical thermodinamic phenomena, like structural phase transitions, including order-disorder transitions and melting, the energetics of defects, e.g. the structure of a dislocation core lie outside the range of thermoelasticity.
In this paper I wish to report a preliminary investigation on an alternative formulation of the thermodynamics of crystals which, in principle at least, should provide a framework for the theoretical description of the fore mentioned phenomena. My approach is tailored on the thermodynamics of dielectric or magnetic materials. A characteristic feature of the thermodynamics of these systems is the appearence of a structural variable, the so called order parameter, e.g. the magnetization vector. How this formulation is applied to the description of phase transitions and to certain non-uniform configurations, like domain walls, is well known · Previous work along similar line of thought has been reported by Parry (1981).
In the next Section we will introduce the concept of ordered medium and will extend the formalism of thermodynamics to deal with energetic problems in these media. In particular we will derive a generalization of the basic equation (1) valid for anisotropic systems. Finally in Sec. III we will introduce a possible characterization of crystals as ordered media.

2 Thermodynamics of ordered media

2.1 The definition of ordered medium

An ordered medium is a thermodymanic system whose particles beside being characterized by their average position x in a region B of the space (B might be called the base space), also have an internal structure which is locally described, on the average, by a field variable $\psi(x)$ (the order parameter) on B.
The order parameter $\psi(x)$ is valued in a space which may be called the order parameter space or perhaps better, the structure space $\mathcal{Y}$. By analogy with the gange theory or elementary particles, the structure space might also be called the isotopic space. This space may have quite varied topological properties.
In the simplest case the structure of the particles is described by a single sealar quantity, e.g. the mass or the charge and as order parameter one can take the local mass or charge density respectively. For a dielectric or magnetic medium, the structure will be described by a polarization vector, like the average magnetization. The one-particle configuration space on the other hand will generally be the ordinary three-dimensional euclidean space E (3), because real thermodynamic systems are always imbedded in the ordinary space, but it has already been proposed (Kléman, 1981) that in certain cases, like in glasses or supercooled liquids, it is more natural to describe the configuration in a riemannian space of constant negative curvature. In any case it is more convenient to adopt general curvilinear coordinates so that the distance between two adjacent points is provided by the metric tensor , field by the well known expression

$$(1) \qquad ds^2 = g_{ij}(x)\, dx^i dx^j$$

where summation over repeated indices is implied.

2.2 Structural and deformation work

In connection with the existence of two spaces one has to introduce two types of reversible changes: changes in the configuration or deformations and structural changes. A continuous, finite deformation is then represented by a continuous path in the configuration space, while a continuous finite structural change will be represented by a continuous path in the structure space $\mathcal{Y}$.
A basic thermodynamic quantity associated with any such a path is the work done by the system. So we distinguish between a structural work W_s, associated with an infinitesimal reversible structural change and a deformation work, W_d.
For the structural work we shall assume that it can be expressed in terms of the order parameter $\psi(x)$ by means of the invariant form

$$(2) \qquad W_s = -\int \sqrt{g}\, \varphi(x)\, \delta\psi(x)\, dx$$

where g is the positive determinant $|g_{ij}|$ of the metrics coefficients.

For example, if the order parameter is simply the mass density $\rho(x)$ we have the familiar expression

$$W_s = -\int \sqrt{g}\, \mu(x)\, \delta \rho(x)\, dx \tag{3}$$

where $\mu(x)$ is the chemical potential, eventually including an external gravitational field.
As a further example we may consider a dielectric medium. In this case if $\vec{E}$ is the electric field inside the dielectric, the structural work can be written

$$W_s = -\frac{1}{4\pi}\int \sqrt{g}\, \vec{E}(x)\cdot \delta\vec{D}(x)\, dx \tag{4}$$

where the displacement vector $\vec{D}$ takes the role of order parameter.
For the deformation work it is no longer possible to maintain the simple expression $p\delta V$, valid for homogeneous systems, because non-uniform systems are also necessarily anisotropic. If the deformation is locally measured by the material strain tensor $u_{ij}(x)$, then the work done by the system during an infinitesimal deformation path is given by

$$W_d = -\int \sqrt{g}\, T^{ij}(x)\, \delta u_{ij}(x)\, dx \tag{5}$$

where T^{ij} are the components of the material stress tensor (also termed the second Piola - Kirchhoff stress tensor).

2.3 The laws of thermodynamics

We are now ready to apply the first law of thermodynamics in the form: the total work done by the system during an infinitesimal, adiabatic reversible change is equal to the corresponding variation of internal energy namely

$$\delta U = -(W_s + W_d) = \int \sqrt{g}\, \varphi(x)\, \delta\psi(x)\, dx + \int \sqrt{g}\, T^{ij}(x)\, \delta u_{ij}(x)\, dx \tag{6}$$

Moreover if the change is not adiabatic, the heat form is defined by

$$q = \delta U + (W_s + W_d) \tag{7}$$

Therefore, like in the thermodynamics of homogeneous systems, q is a linear differential form, but now with an infinite number of terms. It does not seem however that the Caratheodory theorem may be invalidated by this circumstance, so that from the second law of thermodynamics we can introduce the entropy in the usual way, namely

$$q = T\,\delta S \tag{8}$$

Combining Eqs. (7) and (8) we get the fundamental equation

$$\delta U = T\delta S + \int \sqrt{g}\, \varphi(x)\, \delta\psi(x)\, dx + \int \sqrt{g}\, T^{ij}(x)\, \delta u_{ij}(x)\, dx \tag{9}$$

This is a fundamental equation because if we know the internal energy U as a function of S and as a functional of the order parameter ψ and the strain tensor $u_{ij}(x)$, we can get the temperature T, the conjugate field $\varphi(x)$ and the stress tensor T^{ij} by differenziation. We can obtain alternative definitions of these quantities in the usual way, replacing the internal energy U with the free energy F defined by

$$F = U - TS \tag{10}$$

For example

$$\sqrt{g}\, T^{ij}(x) = \left(\frac{\delta U}{\delta u_{ij}(x)}\right)_{S,\psi} = \left(\frac{\delta F}{\delta u_{ij}(x)}\right)_{T,\psi} \tag{11}$$

These equations provide equivalent definitions of the thermodynamic stress tensor, namely the stress tensor associated with the "active tendencies" in the system. They are not however the most appropriate for the actual calculation of the stress because the internal energy or the free energy are in general not known as functional of the strain components, but as functionals of the order parameter and the components of the metric tensor. We shall therefore look for an alternative definition of the thermodynamic stress tensor in analogy with Hilbert's definition of the energy-momentum tensor in field theory.

2.4 The thermodynamic stress tensor (Mistura, 1985)

let x_i $(i = 1, 2, 3)$ denote the coordinates of the position of the particles in the reference configuration and x'_i their coordi nates after the deformation with respect to the same curvilinear coordinate system. Then the material strain tensor $u_{iJ}(x)$ is defined by

$$ds'^2 - ds^2 = 2u_{iJ}(x)\,dx^i\,dx^J \tag{12}$$

On the other hand the strain tensor components may be expressed in terms of the deformation matrix defined by

$$dx'_J = a_{Ji}(x)\,dx_i \tag{13}$$

Indeed

$$ds'^2 = g_{kl}(x')\,a_{ki}(x)\,a_{lJ}(x)\,dx_i\,dx_J \tag{14}$$

and therefore

$$u_{iJ}(x) = \frac{1}{2}\left\{g_{kl}[x'(x)]\,a_{ki}\,a_{lJ} - g_{iJ}(x)\right\} \tag{15}$$

Note in particular that the components of the metric field $g_{iJ}(x)$ are unaffected by the deformation.
We now look at the deformation from an alternative point of view, namely we ask for the coordinate transformation that will pull back the coordinates of all the particles after the deformation to their original values.
It is clear that the coordinates transformation we are looking for is just the inverse of that represented by Eq. (13). As a consequence of the coordinates transformation the components of the metric tensor will now change. The actual law of variation is well known from tensor calculus, namely

$$g'_{iJ}(x) = g_{kl}[x'(x)]\,a_{ki}(x)\,a_{lJ}(x) \tag{16}$$

and comparing with Eq. (16) we get

$$2u_{iJ}(x) = g'_{iJ}(x) - g_{iJ}(x) \tag{17}$$

This relation is essential for the definition of the thermodynamic stress tensor from a given expression of the free energy. Indeed, by taking into account Eq. (17) we can rewrite Eq. (11) in the form

$$\frac{1}{2}\sqrt{g}\,T^{iJ}(x) = \left(\frac{\delta F}{\delta g_{iJ}(x)}\right)_{T,\rho} \tag{18}$$

This is a quite convenient and unambiguous definition of the thermodynamic stress tensor starting from the free energy. Accordingly the components of the metric field play the role of thermodynamic variables.

2.5 An illustrative example: the inhomogeneous dielectric fluid

A non-uniform electric field exerts forces on the particles of a dielectric medium. As a consequence the medium in equilibrium will be inhomogeneous. We will now calculate these ponderomotive forces by taking esplicitly into account the eventual contribution of density gradients.
We postulate that the total free energy can be expressed as the invariant integral of a free energy density, namely

$$F = \int_\Omega \sqrt{g}\, f\, dx \tag{19}$$

In the usual discussion of the thermodynamics of dielectrics the free energy function f (a scalar) is assumed to depend on T, the mass density ρ and the dielectric induction $\vec{D}$. This is a local equilibrium assumption. Moreover cartesian coordinates are adopted, so that no variation of the g_{iJ} is possible. This last circumstance makes the calculation of the stress tensor extremely laborious.
Here we will more generally assume that f depends, beside T, ρ and $\vec{D}$, also on their derivates (a part from T which is necessarily uniform, in equilibrium) and the components of the metric fields g_{iJ}.
However we will assume that there no external charges so that we do not have introduce the derivative of $\vec{D}$, namely $d\vec{D} \equiv \equiv \mathrm{div}\,\vec{D} = 0$. Accordingly we have

$$\delta f = -\sigma\,\delta T + \frac{\partial f}{\partial g_{iJ}}\,\delta g_{iJ} + \frac{1}{4\pi}\,\vec{E}\cdot\delta\vec{D} + \mu\,\delta\rho + \vec{\nu}\cdot\delta d\rho \tag{20}$$

where σ is the entropy density, $\vec{E}$ the electric field inside the dielectric μ the chemical potential, $d\rho \equiv \mathrm{grad}\,\rho$ and $\vec{\nu}$ the field conjugate to $d\rho$.
Our purpose is to arrive at an expression of the stress tensor T^{iJ} in terms of the other thermodynamic variables, by exploiting the fact that f is a scalar, independent of the space coordinates. Ther resulting expression will be the searched generalization for inhomogeneous, anisotropic systems of the fundamental thermodynamic equation (1) giving the pressure for a homogeneous, isotropic system.
The equations determining the equilibrium

configurations of the system are obtained in the familiar way by means of the variational principle: the free energy (19) has to be a minimum with respect to any isothermal, structural change ($\delta T = 0$, $\delta g_{iJ} = 0$) subject to the constraints

$$\int_{\partial\Omega} \vec{D}\cdot d\vec{S} = 4\pi Q \qquad \delta\rho = 0 \quad \partial\Omega$$

$$(21) \quad d\vec{D} = \frac{1}{\sqrt{g}}\frac{\partial}{\partial x^i}\left(\sqrt{g}\, D^i\right) = 0 \qquad \delta M = \delta\int \sqrt{g}\,\rho\, dx = 0$$

Here $\partial\Omega$ denotes the boundary of the space region Ω occupied by the system and Q the charge on this boundary.
Using Logrange's method of multipliers we get the following Euler-Logrange "equation of motion"

$$(22) \quad \mu - \operatorname{div}\vec{\nu} = \mu_0 \qquad \vec{E} = -d\varphi$$

Moreover φ has to be constant on the boundary.
So in a certain sense we can say that at equilibrium the system tends to be as homogeneous as possible $[d(\mu - \operatorname{div}\vec{\nu}) = 0$ and $d\vec{E} = 0]$ compatibly with the constraints.
So far we have considered only structural changes, in order to calculate the stress we must now introduce a deformation. We again assume $\delta T = 0$ and therefore

$$(23) \quad \delta f = \frac{\partial f}{\partial g_{iJ}}\delta g_{iJ} + \frac{1}{4\pi}\vec{E}\cdot\delta\vec{D} + \mu\,\delta\rho + \vec{\nu}\cdot\delta d\rho$$

and

$$(24) \quad \delta F = \int_{\Omega}\left(\sqrt{g}\,\delta f + f\,\delta\sqrt{g}\right)dx$$

Now the terms

$$(25) \quad \int_{\Omega}\sqrt{g}\left(\frac{1}{4\pi}\vec{E}\cdot\delta\vec{D} + \mu\,\delta\rho + \vec{\nu}\cdot\delta d\rho\right)dx$$

which appear in Eq. (24) can be transformed with the help of the equilibrium conditions (22). We thus find

$$(26) \quad \delta F = \int_{\Omega}\left[\sqrt{g}\left(\frac{\partial f}{\partial g_{iJ}}\delta g_{iJ} + \mu_0\delta\rho - \frac{1}{4\pi}d\varphi\,\delta\vec{D}\right) + f\,\delta\sqrt{g}\right]dx$$

But taking further into account the constraints (21) we get

$$(27) \quad \int\sqrt{g}\,\delta\rho\, dx = -\int\rho\,\delta\sqrt{g}\, dx$$
$$\sqrt{g}\,\delta D^i = -D^i\,\delta\sqrt{g}$$

On the other hand

$$(28) \quad \delta\sqrt{g} = \frac{1}{2}\sqrt{g}\, g^{iT}\delta g_{iJ}$$

so that we conlcude, from Eq. (18)

$$(29) \quad T^{iJ} = 2\frac{\partial f}{\partial g_{iJ}} + \frac{1}{2}g^{iJ}\left(f - \mu_0\rho - \frac{1}{4\pi}E_i D^i\right)$$

We can now get an explicit expression of the T^{iJ} in terms of the other thermodynamic variables by exploiting the fact that f is a scalar and therefore independent of the space coordinates. As a consequence if we substitute in the general expression for the variation of f given be Eq. (23), that variation of the structural variables that results from an infinitesimal coordinates transformation, δf must vanish identically. If we write the infitesimal transformation in the form

$$(30) \quad x'_J = x_J + u_J(x)$$

we have $a_{Ji} = \delta_{Ji} + \frac{\partial u_J}{\delta x_i}$ and therefore the transformation laws for the components of the various tensorial quantities are as follows:

31a) $\delta\rho = 0$ because ρ is a scalar

31b) $\delta(d\rho_i) = -\frac{\partial u_J}{\partial x_i}(d\rho)_J$ vector field

31c) $\delta D^i = \frac{\partial u_i}{\partial x_J}D^J$ vector field

31d) $\delta g_{iJ} = -\left(\frac{\partial u_k}{\partial x_i}g_{ki} + \frac{\partial u_k}{\partial x_J}g_{Jk}\right)$

Substituting these expressions into Eq.(23) we get

$$\delta f = -2\frac{\partial u_k}{\partial x_i}\left(\frac{\partial f}{\partial g_{iJ}}g_{kJ}\right) + \frac{1}{4\pi}E_k\frac{\partial u_k}{\partial x_i}D^i +$$
$$(32) \quad -\nu^i\frac{\partial u_k}{\partial x^i}(d\rho)_k = \frac{\partial u_k}{\partial x^i}\left(-2\frac{\partial f}{\partial g_{iJ}}g_{kJ} + \right.$$
$$\left. + \frac{1}{4\pi}E_k D^i - \nu^i(d\rho)_k\right) = 0$$

Due to the arbitrarines of the transformation this is possible only if the expression in brackets yanishes identically.
On the other hand from Eq. (29) we get

$$(33)\quad T^{iJ}g_{KJ} = T^{i}_{K} = 2\frac{\partial f}{\partial g_{iJ}}g_{KJ} + \delta^{i}_{K}(\tilde{f} - \mu_{o}\rho)$$

where $\tilde{f} = f - \frac{1}{4\pi}E_i D^i$, and therefore comparing with Eq. (32)

$$(34)\quad T^{i}_{k} = \frac{1}{4\pi}E_k D^i - \gamma^i \frac{\partial \rho}{\partial x^k} + \delta^i_k(\tilde{f} - \mu_o \rho)$$

This equation represents the appropriate generalization to non-uniform systems of the basic relation (1).
We note in particular that, if the structure of the system is described only by means of scalar quantities, the stress will be isotropic.

3 The order parameter of a crystalline phase

In this Section we will investigate how the thermodynamic of ordered media outlined in the preceding Section can be applied to crystals.
It is clear that the crystal structure cannot be described by a single scalar quantity like in simple fluids. We must take into account that, in crystals, the structural properties of the particles become manifest in their ability to organize themselves in a regular lattice. Thus, following Rogula (1976), we suggest to associate to any particle in a crystal, beside its position in the configuration space E (3), a point in a space of lattices. This space of lattices is a manifold whose elements are in one-to-one correspondence with space lattices. For simplicity we restrict to Bravais lattices whose manifold will be denoted by Λ . Any Bravais lattice is represented by a single point of Λ and a reversible structural change of a crystal will be represented by a continuous path in Λ . From a field theoretical point of view the choice of Λ as the structure space and a Bravais lattice as an order parameter value is admittedly somewhat odd . However a Bravais lattice is completely determined by a point $P \in E(3)$, which may be identified by a position vector $\vec{p}\,(\underline{x})$, and three basic lattice vectors $\vec{a}_1$, $\vec{a}_2$, $\vec{a}_3$. So we have a more traditional representation of the order parameter by identifying it with a moving frame, consisting of the origin P and the triad of basic vectors. In any region of the base space B corresponding to a "good crystal" we can associate to each point a value of the order parameter so that each point x of the base space, which is not in the core of a defect, is lebelled with a triad of basic vectors $\vec{a}_1(x)$, $\vec{a}_2(x)$, $\vec{a}_3(x)$ and a position vector $\vec{p}(x)$ that fix the origin of the lattice. This procedure is reminiscent of that which is usually followed to introduce the tangent space of a manifold, however in the tangent space the origin has an absolute meaning (only homogeneous transformations are allowed) while here the origin can be translated (affine space).
The identification of the order parameter with a moving frame is quite interesting from the theoretical point of view because the moving frame is a basic concept in affine differential geometry which in turn provides the most natural language for the description of Bravais lattices with a continuum distribution of dislocations (Kröner, 1981). In this connection it is appropriate at this point to digress briefly to give an intuitive idea of the relations between the topological properties of the structure space and the structural (as apposite to configurational) defects in ordered media (Rogula, 1976). We started this Section identifying the structure space with the Rogula space Λ , and then introduced a local representation of this space by means of the space of moving frames. However the global properties of the two spaces are different. Indeed, while given a frame a unique Bravais lattice is identified, the converse is not true. Neither the origin, nor the triad of basic vectors are uniquely determined by the lattice. In particular the origin can be any lattice point while the basic vectors are determined apart from a unimodular linear transformation represented by an integral matrix. This situation, although formally more complicated, is conceptually equivalent to that one finds in nematic liquid crystals. In that case the order parameter is locally identified with a unit vector $\vec{n}$ and therefore the order parameter space is homeomorphic to a two-sphere. On the other hand two opposite orientation of $\vec{n}$ are physically equivalent, so that globally the order parameter space is homéomorphic to the projective plane. It is well konwn that it is just to this topological property which is related the existence of structurally stable disclination lines in nematic liquid crystals. Similarly in a crystal, a moving frame may be used to identify a local coordinate system in the

structure space, but to describe correctly its global topological properties we must take into account the identification of different points imposed by the multivaluedness of the representation. By analogy with the nematic case one expects in crystals three types of structural line defects namely dislocations (non-uniqueness of the origin of the moving frame), disclinations (associated with rotations of the basic triad) and also shear defects (associated with equivalent affine transformation of the basic triad). The physical meaning of this prediction, which is due to Rogula, is not yet clear.

Let us now go back to the characterization of the crystal from the energetic point of view.

We will only be able to give some preliminary considerations. We will assume the existence of a free energy density f_c, which however does not coincide with Helmholtz free energy. This is because to study structural transformations in crystals we must imagine the system immersed not only in a thermostat at temperature T, but also in equilibrium with a particle reservoir at chemical potential μ. As a consequence the appropriate free energy f_c will be a function of T and μ. For its dependence on the order parameter let us consider the case $\vec{p} = const$, so that f_c will be independent of $\vec{p}$. With this condition we recover classical thermoelasticity, introducing the notion of a compatible configuration such that

$$d\vec{a}_\ell \equiv rot\,\vec{a}_\ell = 0$$

Then it is natural to assume

$$f_c = f_c(T, \mu, \vec{a}_\ell)$$

which, apart from the explicit appearance of μ, is the Born form of the free energy of classical thermoelasticity. However in our case we don't have to introduce any hypothesis on the relation ships between the order parameter variation and the strain tensor. Indeed if we denote by

$$\vec{\nu}^\ell = \frac{\partial f_c}{\partial \vec{a}_\ell}$$

the thermodynamic variables conjugate to the $\vec{a}_\ell$, at equilibrium we have

$$d\vec{\nu}^\ell \equiv div\,\vec{\nu}^\ell = 0$$

and the stress tensor is obtained, with the procedure outlined above, in the form

$$T^i_J = \nu^{\ell i} a_{\ell J} + \delta^i_J f_c$$

For incompatible configurations, like those occurring in the presence of dislocations, $d\vec{a}_\ell \neq 0$ and therefore

$$f_c = f_c(T, \mu, \vec{a}_\ell, d\vec{a}_\ell)$$

The dependence on the $d\vec{a}_\ell$ is analogous to the dependence on $d\rho$ for non-uniform fluids. There will be corresponding terms in the expression of the stress tensor.

I conlude observing that the language of differential forms and exterior calculus seems to be particularly suited for the mathematical description of the ideas outlined above.

REFERENCES

Kléman, M., Crystallography of amorphous bodies, in: Brulin, O. and Hsieh, R.K.T., (eds.) Continuum models of discrete systems (North Holland Publ. Company 1981).

Kröner, E., Continuum theory of defects, in: Balian, R. et al. (eds.) Les Houches, Session XXXV, 1980, Physics of defects (North Holland Publ. Company 1981).

Mistura, L., The pressure tensor in non-uniform fluids. J. Chem. Phys. (Sept. 1985 to appear).

Parry, G.P., Phase changes in crystal mechanics, in : Brulin O. and Hsieh, R.K.T. (eds.) Continuum models of discrete systems (North Holland Publ. Company 1981).

Parry, G.P., On phase transitions involving internal strain, Int. J. Solid Structures 17 (1981) 361-378.

Rogula, D., Large deformations of Crystals, homotopy and defects, in: Fichera G., (Ed) Trends in Applications of pure mathematics to mechanics (Pitman, London 1976).

5th International Symposium on Continuum Models of Discrete Systems / Nottingham / 14-20 July 1985

Some problems associated with some current non-local theories of elasticity

C.Atkinson
Department of Mathematics, Imperial College of Science & Technology, London, UK

ABSTRACT: In the last few years, non-local elastic theories have been applied to calculate stress and strain fields in the vicinity of defects. The results of these calculations are critically examined from the point of view of their mathematical and physical correctness.

1 INTRODUCTION AND DISCUSSION

It is the purpose of this note to re-appraise the situation concerning the use of certain non-local theories for calculating stress fields associated with defects. In particular we address the claim made by Eringen and co-workers (1977 etc.) that their non-local theory will predict a finite stress at a crack tip. They propose a non-local theory with stress-strain relation of the type

$$t_{ij}(\underline{X}) = \int_V \alpha(|\underline{X}'-\underline{X}|,\tau)\, \sigma_{ij}(\underline{X}')dv(\underline{X}') \quad (1.1)$$

$$\sigma_{ij}(\underline{X}') = \lambda e_{rr}(\underline{X}')\delta_{ij} + 2\, e_{ij}(\underline{X}') \quad (1.2)$$

$$e_{ij}(\underline{X}') = \frac{1}{2}\left[\frac{\partial u_i(\underline{X}')}{\partial X_j'} + \frac{\partial u_j(\underline{X}')}{\partial X_i'}\right] \quad (1.3)$$

where t_{ij} and u_i are the stress tensor and the displacement vector respectively; λ and μ are the Lamé elastic constants and α is the "attenuation function" which depends on the distance $|\underline{X}'-\underline{X}|$ and a parameter τ which denotes the ratio of the internal characteristic length a to the external characteristic length ℓ i.e.

$$\tau = e_0 a/\ell \quad (1.4)$$

For example a may be taken as a lattice parameter and ℓ as the half crack length when discussing cracks in crystals, e_0 is a constant appropriate to a given material.

In addition to the above constitutive equations there are, of course, the equilibrium equations

$$t_{ij,j} = 0 \quad (1.5)$$

together with boundary conditions appropriate to a given problem.

Various "attenuation functions" have been suggested in the literature by matching the dispersion curves of plane waves with those of atomic lattice dynamics (or experiments). Some examples of these are:

(a) One-dimensional moduli

$$\alpha(|x|,\tau) = \frac{1}{\ell\tau}\left(1 - \frac{|x|}{\ell\tau}\right), \quad |x| < \ell\tau$$
$$= 0 \qquad , \quad |x| > \ell\tau \quad (1.6)$$

$$\alpha(|x|,\tau) = \frac{1}{2\ell\tau}\, e^{-|x|/\ell\tau} \quad (1.7)$$

$$\alpha(|x|,\tau) = \frac{1}{\ell\sqrt{\pi\tau}} \exp(-x^2/\ell^2\tau) \quad (1.8)$$

$$\alpha(|x|,\tau) = \frac{1}{\pi}\, \frac{\ell\tau}{(x^2+\ell^2\tau^2)} \quad (1.9)$$

(b) Two-dimensional moduli

$$\alpha(|\underline{x}|,\tau) = (2\pi\ell^2\tau^2)^{-1}\, K_0((\underline{x}.\underline{x})^{\frac{1}{2}}/\ell\tau) \quad (1.10)$$

where K_0 is the modified Bessel function.

$$\alpha(|\underline{x}|,\tau) = (\pi\tau\ell^2)^{-1} \exp(-\underline{x}.\underline{x}/\ell^2\tau) \quad (1.11)$$

and obvious generalisations of these in (a) above.

Similar expressions can be derived for the three dimensional case. Note that equation (1.6) gives a match between the dispersion curve of one-dimensional plane waves based on the above non-local elastic theory and the one-dimensional Born-Karman model of lattice dynamics. According to Eringen (1978) most of the above non-local moduli provide excellent approximations to the atomic dispersion curves for a choice of e_0. All the above expressions have the property that $\alpha(|X|) \to \delta(|X|)$ as $\tau \to 0$.

It is worth noting that for the usual model of a slit crack of length 2ℓ under tension the boundary conditions are taken to be

$$\begin{aligned} t_{yx} &= 0 & y &= 0 & &\forall x \\ t_{yy} &= -t_0 & y &= 0 & |x| &< \ell \\ V &= 0 & y &= 0 & |x| &> \ell \end{aligned} \qquad (1.12)$$

(cf. Eringen (1977) equation (5.8) and Eringen, Speziale and Kim (1977)).

For a line crack subject to antiplane shear (Eringen (1979) equation (3.7)), the boundary conditions are taken to be

$$\begin{aligned} w(x,0) &= 0 & &\text{for} & |x| &> \ell \\ t_{yz}(x,0) &= \tau_0 & &\text{for} & |x| &< \ell \\ w(x,y) &= 0 & &\text{as} & (x^2+y^2)^{\frac{1}{2}} &\to \infty. \end{aligned} \qquad (1.13)$$

In the above t_0 and τ_0 are constants. Similar boundary conditions are taken for the plane strain shear problem in Eringen (1978). The notation in the above papers corresponds with that of equations (1.1) to (1.3) in a fairly straightforward way, i.e. (x_1,x_2,x_3) corresponds to (x,y,z), (u_1,u_2,u_3) with (u,v,w) etc.

An attempt was made by Eringen, Speziale and Kim (1977) to solve numerically the problem posed by the above boundary conditions. They concluded from their numerical results that a good approximation could be obtained by taking the classical elastic solution and feeding this into their integral equation formulation of the problem.

This procedure was questioned by Atkinson (1980) where it was shown that the approximation was non-uniform. This meant that a new co-ordinate system could be defined in the neighbourhood of the crack tip for which as the parameter τ tended to zero the solution did not converge to the original boundary conditions (1.12), (1.13), etc. Moreover, in Atkinson (1980a, b) non-existence theorems are sketched which state that no solution exists for the problem as posed with boundary conditions (1.12) or (1.13) and the field equations (1.1) through (1.5) together with the condition of non-infinite crack opening displacements. In addition to this we have derived in Atkinson (1980b) and (1980c) results when the "attenuation function" α has a delta function component so

$$\alpha(|\underline{x}|) = b\delta(|\underline{x}|) + (1-b)\,\alpha_1(|\underline{x}|) \qquad (1.14)$$

where $\alpha_1(|\underline{x}|)$ is of a form similar to that defined in equations (1.10) and (1.11) if two-dimensional problems are considered. With $\alpha(|\underline{x}|)$ defined as in (1.14) and boundary conditions like (1.12) or (1.13) we have shown that for $\tau \ll 1$, the solution for the crack tip stress and displacement fields tend to the classical crack tip elastic displacement field multiplied by $b^{-\frac{1}{2}}$ and the classical crack tip stress field multiplied by $b^{\frac{1}{2}}$. These results agree with numerical results given in Atkinson (1980c) and support the non-existence theorems mentioned above since b=0 for those cases. A rigorous non-existence proof is given in Atkinson (1980c), we reproduce this here in the Appendix to illustrate the difficulty with the theory.

Eringen (1981) has replied to the above criticisms by stating that in Atkinson (1980a) incorrect boundary conditions were used for the one-dimensional model problem discussed there. However, the boundary conditions used were precisely those used in the treatment of Eringen et al. (1977). Furthermore, the changes in the boundary conditions suggested by Eringen (1981) are somewhat contrived and seem to be designed to justify the approximation scheme he had used previously and which we had shown had a non-uniform character. Even with the changed conditions, however, his conclusions are not all correct. For example with the one-dimensional modulus (1.7), with $\beta = \frac{1}{\ell\tau}$ in the notation of Eringen (1981) he presents the solution (his equation (4.5))

$$\frac{t_{yy}}{t_0} = -1 + e^{-\beta\ell}\cosh\beta x \quad |x| < \ell \qquad (1.15)$$

and then states that "using the reasoning of Atkinson (1980a) we set $\beta x = X$ then

$$P_c(x,\beta) = (t_{yy}+t_0)/t_0 = e^{-\beta\ell}\cosh X\text{"} \qquad (1.16)$$

He then concludes that this tends to zero uniformly in β as $\beta \to \infty$ for $x < \ell$. This is not in fact the reasoning of Atkinson (1980a), the correct reasoning would rescale the coordinates near the crack tips (e.g $x = \ell$) then writing

$$\beta(x-\ell) = U, \qquad U < 0 \tag{1.17}$$

one gets

$$(t_{yy}+t_0)/t_0 = \frac{1}{2}(e^U+e^{-U}e^{-2\beta\ell}) \tag{1.18}$$

and this does not tend to zero uniformly as $\beta \to \infty$. The one-dimensional model is, of course, meant to be merely illustrative so does not perhaps merit too much detailed attention. Our point here is that fundamental difficulties do not necessarily go away even in this simple model situation.

To summarise the conclusion of the above discussion the results of Atkinson (1980a, b) and (1980c) suggest that with $\alpha(|\underline{X}|)$ defined as in (1.14) (i.e. with a delta function component) solutions exist to problems defined by boundary conditions such as (1.12) but in this case the stress field has the characteristic stress singularity of the classical elastic solution. Constitutive equations of this type were suggested by Kroner (1970). If b in equation (1.14) is set equal to zero then no solution exists if the crack opening displacement is to be non-infinite.

The detailed results presented in the above mentioned papers do not consider the particular non-local modulus given in (1.10) above. This modulus has been considered recently by Ari and Eringen (1983) and Eringen (1983a) and (1983b). In Ari and Eringen (1983) boundary conditions identical to those of (1.12) are taken (their equations (3.1)) in a medium with non-local modulus given by (1.10). They reduce the boundary conditions to identical ones on the tensor σ_{ij} given in (1.2) thus the problem for σ_{ij} looks like the classical elastic problem. The results are complicated but it seems to us that this amounts to specifying the shape of the crack as that appropriate to the classical elastic problem and then substituting this into (1.1) to evaluate the stresses t_{ij}. The stress field on the crack line $|x| < \ell$, $y = 0$ will not then satisfy the boundary conditions (1.12) uniformly (this is the point made in the papers of Atkinson cited above). Indeed this appears to be the case in Fig. 2. of Ari and Eringen's paper. This problem is readdressed by yet a different method in the papers of Eringen (1983) where the case of a screw dislocation is re-examined and earlier results rederived by this new method. We have checked some of these results for the dislocation but have not yet rationalized the results for the crack problem.

To conclude, the major claim of the non-local theory is that it predicts finite stresses at crack tips and at dislocation cores where classical elasticity will lead to infinite stresses. Such a comment on classical elasticity is not of course quite true, finite stresses at a dislocation core can be achieved if instead of a volterra dislocation model the dislocation is modelled by infinitesimal somigliana dislocations in just the right proportion near the dislocation core, whereas finite stresses at a crack tip can be achieved by non-uniform loading of the crack in an appropriate manner (e.g. the Dugdale model or the model of Barenblatt). A non-local constitutive law such as (1.1) does have obvious potential for producing finite stresses since if $\alpha(|\underline{X}|)$ is of a form such as those of (1.6) to (1.11) singularities in σ_{ij} will be smoothed out when t_{ij} is computed. However, one consequence of such smoothness is that reasonable t_{ij}'s could be computed from a σ_{ij} which is derivable from a displacement field u_i which is unphysical (e.g. singular). If the problem is specified only in terms of displacements such as the dislocation problem then there may be no difficulties. If the problem is specified in terms of stresses such as equations (1.12) then there may be problems such as we have indicated. The calculations made by Eringen et al. for the crack problem can perhaps be most accurately interpreted as having been made for a crack opening displacement of a specified shape, i.e. an elliptical crack opening, the stress on the crack faces is then non-uniform depending on the "attenuation factor" α and the parameter τ. To compare the stress concentrating character of such a crack for different materials would then be possible but is not so useful as the case of a uniformly loaded crack. Finally, we should mention that other more general non-local continuum theories have been suggested (e.g. Kunin (1982)). We are not at present qualified to comment on these, the issue discussed here is contentious enough.

REFERENCES

Ari, N. and Eringen, A.C. 1983. Non-local stress field at Griffith crack. Cryst. Latt. Def. and Amorph. Mat. 10:33-38.

Atkinson, C. 1980a. On some recent crack tip stress calculations in non-local elasticity. Arch. Mech. 32: 317-328.

Atkinson, C. 1980b. Crack problems in non-local elasticity. Arch. Mech. 32: 597-614.
Atkinson, C. 1980c. Non-local elasticity and the line crack problem: in Proc. Int. Symposium on Defects & Fracture. Poland (13-17)
Eringen, A.C., C.G. Speziale and B.S. Kim. 1977. Crack tip problem in non-local elasticity, J. Mech. Phys. Solids 25: 339-356.
Eringen, A.C. 1978. Nonlinear equations of physics and mathematics. p.271, A.O. Barut (ed.) Dordrecht-Holland: Reidel.
Eringen, A.C. 1978. Line crack subject to shear. Int. J. Frac. 14: 367-379.
Eringen, A.C. 1979. Line crack subject to anti-plane shear. Engng. Frac. Mechs. 12: 211-219.
Eringen, A.C. 1981. On the nature of boundary conditions for crack tip stress. Arch. Mech. 33: 937-947.
Eringen, A.C. 1983a. On differential equations of non-local elasticity and solutions of screw dislocation and surface waves. J. Appl. Phys. 54: 4703-4710.
Eringen, A.C. 1983b. Interaction of a dislocation with a crack. J. Appl. Phys. 54: 6811-6817.
Kroner, E. The problem of non-locality in the mechanics of solids. Review on present status in Fundamental Aspects of Dislocation Theory. J.A. Simmons (ed.) et al. NBS Special Publ. 317: II: (1970) 729-736.
Kunin, I.A. 1982-83. Elastic Media with Microstructure. I. One-dimensional Models. II. Three-dimensional Models (1983) (Springer Series in Solid State Sciences). 26 Springer-Verlag, Berlin, New York (1982).

APPENDIX

We paraphrase here a non-existence theorem given in Atkinson (1980c) other theorems are sketched in Atkinson (1980a) and (1980b). The purpose of including such a theorem here is to indicate the kind of difficulties we think occur in theories such as proposed by equation (1.1).

Consider the integral equation

$$1 = \int_{-\ell}^{\ell} K(x-\xi) f(\xi)\, d\xi \quad , \quad |x| \leqslant \ell \qquad (A.1)$$

where $$U(x) = \int_{-\ell}^{x} f(\xi) d\xi \quad |x| \leqslant \ell \qquad (A.2)$$

and

$$\int_{-\ell}^{\ell} f(\xi) d\xi = 0 \qquad (A.3)$$

The kernel K is defined as

$$K(x) = \frac{x}{x^2+a^2} \qquad (A.4)$$

a is a physical constant not equal to zero. Note that if a is zero (A.1) reduces to the well known Cauchy integral with solution $f(\xi) = \xi/(1-\xi^2)^{\frac{1}{2}}$. The reader familiar with dislocation models of fracture will recognise $f(\xi)$ as a dislocation density and $U(x)$ as proportional to the crack opening. Since $U(x)$ is a physical displacement we impose the condition on $f(\xi)$ that it must be integrable. With this condition we now prove that (A.1) does not have a solution with $a \neq 0$. (Note equation (A.1) can be derived for a problem with a two-dimensional modulus a generalisation of (1.9)).

Proof

For complex z, define

$$\varphi(z) = \int_{-\ell}^{\ell} K(z-\xi) f(\xi) d\xi \quad . \qquad (A.5)$$

This makes sense in the strip $|\text{Im} z| < a$ and is analytic in z. Furthermore, $\varphi(z) = 1$ for $z = x$, $|x| \leqslant \ell$, hence $\varphi(z) \equiv 1$ in the whole region $|\text{Im} z| < a$, and in particular for $z = x$ (real) for all x. (Here the theorem used is that if $\psi(z)$ is analytic in region D, and $\psi(z) = 0$ at a sequence of points $z = z_n$ with $z_n \to z^*$ in D then $\psi(z) = 0$ in D). However, for real $x > \ell$ and $|\xi| \leqslant \ell$

$$|K(x-\xi)| \leqslant \frac{1}{|x-\xi|} \leqslant \frac{1}{x-\ell} \qquad (A.6)$$

so $$|\varphi(x)| \leqslant \frac{1}{(x-\ell)} \int_{-\ell}^{\ell} f(\xi) d\xi$$

this gives $\varphi(x) \to 0$ as $x \to +\infty$ contradicting $\varphi(x) \equiv 1$.

Note that the integrability of $f(\xi)$ is used in the demonstration of the analyticity of $\varphi(z)$ above.

5th International Symposium on Continuum Models of Discrete Systems / Nottingham / 14-20 July 1985

Asymmetric swirling flows and related problems in non-linear elasticity

K.R.Rajagopal
Department of Mechanical Engineering, University of Pittsburgh, Pa., USA

ABSTRACT: Few boundary value problems in fluid mechanics can match the attention which has been accorded to the flow of the classical linearly viscous fluid between parallel rotating disks. A new and interesting feature which has been introduced recently is the existence of solutions that are not symmetric. In this article we review the existence of such asymmetric solutions in the case of the linearly viscous fluid. Related problems in the instance of general non-Newtonian fluids and non-Linear elasticity are also discussed.

1 INTRODUCTION

Karman (1) used a similarity transformation to study the steacy axially symmetric swirling flow of the classical incompressible linearly viscous fluid, induced by the rotation of an infinite disk. Later, Batchelor (2) showed that such a similarity transformation would be appropriate for studying the flow of a linearly viscous fluid between two infinite parallel disks, rotating with constant but differing angular velocities, about a common axis. These two works have been followed by extensive studies which can be matched by few boundary value problems in fluid mechanics.

Recently, breaking away from the approaches of Karman (1) and Batchelor (2) which assumed axial symmetry, Berker (3) considered the possibility of solutions that are not necessarily axially symmetric and established a one-parameter family of solutions for the flow of the classical linearly viscous fluid between two plane parallel disks rotating about a common axis with the same angular speed. The only axially symmetric solution in this family is the rigid body motion; the only solution that would follow from the classical assumptions of Karman. In the light of Berker's work, Parter and Rajagopal (4) re-examined the problem of flow between parallel disks rotating about a common axis with differing angular speeds. Parter and Rajagopal (4) rigorously prove that the axially symmetric solutions are never isolated when considered within the full scope of the Navier-Stokes equation. Similar results apply in the case of the flow due to a single rotating disk and flow due to rotating disks subject to suction or injection at the disk. We shall discuss the implications of these asymmetric solutions and several interesting related problems: the possibility of asymmetric solutions in the case of viscoelastic fluids due to parallel disks rotating about a common axis, and the possibility of asymmetric solutions in the case of twisting of an infinite slab of non-linearly elastic materials.

The results of Berker (3) have relevance to another very interesting application in non-Newtonian fluid dynamics, the flow occurring in the orthogonal rheometer. The apparatus consists of two parallel disks which rotate with the same constant angular speed about two parallel but different axis. The fluid to be tested fills the space between the plates. If the fluid is non-Newtonian, then normal stresses develop and measuring these will help in characterizing the fluid that fills the appartus.

Rajagopal* (5) recognized that a velocity field similar to that used by Berker (3) can be employed in this problem and that the velocity field assumed by Berker (3) is a motion with constant principal relative stretch history. He used this fact to show that the flow of any homogeneous incompressible simple fluid in such a configuration is governed by a second order partial differential equation. Thus, unlike other boundary value problems in which one might

*Goddard (6) independently established results which are essentially the same. later.

require additional boundary conditions for specific non-Newtonian fluid models (of the differential type), the adherence boundary condition is sufficient for determinacy.

2 FLOW BETWEEN PARALLEL DISKS ROTATING ABOUT A COMMON AXIS

We shall start with a brief discussion of the axially symmetric flow of a Navier-Stokes fluid due to two infinite parallel disks rotating with differing angular speeds about a common axis. An up-to-date review of the numerical and mathematical work on the axially symmetric solutions for a Navier-Stokes fluid can be found in the review article by Parter (7).

Karman assumed an axially symmetric velocity field of the form:

$$v_r = \frac{r}{2} H'(z), \quad v_\theta = \frac{r}{2} G(z) \text{ and } v_z = -H(z). \tag{2.1}$$

Here v_r, v_θ and v_z denote the components of the velocity in the r, θ and z directions, respectively. Notice that the velocity field (2.1) automatically satisfies the constraint of incompressibility. Substituting (2.1) into the Navier-Stokes equations yields

$$\varepsilon H^{iv} + HH''' + GG' = 0 \;, \tag{2.2}$$

and

$$\varepsilon G'' + HG' - H'G = 0 \;, \tag{2.3}$$

where $\varepsilon \equiv \frac{\mu}{\rho}$.

If we are interested in the flow due to the two rotating disks at $z = h$ and $z = -h$, then (2.2) and (2.3) would be valid in the interval $-h \le z \le h$. The appropriate boundary conditions for the problem are

$$H(-h,\varepsilon) = H(h,\varepsilon) = 0 \text{ (no penetration)} \tag{2.4}$$

$$H'(-h,\varepsilon) = H'(h,\varepsilon) = 0 \text{ (adherence)} \tag{2.5}$$

$$G(-h,\varepsilon) = 2\Omega_{-h}, \quad G(h,\varepsilon) = 2\Omega_{+h} \tag{2.6}$$

where Ω_{-h} and Ω_{+h} are the angular speeds of the disks at $z = -h$ and $z = h$, respectively.

Let us consider the case $\Omega_h = \Omega_{-h} \neq 0$, within the context of the Karman equations (2.2)-(2.6). Then, the only solution to equations (2.2)-(2.6) is the rigid body solution:

$$G \equiv 2\Omega \text{ and } H \equiv 0 \;, \tag{2.7}$$

which is isolated and stable. By isolated we mean there is a neighborhood of this solution wherein there are other solutions, and by stable we mean there is no bifurcation from this solution, in particular the linearized problem at this solution is nonsingular.

Recently, Berker (3) established a truly remarkable result for the problem of two infinite parallel disks rotating with the same angular velocity Ω, about a common axis. He sought solutions of the form

$$v_x = -\Omega[y - f(z)] \;, \tag{2.8}$$

$$v_y = \Omega[z - g(z)] \;, \tag{2.9}$$

$$v_z = 0 \;, \tag{2.10}$$

where v_x, v_y and v_z are the components of the velocity in the x, y and z directions, respectively. Such a velocity field corresponds to a flow wherein streamlines in any z = constant plane are concentric circles, with no flow across any such plane; the locus of the centers of these circles as the z = constant plane shifts from $z = -h$ to $z = h$ being a curve in space described by $x = f(z)$ and $y = g(z)$. Notice that the velocity field (2.8)-(2.10) automatically satisfy the constraint of incompressibility.

Substituting (2.8)-(2.10) into the Navier-Stokes equation, we obtain

$$\mu f''' + \rho\Omega g' = 0 \;, \tag{2.11}$$

$$\mu g''' - \rho\Omega f' = 0 \;, \tag{2.12}$$

The appropriate boundary conditions are

$$f(h) = 0, \; f(-h) = 0, \; g(-h) = 0, \; g(h) = 0. \tag{2.13}$$

Since the plane $z = 0$ and the locus of the centers of the rotation intersect at some point $(\ell,0,0)$ (we can always pick such a Cartesian co-ordinate system)

$$f(0) = \ell \text{ and } g(0) = 0 \;. \tag{2.14}$$

Berker (3) showed that the equations (2.11)-(2.13) have a one-parameter family of solutions.

When $\ell = 0$, we obtain the rigid body solution. However, this is but one of an infinity of solutions that are possible. Moreover, when $\ell \neq 0$, the solutions are not axially symmetric. Thus, in the special case when $\Omega_h = \Omega_{-h}$, the axially symmetric solution of the Karman equation is imbedded in a much larger class of solutions. This naturally leads us to ask the question whether the axially symmetric solutions to the Karman equations are imbedded in a larger class of solutions when $\Omega_h \neq \Omega_{-h}$? This question has been recently answered by Parter and Rajagopal (4).

Parter and Rajagopal (4) assumed a velocity field of the form

$$v_x = \frac{x}{2} H'(z) - \frac{y}{2} G(z) + g(z) , \qquad (2.15)$$

$$v_y = \frac{y}{2} H'(z) + \frac{x}{2} G(z) - f(z) , \qquad (2.16)$$

and

$$v_z = -H(z) . \qquad (2.17)$$

The above velocity field in cylindrical coordinates has the form

$$v_r = \frac{r}{2} H'(z) + g(z) \cos\theta - f(z) \sin\theta , \qquad (2.18)$$

$$v_\theta = \frac{r}{2} G(z) - g(z) \sin\theta - f(z) \cos\theta , \qquad (2.19)$$

$$v_z = -H(z) .$$

Notice that when $f \equiv 0$, $g \equiv 0$, we recover the velocity field assumed by Karman. When $H \equiv 0$ and $G \equiv 2\Omega$, we obtain the velocity field assumed by Berker.

Substituting equations (2.21)-(2.23) into the Navier-Stokes equations yields

$$\varepsilon H^{iv} + HH''' + GG' = 0 , \qquad (2.21)$$

$$\varepsilon G'' + HG' - H'G = 0 , \qquad (2.22)$$

$$\varepsilon f''' + Hf'' + \frac{1}{2} H'f' - \frac{1}{2} H''f + \frac{1}{2} (Gg)' = 0 , \qquad (2.23)$$

$$\varepsilon g''' + Hg'' + \frac{1}{2} H'g' - \frac{1}{2} H''g - \frac{1}{2} (Gf)' = 0 . \qquad (2.24)$$

The appropriate boundary conditions are

$$H (-h,\varepsilon) = H(h,\varepsilon) = 0 , \qquad (2.25)$$

$$H'(-h,\varepsilon) = H'(h,\varepsilon) = 0 , \qquad (2.26)$$

$$G (h,\varepsilon) = 2\Omega_{-h} , \quad G(h,\varepsilon) = 2\Omega_h , \qquad (2.27)$$

$$f(-h,\varepsilon) = f(h,\varepsilon) = 0 , \qquad (2.28)$$

$$g(-h,\varepsilon) = 0 , \quad g(h,\varepsilon) = 0 . \qquad (2.29)$$

The system of equations (2.21)-(2.24) and boundary conditions (2.25)-(2.29) are undetermined and as before we can augment the system be requiring

$$f(0, \varepsilon) = \ell_1 , \quad g(0, \varepsilon) = \ell_2 . \qquad (2.30)$$

The above system has a very interesting feature. The equations (2.21), (2.22) and the boundary conditions (2.25)-(2.27) are precisely the same as those that govern the axially symmetric problem. More importantly, these are the only equations which are non-linear. The equations for f and g are linear with coefficients which are solutions to the non-linear axially symmetric problem. We are now in a position to answer the following question: whenever there is a solution (H(z), G(z)) to the system (2.21), (2.22), (2.25)-(2.27), can we find a one-parameter family of solutions (f,g) to the system (2.23), (2.24), (2.28)-(2.30)? Parter and Rajagopal (4) answer the question in the affirmative. Thus, none of the axially symmetric solutions are isolated when considered within the full scope of the Navier-Stokes equation. Details of the proof of the existence of a one-parameter family of solutions (f,g) and a detailed discussion of related questions can be found in the paper by Parter and Rajagopal (4). The most interesting feature of these solutions is their existence at even low Reynolds numbers.

A detailed numerical study of the asymmetric flow between parallel rotating disks has been carried out recently by Lai, Rajagopal and Szeri (8).

We next turn our attention to discussing solutions that are not axially symmetric in the case of non-Newtonian fluids. In the special case $\Omega_h = \Omega_{-h}$, one-parameter family of solutions that are not axially symmetric have been analytically established by Rajagopal and Gupta (8) in the case of the incompressible fluid of second grade*, by Rajagopal and Wineman (11) in the case of a special subclass of the K-BKZ fluid.

When $\Omega_h \neq \Omega_{-h}$, Huilgol and Rajagopal (12) show that a situation similar to that considered by Parter and Rajagopal (4) obtains. In the case of an Oldroyd fluid Huilgol and Rajagopal (12) assume a velocity field of the form (2.21)-(2.23) and show that the problem is governed by four coupled equations for the four functions H, G, f and g. Similar to the situation in the case of the Navier-Stokes fluid the functions H and G

*Earlier, Drouot (10) extended Berker's analysis, and her work clearly implies the possibility of exact solutions in the case of an incompressible homogeneous fluid of second grade. However, she did not solve the specific boundary value problem.

are governed by two coupled non-linear ordinary differntial equations. The other two equations involve the four functions H, G, f and g. Having established the existence of the solution (H, G) we can once again proceed to ask the question whether a one-parameter family of solutions (f, g) exist?

3 FLOW BETWEEN PARALLEL DISKS ROTATING ABOUT DIFFERENT AXES

3.1 $\Omega_h = \Omega_{-h}$

We shall first discuss the special case when $\Omega_h = \Omega_{-h} = \Omega \neq 0$. Let 'a' denote the distance between the parallel but distinct axes. In the case of Navier-Stokes fluid, Abbot and Walters (13) restricted themselves to solutions which possess midplane symmetry and exhibited an exact solution to that problem, the result being valid for arbitrary values of the offset. However, if we relax the requirement of midplane symmetry, it is easy to show that the problem under consideration possesses a one-parameter family of solutions (cf. Berker (14)). As we mentioned earlier, the above flow has relevance to the flow occurring in an orthogonal rheometer. The analysis of Drouot (10) once again implies the possibility of exact solutions, similar to Berker's in the case of the fluid of second grade. Rajagopal and Gupta (15) and Rajagopal (16) have since established these exact solutions.

We shall now discuss briefly the flow of an incompressible simple fluid in the orthogonal rheometer. We shall assume that the motion occurring in the orthogonal rheometer has the form (2.8)-(2.10). We shall first show that such a flow is a motion with constant principal relative stretch history (cf. Noll (17)). Let $\underset{\sim}{\xi} = (\xi, \eta, \zeta)$ denote the position occupied by a particle $\underset{\sim}{X} = (X,Y,Z)$ at time τ. Let $\underset{\sim}{x} = (x,y,z)$ denote the position occupied by the same particle $\underset{\sim}{X}$ at time t. It follows from (2.8)-(2.10) that

$$\dot{\xi} = -\Omega(\eta - g(\zeta)) \,, \tag{3.1}$$

$$\dot{\eta} = \Omega(\xi - f(\zeta)) \,, \tag{3.2}$$

$$\dot{\zeta} = 0 \,, \tag{3.3}$$

with

$$\xi(t) = x, \ \eta(t) = y, \text{ and } \zeta(t) = z \,. \tag{3.4}$$

It follows from a result due to Wang (18), and the assumed form for the motion, that

$$\underset{\sim}{T} = -p\underset{\sim}{1} + \underset{\sim}{f}(\underset{\sim}{A}_1, \underset{\sim}{A}_2) \,, \tag{3.5}$$

where $\underset{\sim}{A}_1$ and $\underset{\sim}{A}_2$ are the first two Rivlin-Ericksen tensors (cf. Rivlin and Ericksen (19)).

If the body force field $\underset{\sim}{b}$ is conservative and hence derivable from a potential, then $\underset{\sim}{b} = -\text{grad}\,\phi$, where ϕ is a scalar field. It then follows that the balance of linear momentum has the form:

$$\frac{1}{\rho}\frac{\partial p}{\partial x} = -\frac{\partial\phi}{\partial x} + \Omega^2[x - f] + \frac{1}{\rho} h_1(f', g', f'', g'') \,, \tag{3.6}$$

$$\frac{1}{\rho}\frac{\partial p}{\partial y} = -\frac{\partial\phi}{\partial y} + \Omega^2[y - g] + \frac{1}{\rho} h_2(f', g', f'', g'') \,, \tag{3.7}$$

$$\frac{1}{\rho}\frac{\partial p}{\partial z} = -\frac{\partial\phi}{\partial z} + \frac{1}{\rho} h_3(f', g', f'', g'') \,. \tag{3.8}$$

A straightforward computation then yields the pressure:

$$\frac{p}{\rho} = -\phi + \frac{\Omega^2}{2}(x^2 + y^2) + (\frac{s}{\rho}x + \frac{q}{\rho}y) + h(z) + c \tag{3.9}$$

where s, q and c are constants and

$$h(z) = \frac{1}{\rho}\int h_3\, dz \,. \tag{3.10}$$

The specific constitutive equations determine the functions h_1, h_2, and h_3. This can then be substituted into (3.6)-(3.8) and the appropriate partial differential equations analyzed. Notice that the equations (3.6)-(3.8) are of second order and hence the no slip boundary conditions are sufficient for determinancy.

The appropriate boundary conditions for the velocity field are

$$u = \frac{\Omega a}{2} - \Omega y, \ v = \Omega x, \ w = 0 \quad \text{at } z = h \,, \tag{3.11}$$

$$u = -\frac{\Omega a}{2} - \Omega y, \ v = \Omega x, \ w = 0 \quad \text{at } z = -h \,, \tag{3.12}$$

and

$$u \to \mp \infty, \; v \to \pm \infty \text{ as } x,y \to \pm \infty . \qquad (3.13)$$

It follows from (3.11), (3.12) and (3.13) that

$$f(h) = f(-h) = 0 , \qquad (3.14)$$

$$g(h) = \frac{a}{2} , \; g(-h) = -\frac{a}{2} . \qquad (3.15)$$

We can eliminate the pressure gradient which occurs in the equations of motion (3.25)-(3.27), in the usual manner, to obtain

$$h_2' = \rho\Omega^2 g' , \qquad (3.16)$$

and

$$h_1' = \rho\Omega^2 f' , \qquad (3.17)$$

In eliminating the pressure field we have raised the order of the equations. As before, we augment the number of boundary conditions by recognizing that the locus of the centers of rotation cut plane z = 0 at some point, in this case (ℓ_1, ℓ_2). Thus

$$f(0) = \ell_1 \; , \quad g(0) = \ell_2 . \qquad (3.18)$$

We shall next discuss a non-trivial example, namely the problem associated with the flow of a K-BKZ fluid in an orthogonal rheometer.

The Cauchy stress $\underset{\sim}{T}$ in the K-BKZ fluid has the structure (cf. Kaye (20), Bernstein, Kearsley and Zapas (21)):

$$\underset{\sim}{T} = -p\underset{\sim}{1} + 2\int_{-\infty}^{t} \{U_1 \underset{\sim}{C}_t^{-1}(\tau) - U_2 C_t(\tau)\} d\tau , \qquad (3.19)$$

where

$$\underset{\sim}{C}_t(\tau) = \underset{\sim}{F}_t^T(\tau)\, \underset{\sim}{F}_t(\tau) . \qquad (3.20)$$

In equation (3.19), U denotes the strain energy function for the viscoelastic fluid of the principal invariants of $\underset{\sim}{C}_t(\tau)$ and $\underset{\sim}{C}_t^{-1}(\tau)$:

$$U = U(I_1, I_2, t - \tau) \qquad (3.21)$$

$$I_1 = \mathrm{tr}\, \underset{\sim}{C}_t(\tau) \; , \; I_2 = \mathrm{tr}\, \underset{\sim}{C}_t^{-1}(\tau) \qquad (3.22)$$

and

$$U_i = \frac{\partial U}{\partial I_i} , \; i = 1, 2 . \qquad (3.23)$$

For the motion under consideration a lengthy but straightforward computation yields

$$\underset{\sim}{C}_t(\tau) = \underset{\sim}{1} - \frac{s}{\Omega} \underset{\sim}{A}_1 + \frac{(1-c)}{\Omega^2} \underset{\sim}{A}_2 , \qquad (3.24)$$

and

$$\begin{aligned} \underset{\sim}{C}_t^{-1}(\tau) = {} & 1 + \frac{s}{\Omega} [1 + 2(1-c)(f'^2 + g'^2)]\underset{\sim}{A}_1 \\ & - \frac{(1-c)}{\Omega^2} [1 + 2(1-c)(f'^2 + g'^2)]\underset{\sim}{A}_2 \\ & + \frac{s^2}{\Omega^2} \underset{\sim}{A}_1^2 + \frac{(1-c)^2}{\Omega^4} \underset{\sim}{A}_2^2 \\ & + \frac{s(1-c)}{\Omega^3} (\underset{\sim}{A}_1\underset{\sim}{A}_2 + \underset{\sim}{A}_2\underset{\sim}{A}_1) , \end{aligned} \qquad (3.25)$$

where

$$s \equiv \mathrm{Sin}\, \Omega(t - \tau), \; c \equiv \mathrm{Cos}\, \Omega(t - \tau) \qquad (3.26)$$

Also, notice that

$$\begin{aligned} I_1(t,\tau) = I_2(t,\tau) &= 3 + 2(1-c)(f'^2 + g'^2) \\ &\equiv I(\Omega(t - \tau), z) . \end{aligned} \qquad (3.27)$$

It follows from (3.6), (3.7) and (3.8) that

$$\frac{d}{dz} \{f'B(\kappa) + g'A(\kappa)\} = \rho\Omega^2 f , \qquad (3.28)$$

$$\frac{d}{dz} \{-f'A(\kappa) + g'B(\kappa)\} = \rho\Omega^2 g , \qquad (3.29)$$

where

$$\kappa \equiv (f'^2 + g'^2)^{1/2} \qquad (3.30)$$

and

$$A(\kappa) = 2 \int_0^\infty \tilde{U}[3+2(1-\cos\Omega\alpha)\kappa^2, \alpha] \sin\Omega\alpha \, d\alpha , \qquad (3.31)$$

$$B(\kappa) = 2 \int_0^\infty \tilde{U}[3+2(1-\cos\Omega\alpha)\kappa^2, \alpha](1-\cos\Omega\alpha) d\alpha , \qquad (3.32)$$

$$\tilde{U}(I,\alpha) \equiv U_1(I,I,\alpha) + U_2(I,I,\alpha) . \qquad (3.33)$$

Let us consider a special subclass of K-BKZ fluids wherein U_1 and U_2 are independent of I_1 and I_2, i.e

$$U_1 = U_1(t - \tau) \ , \ U_2 = U_2(t - \tau) \ . \qquad (3.34)$$

It follows from (3.60) and (3.61) that

$$A(\kappa) = G_2(\Omega) \text{ and } B(\kappa) = G_1(\Omega) \ , \qquad (3.35)$$

where $G_1(\Omega)$ and $G_2(\Omega)$ denote the real and imaginary parts of the complex shear modulus of linear viscoelasticity. If follows from (3.28) and (3.19) that

$$G_1(\Omega)f'' + G_2(\Omega)g'' - \rho\Omega^2 f = q_1 \ , \qquad (3.36)$$

$$-G_2(\Omega)f'' + G_1(\Omega)g'' - \rho\Omega^2 g = q_2 \ , \qquad (3.37)$$

where q_1 and q_2 are constants. The appropriate boundary conditions are once again (3.14) and (3.15). The above problem can be solved exactly (cf. Rajagopal and Wineman (22)). Detailed numerical solutions have also been carried out in the case of a special subclass of K-BKZ fluids, namely the Wagner (cf. (23)) and Currie (cf. (24)) models by Bower, Rajagopal and Wineman (25) and by Rajagopal, Renardy, Renardy and Wineman (26).

3.2 $\Omega_h \neq \Omega_{-h}$

When $\Omega_h \neq \Omega_{-h}$, the flow of a Navier-Stokes fluid between plates rotating about different axes is governed by the same system of differential equations (2.21)-(2.24). The only difference in the boundary value problem from that governing the flow about a common axis occurs in the specification of boundary conditions. The boundary conditions $(2.28)_1$ and $(2.29)_1$ are replaced by

$$g(-h,\varepsilon) = -\frac{a\Omega_{-h}}{2} \ , \ g(h,\varepsilon) = \frac{a\Omega_{+h}}{2} \qquad (3.38)$$

Parter and Rajagopal (4) have discussed questions regarding the existence of solutions to the system (2.21)-(2.24), subject to (2.25) - (2.30). Numerical solutions of the above system have been carried out by Lai, Rajagopal and Szeri (8). In this case, the locus of the stagnation points is far from simple and once again does not possess any mid-plane symmetry. Not much has been done with regard to non-Newtonian fluids.

4 ASSOCIATED PROBLEMS IN NON-LINEAR ELASTICITY

We conclude this article by mentioning that interesting asymmetric solutions exist to the boundary value problems corresponding to those which have been discussed earlier, in non-linear elasticity (cf. Rajagopal and Wineman (27)), that is in the case of the twisting of a non-linear elastic slab about a common axis, or about distinct axes.

Consider the problem of rotation the top and bottom layers of an elastic slab about a common axis or that about non-coincident axes. We shall assume that the corresponding motion has the form (cf. Rajagopal and Wineman (27)).

$$x = [x-f(Z)]\cos\Omega(Z)-[Y-g(Z)]\sin\Omega(Z) + f(Z), \qquad (4.1)$$

$$y = [x-f(Z)]\sin\Omega(Z)-[Y-g(Z)]\cos\Omega(Z) + g(Z), \qquad (4.2)$$

$$z = Z \ , \qquad (4.3)$$

where X, Y, Z and x, y, z represent the reference and current co-ordinates of the same material point. This represents a deformation in which material points which lie in any plane parallel to the top and bottom layers continue to remain in the plane, the plane rotating about a point by an amount $\Omega(Z)$. A simple computation verifies that $\det \underset{\sim}{F} = 1$. In the case of rotation about a common axis, Rajagopal and Wineman (27) have exhibited a one-parameter family of solutions to the equations of equilibrium in the case of the Neo-Hookean and Mooney-Rivlin materials. Only one member of this family is the classical symmetric solution, all other solutions being asymmetric. As in the fluid problem, rotation about non-coincident axes, possesses a two parameter family of solutions.

Next, we shall consider the finite extension and twisting of an infinite non-linearly elastic slab. Let us suppose that a finite extension λ of an infinite sandwich is followed by twisting the top plate by an amount Ω. In this case we shall assume a deformation of the following type

$$x = \frac{1}{\sqrt{\lambda}}[X \cos\Omega(\lambda Z) - Y \sin\Omega(\lambda Z)] + f(\lambda Z) \qquad (4.4)$$

$$x = \frac{1}{\sqrt{\lambda}} [X \sin\Omega(\lambda Z) + Y \cos\Omega(\lambda Z)] + g(\lambda Z) \qquad (4.5)$$

$$z = \lambda Z \ , \qquad (4.6)$$

where λ is a constant. Rajagopal and Wineman (27) show that similar to the preceding problem, in the case of Neo-Hookean and Mooney-Rivlin materials, one or two parameter family of solutions exist, depending on whether the rotation is about a common axis or about non-coincident axes.

We conclude this presentation with a discussion of a problem related to the previous boundary value problem, which does not involve any rotation, namely the problem of non-uniform uniaxial extension of an infinite slab of incompressible non-linear elastic materials.

In this case we consider a deformation of the form

$$x = \frac{1}{\sqrt{\lambda'(Z)}}\,X, \quad Y = \frac{1}{\sqrt{\lambda'(Z)}}\,Y, \quad z = \lambda(Z) \,. \tag{4.7}$$

A lengthy but straightforward manipulation yields the following equation of equilibrium

$$\lambda''' - \frac{3}{2}\frac{(\lambda'')^2}{\lambda'} = C\lambda_1 \,, \tag{4.8}$$

where C is a constant. The appropriate boundary conditions are

$$z(0) = 0 \,, \tag{4.9}$$

$$z(H) = h \,, \tag{4.10}$$

where h is the current thickness of the slab of original thickness H.

The above problem can also be solved exactly (cf. Rajagopal and Wineman (27)) and the form of the solution depends on whether C is positive, negative or zero. When $C > 0$,

$$\lambda'(Z) = \frac{1}{\{A_1 \sin\sqrt{(\frac{C}{2})}\,Z + B_1 \cos\sqrt{\frac{C}{2}}\,Z\}^2} \tag{4.11}$$

when $C < 0$,

$$\lambda'(Z) = \frac{1}{\{A_2 e^{\sqrt{\frac{C}{2}}\,Z} + B_e{}^{\sqrt{\frac{C}{2}}\,Z}\}^2} \tag{4.12}$$

and when $C = 0$

$$\lambda' = \text{const} \,, \tag{4.13}$$

which corresponds to the classical solution.

REFERENCES

Karman, T. von. 1921. Uber laminare und turbulente Reibung, Z. Angen. Math. Mech. 1:232-252.

Batchelor, G.K. 1951. Note on a class of solutions of the Navier-Stokes equations representing steady rotationally-symmetric flow. Quart. J. Mech. Appl. Math. 4:29-41.

Berker, R. 1979. A new solution of the Navier-Stokes equation for the motion of a fluid contained between two parallel planes rotating about the same axis. Archiwum Mechaniki Stosowanej. 31:265-280.

Parter, S.V. & K.R. Rajagopal 1984. Swirling flow between rotating plates. Arch. Ratl. Mech. Anal. 86:305-315.

Rajagopal, K.R. 1982. On the flow of a simple fluid in an orthogonal rheometer. Arch. Ratl. Mech. Anal. 79:29-47.

Goddard, J.D. 1983. The dynamics of simple fluids in steady circular shear. Quart. Appl. Math. 31:107-118.

Parter, S.V. 1982. On the swirling flow between rotating co-axial disks: A survey, pages 258-280 in Theory and Applications of Singular Perturbations, Proc. of a Conf. Oberwolfach, 1981, Edited by W. Eckhaus and E.M. DeJager. Lecture Notes in Mathematics #942, Springer-Verlag; Berlin-Heidelberg-New York.

Lai, C.Y., K.R. Rajagopal & A.Z. Szeri 1984. Asymmetric flow between parallel rotating disks. J. Fluid Mech. 146:203-225.

Rajagopal, K.R. & A.S. Gupta 1981. Flow and stability of second grade fluids between two parallel rotating plates, Archiwum Mechaniki Stosowanej. 33:663-674.

Drouot, R. 1967. Sur un cas d'integration des equations du mouvement d'un fluide incompressible du deuxieme ordre, C.R. Acad. Sc. Paris. 265:300-304.

Rajagopal, K.R. & A.S. Wineman 1983. A class of exact solutions for the flow of a viscoelastic fluid, Archiwum Mechaniki Stosowanej. 35:747-752.

Huilgol, R.R. & K.R. Rajagopal Non-axisymmetric flow of a viscoelastic fluid between rotating disks, submitted for publication.

Abbot, T.N.G. & K. Walters 1970. Rheometrical flow systems, Part 2, Theory for the orthogonal rheometer, including an exact solution for the Navier-Stokes equations. J. Fluid Mech. 40:205-213.

Berker, R. 1982. An exact solution of the Navier-Stokes equation, the vortex with curvilinear axis. Intl. J. Eng. Science. 20:217-230.

Rajagopal, K.R. & A.S. Gupta 1985. Flow and stability of a second grade fluid between two parallel rotating plates about non-coincident axes. Intl. J. Eng. Science. 19:1401-1409.
Rajagopal, K.R. 1981. The flow of a second order fluid between rotating parallel. J. of Non-Newtonian Fluid Mech. 9:185-190.
Noll, W. 1962. Motions with constant stretch history. Arch. Rational Mech. Anal. 11:97-105.
Wang, C.C. 1965. A representation theorem for the constitutive equation of a simple material in motions with constant stretch history. Arch. Rational Mech. Anal. 20: 329-340.
Rivlin, R.S. & J.L. Ericksen 1955. Stress deformation relations for isotropic materials. J. Rational Mech. Anal. 4:323-425.
Kaye, A. 1962. Note No. 134, College of Aeronautics Cranfield Institute of Technology.
Bernstein, B., E.A. Kearsley & L.J. Zapas 1963. A study of stress relaxation with finite strain. Trans. Soc. Rheol. 7:391-410.
Rajagopal, K.R. & A.S. Wineman 1983. Flow of a BKZ fluid in an Orthogonal Rheometer. Journal of Rheology. 27:509-516.
Wagner, M.H. 1976. Analysis of time dependent non-linear stress growth data for elongational flow of a low density branched polyethylene line melt. Rheologica Acta. 15:133
Currie, P.K. 1982. Constitutive equations for polymer melts predicted by Doi-Edwards and Curtiss-Bird Kinetic theory models. J. Non-Newtonian Fluid Mechanics. 11:53-68.
Bower, M., K.R. Rajagopal & A.S. Wineman A numerical study of the inertial effects of the flow of a shear thinning K-BKZ fluid between rotating parallel plates, submitted for publication.
Rajagopal, K.R., M. Renardy, Y. Renardy & A.S. Wineman. Flow of viscoelastic fluids between plates rotating about distinct axes, In Press, Rheologica Acta.
Rajagopal, K.R. & A.S. Wineman 1985. New exact solutions in non-linear elasticity Int. J. Eng. Science. 23:217-234.

5th International Symposium on Continuum Models of Discrete Systems / Nottingham / 14-20 July 1985

A virial stochastic solution for a random suspension of spheres

K.Z.Markov
University of Sofia, Bulgaria

ABSTRACT: The heat conduction problem for a random suspension of equisized nonoverlapping spheres is considered. The random temperature field $T(x)$ is expanded in a functional series, generated by the given random field of thermal conductivity, with certain nonrandom kernels. The series is rendered virial, i.e. its truncation after the p-tuple term yields results for the averaged statistical characteristics, which are correct to order c^p, c being the volume fraction of the spheres. The procedure of identification of the kernels is considered in detail for $p=2$. The respective kernels are explicitly found and connected to the temperature fields in an unbounded matrix, containing one or two spherical inhomogeneities, and undergoing constant temperature gradient. In this way the full stochastic solution to the problem of heat conduction through the suspension is obtained to order c^2 in a closed form for an arbitrary random distribution of spheres.

1 INTRODUCTION

Consider an infinite heterogeneous medium whose coefficient of thermal conductivity is a statistically homogeneous and isotropic random field $k(x)$. The temperature field in the medium, at the absence of body sources, is governed by the equations

$$(1.1a)\quad \nabla \cdot q(x)=0, \quad q(x)=k(x)\nabla T(x),$$

where $q(x)$ is, for convenience, the opposite heat flux vector. We prescribe also the mean value of the temperature gradient to be constant

$$(1.1b)\quad \langle \nabla T(x)\rangle = G$$

which plays the role of a boundary condition for the random equation (1.1a); hereafter the brackets $\langle\cdot\rangle$ denote ensemble averaging.

The random problem (1.1) represents well the typical problem one encounters when calculating the bulk properties for composite materials of random constitution, see e.g. Beran (1968). This is a stochastically nonlinear problem, which is the source of all the troubles in this field.

A general approach to the problem (1.1) was recently proposed (Markov 1984, Christov & Markov 1985a,b). It is based on the obvious remark that the problem (1.1) defines implicitly a nonlinear operator which transforms the random conductivity field $k(x)$ - the "input" - into the random temperature field - the "output"

$$T(\cdot) = \mathfrak{T}[k(\cdot)].$$

Following the general idea of Volterra and Wiener, it is then fully natural to expand the operator $\mathfrak{T}$ in a functional series, generated by the input $k(x)$

$$(1.2)\quad T(x) = K_0(x) + \int K_1(x-y)k(y)dy + \iint K_2(x-y_1,x-y_2)k(y_1)k(y_2)\,dy_1dy_2 + \dots$$

with certain nonrandom kernels K_0, K_1, ..., and to truncate the series afterwards. (Hereafter, if the integration domain is not explicitly indicated, the integrals are taken over the whole $\mathbb{R}^3$.)

Two kinds of applications of such truncated series may be envisaged here. The first and the more general is to try to specify the first few kernels T_p in order to get approximate, in a certain sense, solutions to the stochastic problem (1.1). This idea was worked up by Christov & Markov (1985a, b) and Markov (1984) who derived and examined the respective equations for the kernels in certain special cases of random constitution. The second consists in employing the series as classes of trial functions for the classical variational principles, corresponding to the problem (1.1), and thus to obtain bounds on the bulk properties of the medium. This possibility was indicated by Markov (1985) who was thus able to derive, in particular, some new bounds on the bulk conductivity for an arbitrary random suspension of spheres.

The aim of this lecture is to outline the way how the full stochastic solution to the problem (1.1) can be found in a virial form for a random suspension of spheres by means of functional series of the type (1.2). As an illustration the full stochastic solution and, in particular, the bulk conductivity for the suspension will be given to order c^2 in a closed form.

2 STATISTICAL DESCRIPTION OF THE SUSPENSION

Consider a rigid suspension of spheres, each of radius a and thermal conductivity k_f, randomly distributed in an unbounded matrix of conductivity k_m. The volume fraction of the spheres is c, so that the mean number of spheres per unit volume is $n = c/V_a$, $V_a = 4\pi a^3/3$.

Let x_j be the random point system which comprises the centres of the spheres. A full, and convenient, statistical description of the suspension is provided by the set of multipoint distribution densities $f_p(y_1, \dots, y_p)$, $p = 1,2, \dots$, which give the probability dP to find a point from x_j per each of the volumes $y_i \le y \le y_i + dy_i$, $i = 1, \dots, p$, to be

$$(2.1) \qquad dP = f_p(y_1,\dots,y_p)dy_1\dots dy_p$$

(Stratonovich 1963).

We assume that the suspension is statistically homogeneous and isotropic. Then, in particular, $f_1 = n$ and $f_p(y_1,\dots,y_p) = f_p(y_{21},\dots,y_{p1})$, $y_{ij} = y_i - y_j$. Let us recall that if $f_p = n^p$, p=1,2,..., we have the so-called Poisson sytem of points (Stratonovich 1963).

Since the spheres are forbidden to overlap, the functions $f_p(Y_p)=0$, if $|y_i - y_j| < 2a$ for a pair i, j. This means that

$$(2.2) \qquad f_p(Y_p) = f_p(Y_p)\prod_{\substack{i,j=1\\ i\neq j}}^{p} Q_{2a}(y_i - y_j),$$

where $Q_{2a}(y)$ equals 0 at $|y| < 2a$ and 1 otherwise; $Y_p = (y_1,\dots,y_p)$.

After Markov (1985), we restrict the analysis to the class of suspensions which possess the property

(2.3) The distance between the nearest spheres tends to infinity at $n \to 0$.

A simple analysis (Markov 1985) shows that (2.3) yields

$$(2.4) \qquad f_p(Y_p) = n^p f_{p,p}(Y_p) + o(n^p),$$

p = 1,2,... .

The assumption (2.3) means that the manifacturing process, through which the class of suspension under consideration is produced, does not put spheres in rigid complexes, say, in dumb-bells. However, (2.3) does not exclude the possibiliy for the spheres to form groups (clusters) at higher values of n. The only thing we assume here is for those groups to fall apart at $n \to 0$ which seems perfectly natural when speaking of suspensions of spheres. (Otherwise we should have spoken of suspensions of dumb-bells or other more complicated sphere complexes.) For more details see again Markov (1985).

Note that $f_{p,p} = 1$, p=1,2,..., in (2.4) corresponds to the so-called perfect disorder of spheres (Christov & Markov 1985 a,b), which combines the definition of the Poisson system with the condition (2.2) of nonoverlapping of the spheres. If only $f_{2,2} = 1$, i.e.

$$(2.5) \qquad f_2(y_1,y_2) = n^2 Q_{2a}(y_1 - y_2)$$

we have the popular case of well-stirred suspension.

3 FUNCTIONAL EXPANSION

After Stratonovich (1963) let us introduce the so-called random density field

$$(3.1)\quad f(x) = \sum_{J} \delta(x-x_j)\ ,$$

generated by the random set of spheres centres. The random conductivity field $k(x)$ of the suspension can be then written down as

$$(3.2)\quad k(x) = \langle k\rangle + [k]\int h(x-y) f'(y)dy.$$

Here $\langle k\rangle = ck_f + (1-c)k_m$, $[k] = k_f - k_m$, $f'(y) = f(y) - n$ is the centered field f so that $\langle f'(y)\rangle = 0$ and $h(x)$ is the characteristic function for a single sphere located at the origin.

The representation (3.2) suggests that the temperature field in the suspension be seeked, instead of (1.2), as a functional series generated by the random density function $f(x)$.

Let us recall now that when employing functional series for random problems a central question to be solved,if we want to get rigorous results, consists in rendering the series orthogonal in stochastical sense. The importance of such an orthogonalization was acknowledged by Wiemer himself (see Wiener 1958, and also Schetzen 1980 for a comprehensive survey of the related problems and approaches), who chose the input as a Gaussian white noise and showed that the respective orthogonal functionals are generated by the multivariate Hermite polynomials. However, as argued by Christov (1981) and Christov & Markov (1985a, b), the Gaussian input is badly fitted for problems, connected with random point systems for which the Poisson point system represents much more plausible an approximation.

It is known (Ogura 1972) that if the input is a Poisson point system, the orthogonal functionals are generated by the multivariate Charlier polynomials of the random density function $f(x)$, the first few of them being

$$(3.3)\quad \begin{aligned} C_f^{(0)} &= 1\ ,\\ C_f^{(1)}(y) &= f'(y) = f(y) - n\ ,\\ C_f^{(2)}(y_1,y_2) &= f(y_1)(f(y_2) - \delta(y_{12}))\\ &\quad - n(f(y_1) + f(y_2)) + n^2\ , \ \ldots\ . \end{aligned}$$

It appears (Christov & Markov 1985a) that the same result holds as well for a perfectly disordered suspension of spheres (see Sec. 2), provided the kernels in the expansion comply with the condition (3.11) below. For an arbitrary point system x_j, however, there is no procedure hdw to construct the orthogonal functionals; moreover, it is not clear whether they exist at all.

To overcome this difficulty we argue as follows. Note that in the problem (1.1) for a suspension there exists a small parameter c - the volume fraction of the spheres. This suggests to look for the solution to (1.1) in the usual virial form, i.e. as a certain series whose truncation after the p th order term brings results valid to order c^p, for any p prescribed. Then, if we want to solve the problem (1.1) to order c^p, it is not necessary to have the respective functionals orthogonal, but only virial-orthogonal, in the sense that their moments are to be small of order c^{p+1}.

We shall state now our central result, namely, it appears that for an arbitrary statistics of spheres, which satisfies (2.3) and thus (2.4), such a virial-orthogonal system of functionals can be constructed on the base of the Charlier polynomials by means of a kind of orthogonalization process. In general, the construction is rather tedious and we shall consider here only the case $p = 2$, for which we introduce the modified Charlier polynomials

$$(3.4)\quad \begin{aligned} D_f^{(m)}(Y_m) &= C_f^{(m)}(Y_m),\ m \neq 2,\\ D_f^{(2)}(y_1, y_2) &= C_f^{(2)}(y_1,y_2)\\ &\quad - nR(y_{12})(C_f^{(1)}(y_1) + C_f^{(1)}(y_2))\\ &\quad + n^2 R(y_{12})\ , \end{aligned}$$

where

$$(3.5)\quad \begin{aligned} R(y) &= 1 - Q(y),\\ Q(y) &= f_2(y)/n^2 = f_{2,2}(y)\ . \end{aligned}$$

Due to (2.2), we have

$$(3.6)\quad Q(y) = 0,\ R(y) = 1 \text{ at } |y| < 2a.$$

In particular, for a well-stirred suspension of spheres, see (2.5),

$$(3.7) \quad Q(y) = Q_{2a}(y), \ R(y) = h_{2a}(y),$$

where $h_{2a}(y)$ is the characteristic function of a sphere of radius 2a, located at the origin.

For the moments of the quantities $D_f^{(p)}$ we shall need the following formulas, valid to order c^2 (i.e. n^2, since c and n are proportional):

$$(3.8) \quad \begin{aligned} &\langle D_f^{(1)}(y_1)\, D_f^{(1)}(y_2)\rangle = n\,\delta_{12} - n^2 R_{12}, \\ &\langle D_f^{(1)}(y_1)\, D_f^{(1)}(y_2)\, D_f^{(1)}(y_3)\rangle \\ &= n\,\delta_{12}\delta_{13} - 3\{\delta_{12} R_{13}\}_s\, n^2, \\ &\langle D_f^{(2)} D_f^{(1)}\rangle = o(n^3), \\ &\langle D_f^{(2)}(y_1,y_2) D_f^{(1)}(y_3) D_f^{(1)}(y_4)\rangle \\ &= \langle D_f^{(2)}(y_1,y_2) D_f^{(2)}(y_3,y_4)\rangle \\ &= n^2(1-R_{12})(\delta_{13}\delta_{24} + \delta_{14}\delta_{23}), \\ &\langle D_f^{(2)}(y_1,y_2) D_f^{(2)}(y_3,y_4) D_f^{(1)}(y_5)\rangle \\ &= n^2(1-R_{12})(\delta_{15}+\delta_{25})(\delta_{13}\delta_{24} \\ &+ \delta_{14}\delta_{23}); \end{aligned}$$

For brevity, $\delta_{pq} = \delta(y_{pq})$, $R_{pq} = R(y_{pq})$, $N\{\cdot\}_s$ denotes symmetrization with respect to indexes listed in the brackets, N is the number of terms in the full-scale expression. The formulas (3.8) follow from the known relations between the moments of the random density function f(x) and the multipoint functions f_p of the random set x_j (Stratonovich 1967).

Let us introduce the so-called factorial moments of the random function f(x)

$$\Delta_f(y_1,\ldots,y_p) = f(y_1)(f(y_2) - \delta(y_{12})) \ldots (f(y_p) - \delta(y_{1p}) - \ldots - \delta(y_{p-1p})).$$

As seen from (3.3) and (3.4)

$$D_f^{(p)}(Y_p) = \Delta_f(Y_p) + O(n), \ p=1,2,\ldots.$$

On the other hand, it was recently proved (Christov 1985) that

$$\langle \Delta_f(Y_p)\rangle = f_p(Y_p), \ p=1,2,\ldots.$$

On the base of the last two relation together with (2.4), it is readily shown that

$$(3.9) \quad \begin{aligned} &\langle D_f^{(p)}\rangle, \langle D_f^{(p)} D_f^{(1)}\rangle, \langle D_f^{(p)} D_f^{(1)} D_f^{(1)}\rangle, \\ &\langle D_f^{(p)} D_f^{(2)}\rangle, \langle D_f^{(p)} D_f^{(2)} D_f^{(1)}\rangle, \ldots = o(n^p), \\ &\langle D_f^{(p)} D_f^{(q)}\rangle, \langle D_f^{(p)} D_f^{(q)} D_f^{(1)}\rangle, \langle D_f^{(p)} D_f^{(q)} D_f^{(2)}\rangle, \\ &\langle D_f^{(p)} D_f^{(q)} D_f^{(1)} D_f^{(1)}\rangle, \langle D_f^{(p)} D_f^{(q)} D_f^{(1)} D_f^{(2)}\rangle, \\ &\ldots = o(n^r), \end{aligned}$$

at $p, q \geq 3$, $r = \max(p,q)$.

Let us represent now the temperature field T(x) as a functional series, generated by the introduced system of functions (3.4)

$$(3.10) \quad \begin{aligned} T(x) &= G\cdot x + \int T_1(x-y) D_f^{(1)}(y)dy \\ &+ \iint T_2(x-y_1, x-y_2) D_f^{(2)}(y_1,y_2)dy_1 \\ &dy_2 + \sum_{p=3}^{\infty} \int\ldots\int T_p(x-y_1,\ldots, \\ &x-y_p) D_f^{(p)}(y_1,\ldots,y_p)dy_1\ldots dy_p. \end{aligned}$$

Due to (3.4) and (3.8), $D_f^{(1)}$ and $D_f^{(2)}$ are centered, so that the random field (3.10) satisfies (1.1b). Note also that since there are no points x_j located closer than 2a, the kernels T_p in (3.10) are defined only if the distances between their arguments exceed 2a. That is why we can always assume after Christov & Markov (1985a) that

$$(3.11) \quad \begin{aligned} &T_p(y_1,\ldots,y_p) = 0 \ \text{if} \\ &|y_i - y_j| < 2a \ \text{for a pair } i,j. \end{aligned}$$

The relations (3.8) and (3.9) ensure that the functionals, generated by the functions (3.4), form a virial-orthogonal system to order c^2. Thus a full stochastic solution of the problem (1.1) can be obtained from the series (3.11) to the same order c^2, provided the latter is truncated after the two-tuple term. Then it remains to specify the kernels T_1 and T_2 only which will be done in the next Section. In what follows the remainder $\sum_{p=3}^{\infty}$ will be omitted as contributing quantities of order $o(c^2)$.

4 IDENTIFICATION OF THE KERNELS T_1 AND T_2

To specify the kernels T_1, T_2 we argue as usual (Christov&Markov 1985a,b, Markov 1984): insert (3.10) and (3.2) in (1.1a), multiply by $D_f^{(1)}(0\cdot)$ and by $D_f^{(2)}(0,z)$, and average the results. In virtue of (3.8), this procedure yields the pair of equations

$$k_m \Delta S + [k]\nabla\cdot\Big\{h(x)(G+\nabla S) - nF(x)G + n(V_a - F(x))\nabla S - n\int \nabla S(x-y)h(x-y)R(y)\,dy + 2nI_2(x)\Big\} = 0; \quad (4.1)$$

$$(1-R(z))\nabla\cdot\Big\{2(k_m + [k](h(x) + h(x-z)))\nabla T_2(x,x-z) + [k](h(x)\nabla T_1(x-z) + h(x-z)\nabla T_1(x)\Big\} = 0, \quad (4.2)$$

where

$$S(x) = T_1(x) - n\int T_1(x-y)R(y)\,dy, \quad (4.3)$$

$$F(x) = \int h(x-y)R(y)\,dy, \quad (4.4)$$

$$I_2(x) = \int (1-R(y))h(x-y)\nabla T_2(x-y,x)\,dy, \quad (4.5)$$

hereafter $\nabla = \nabla_x$. Since everywhere in the averaged relations T_1 appears scaled by n, and T_2 by n^2, cf. (3.8), we seek the solution of (4.1) and (4.2) to order n^2 in the form

$$T_1(x) = T_{10}(x) + nT_{11}(x), \quad T_2(x) = T_{20}(x), \quad (4.6)$$

so that

$$S(x) = S_0(x) + nS_1(x), \quad (4.7)$$

$$S_0(x) = T_{10}(x), \quad (4.8)$$

$$S_1(x) = T_{11}(x) - \int T_{10}(x-y)R(y)\,dy. \quad (4.9)$$

Moreover, since $R(z) = 1$ at $|z| < 2a$, cf. (3.6), eqn (4.2) should be considered only in the region $|z| > 2a$, in which $1 - R(z) = f_2(z)/n^2 > 0$. Thus we can omit the multiplier $1-R$.

Upon inserting (4.6) and (4.7) in (4.1) and (4.2), we get in usual way

$$k_m \Delta S_0 + [k]\nabla\cdot\{h(x)(G + \nabla S_0)\} = 0, \quad (4.10)$$

$$k_m \Delta S_1 + [k]\nabla\cdot\Big\{h(x)\nabla S_1 + (V_a - F(x))\nabla S_0 - \int \nabla S_0(x-y)h(x-y)R(y)\,dy + 2I_{20}(x)\Big\} = 0, \quad (4.11)$$

$$\nabla\cdot\Big\{2(k_m + [k](h(x) + h(x-z))\nabla T_{20}(x,x-z) + [k](h(x)\nabla T_{10}(x-z) + h(x-z)\nabla T_{10}(x))\Big\} = 0, \quad (4.12)$$

where

$$I_{20}(x) = \int (1-R(y))h(x-y)\nabla T_{20}(x-y,x)\,dy. \quad (4.13)$$

Eqn (4.10) is nothing but the equation for the disturbance to the temperature field introduced by a single spherical inclusion, when the temperature gradient at infinity equals G. Its solution is thus well known

$$T_{10}(x) = \begin{cases} -\beta G\cdot x & \text{at } |x| \leq a \\ \beta a^3 G\cdot\nabla 1/|x| & \text{at } |x| > a \end{cases}, \quad (4.14)$$

$\beta = [k]/(k_f + 2k_m)$, so that T_{10} does not depend on the statistics of the spheres in the suspension.

Consider next eqn (4.12). Again it does not depend on the statistics of the spheres and therefore it coincides with that derived by Christov & Markov (1985a) for the case of a perfectly disordered suspension. Thus, as noted there, the solution to eqn (4.12) has the form

$$2T_{20}(x,x-z) = T^{(2)}(x;z) - T_{10}(x) - T_{10}(x-z), \quad (4.15)$$

where $T^{(2)}(x;z)$ is the disturbance to temperature field in a homogeneous unbounded matrix, introduced by a pair of identical spherical inclusions whose centres are at the origin and at the point z, when the temperature gradient at infinity equals G. Each of these inclusions, if it were alone, would disturb the temperature field $G\cdot x$ in the matrix by $T_{10}(x)$ and $T_{10}(x-z)$, respectively. Thus the kernel $T_{20}(x,x-z)$ is the field which should be added to the single-inclusion disturbances $T_{10}(x)$, $T_{10}(x-z)$ in order to obtain the double-inclusion disturbance $T^{(2)}(x;z)$.

It is important to note that to the order $o(|z|^{-3})$ $T_{20}(x,x-z)$ has the following asymptotics at $|z|\gg 1$:

$$(4.16)\qquad \begin{cases} -\beta^2/2\, G\cdot\nabla_z \nabla_z 1/|z|\cdot x & \text{at } |x|\leq a \\ \beta^2/2\, a^3 G\cdot\nabla_z \nabla_z 1/|z|\cdot 1/|x| & \text{at } |x|>a \end{cases}$$

so that, at $|x|\leq a$,

$$(4.17)\qquad \begin{aligned} &2\nabla T_{20}(x,x-z) \\ &= \beta\nabla T_1(x-z) + o(|z|^{-3}),\ |z|\gg 1 \end{aligned}$$

The field $T^{(2)}(x;z)$ can be found, e.g. by means of the method of twin expansions (Jeffrey 1973 et al.) and therefore we shall think it known. Thus it remains to find S_1, i.e. T_{11}, which is the only one depending on the statistics of the suspension.

In virtue of (4.8) and (4.14), eqn (4.11) becomes

$$(4.18)\qquad \begin{aligned} &k_m\Delta S_1 + [k]\nabla\cdot\{h(x)\nabla S_1(x) \\ &+ (V_a - F(x))\nabla T_{10}(x) \\ &+ (\beta-1)F(x)G + 2I_{20}(x)\} = 0. \end{aligned}$$

A straightforward calculation of $I_{20}(x)$ by means of $T^{(2)}(x;z)$ seems unsurmountable. We shall circumvent this difficulty in the following manner.

Let us multiply (4.12) by $1-R(z)$ and rewrite the result in the form

$$(4.19)\qquad \begin{aligned} &(1-R(z))\nabla\cdot\{2[k]\,h(x-z) \\ &\nabla T_{20}(x,x-z)\} = -(1-R(z)) \\ &\nabla\cdot\{2(k_m+[k]h(x))\nabla T_{20}(x,x-z) \\ &+[k](h(x)\nabla T_1(x-z) \\ &+ h(x-z)\nabla T_1(x)\}. \end{aligned}$$

Due to the presence of $h(x-z)$, the l.h.side of (4.19) is absolutely integrable with respect to z in the region $Z_{2a}=\{z \mid |z|>2a\}$ and thus the same holds for the r.h.side. (let us remind that the differentiation in (4.19) is with respect to x, which commutates with the integration over z.) That is why we can choose the mode of integration at our ease. We choose to take the integrals in the sense

$$(4.20)\qquad \int_{Z_{2a}} = \lim_{R\to\infty}\int_{Z_{2a,R}},$$

where $Z_{2a,R}=\{z \mid R>|z|>2a\}$, and in the region $Z_{2a,R}$ we first integrate with respect to the angular coordinates and then with respect to the radial coordinates.

Let us introduce the functions

$$(4.21)\qquad \begin{aligned} &L_{20}(x) = \int_{Z_{2a}} (1-R(z)) \\ &T_{20}(x,x-z)\,dz, \end{aligned}$$

$$(4.22)\qquad \begin{aligned} &L_{10}(x) = \int_{Z_{2a}} (1-R(z)) \\ &\nabla T_{10}(x-z)\,dz. \end{aligned}$$

As seen from (4.14) and (4.16) both integrals in (4.21) and (4.22) are not absolutely convergent. However, they exist in the sense (4.20) since when integrating first with respect to the angular coordinates the contribution of the leading term $|z|^{-3}$ in the asymptotics at $|z|\gg 1$ vanishes, cf. again (4.14), (4.16), and the remaining terms $O(|z|^{-m})$, $m\geq 4$, are already absolutely integrable. Note that the existence of the integrals (4.21) and (4.22) is the reason why the mode of integration (4.20) was chosen.

Let us integrate now (4.19) with respect to z, taking the integrals in the sense (4.20) when needed

$$(4.23)\qquad \begin{aligned} &2k_m\Delta L_{20}(x) + [k]\nabla\cdot\{2h(x)\nabla L_{20}(x) \\ &+ 2I_{20}(x) + h(x)L_{10}(x) + \nabla T_{10} \\ &(x)\int_{Z_{2a}}(1-R(z))h(x-z)\,dz\} = 0. \end{aligned}$$

Thus, having evaluated the conditionally convergent integrals (4.21) and (4.22), we combine them in (4.23) in a way which cancels out the contributions of the non-convergent parts of the two integrals, in order to obtain the absolutely convegent integral $I_{20}(x)$. A similar combination of conditionally convergent integrals was employed by Einstein (1906) and we just repeated the remark of Jeffrey (1974), concerning this work of Einstein.

A simple analysis shows that

$$h(x)\, L_{10}(x) = 0 , \tag{4.24}$$

$$\int_{Z_{2a}} (1 - R(z))h(x-z)dz = V_a - F(x), \tag{4.25}$$

cf. (4.4). Upon subtracting (4.23) from (4.18) and taking into account (4.24) and (4.25), we get

$$k_m \Delta H(x) + [k]\nabla\cdot\{h(x)\nabla H(x) + (\beta - 1)F(x)G\} = 0 , \tag{4.26}$$

where

$$H(x) = S_1(x) - 2L_{20}(x) . \tag{4.27}$$

Thus we have eliminated $I_{20}(x)$, replacing it by the integral $L_{20}(x)$ which, though conditionally convergent only, is much easier to be calculated (see next Section). Moreover eqn (4.26) can be readily solved now, yielding

$$H(x) = \begin{cases} \beta(1-\beta)V_a G\cdot x & \text{at } |x| \leq a \\ (-\beta^2 a^3 V_a + 3\beta \int_0^{|x|} t^2 F(t)dt)\, G\cdot x/|x|^3 & \text{at } |x| > a \end{cases} \tag{4.28}$$

Finally, with $S_1(x)$ and $T_{10}(x)$ known, we find $T_{11}(x)$ by means of (4.9) and thus the kernels $T_1(x)$ and $T_2(x)$ in the series (3.10) are obtained in an integral form through the solutions of one- and two-sphere problems $T_{10}(x)$ and $T^{(2)}(x;z)$, respectively. In this way we can calculate explicitly all the moments of the temperature field $T(x)$ and the joint moments of $T(x)$ and $k(x)$ to order c^2, which just means that the full stochastic solution to order c^2 is found in the course of the foregoing reasoning for the random problem (1.1).

Eqns (4.27) and (4.28) reveal the reason why non-absolutely convergent integrals have repeatedly appeared in the analysis of bulk properties of suspensions to order c^2 (Peterson & Hermans 1969, Batchelor & Green 1972, Jeffrey 1973 et al.): simply because the problem (1.1) has in this case a solution which includes a conditionally convergent integral.

5 BULK CONDUCTIVITY OF THE SUSPENSION TO ORDER c^2

To illustrate the performance of the obtained solution to the problem (1.1), let us calculate the simplest and most popular statistical characteristics of the rigid suspension - its bulk (or effective) conductivity k^*. In virtue of (3.2), (3.8), (3.10), (4.3), and (4.7), we find

$$\begin{aligned} k^* G &= \langle k(x)\nabla T(x)\rangle \\ &= \langle k\rangle G + [k]n\int_{V_a} \nabla S(x)dx \\ &= \langle k\rangle G + [k]n\int_{V_a} \nabla S_0(x)dx \\ &\quad + n^2[k]\int_{V_a} \nabla S_1(x)dx . \end{aligned} \tag{5.1}$$

Thus only the values of $S(x)$ within the sphere $V_a = \{ x \mid |x| \leq a \}$ are needed when calculating k^*.

Eventually, upon taking into account (3.5), (4.8), (4.14), (4.15), (4.21), (4.24), (4.27), and (4.28), we get for the bulk conductivity

$$\begin{aligned} k^* G &= k_m (1 + 3\beta c + 3\beta^2 c^2)\, G \\ &\quad + [k]c^2 \int_{V_a} dx \int_{Z_{2a}} dz\, f_{2,2}(z) \\ &\quad (\nabla T^{(2)}(x;z) - \nabla T_{10}(x)) . \end{aligned} \tag{5.2}$$

It could be shown that eqn (5.2) leads to the same result as that given by Jeffrey (1973), provided a "renormalization" is performed. (The idea of renormalization in the theory of suspensions was proposed by Batchelor (1972) and Jeffrey (1973), as a method to avoid the conditionally convergent integrals.) It is easy to perform it within the frame of our analysis. Indeed, taking into account (4.24), we rewrite the integral in (5.2) in the form, given, as a matter of fact, by Jeffrey (1973)

$$\int_{V_a} dx \int_{Z_{2a}} dz\, f_{2,2}(z)\, (\nabla T^{(2)}(x;z) - \nabla T_{10}(x) - \beta\nabla T_1(x-z)) .$$

Due to (4.15) and (4.17) the integrand here is already of order $o(|z|^{-3})$ at $|z| \gg 1$ and thus it is absolutely integrable. (A more detailed analysis of the field $T^{(2)}(x;z)$ yields that the integrand is of order $O(|z|^{-6})$, see Jeffrey(1973).)

It is clear, however, that the above renormalization is by no means necessary in our analysis, for (4.20) defines unambiguously the mode of integration in (5.2). The renormalization here could be viewed only as a computational trick which may help evaluating the respective integrals. But even as a trick it is not needed in certain cases, e.g. for the two-dimensional case considered below.

It is to be noted also that for a well-stirred suspension, cf. (2.5), eqn (5.2) coincides with that proposed by Peterson & Hermans (1969) who, without giving any reason, tacitly adopted the mode of integration (4.20). They proposed as well a useful computational technique for evaluation of the conditionally convergent integral in (5.20), which does not need a renormalization and is easily extandable to an arbitrary statistics of spheres. This technique allows to obtain explicit results for k^* for the two-dimensional counterpart of the suspension under study, i.e. for a fibre-reinforced material: an array of equisized parallel cylinders under a temperature gradient orthogonal to the cylinders' axes. Then k^* is just the transverse bulk conductivity of the material and its eventual form reads

$$(5.3)\quad \begin{aligned} k^*/k_m &= 1 + 2\beta c + 2\beta^2 c^2 (1 + 2\beta M(\beta)) + o(c^2), \\ M(\beta) &= 16 \sum_{s=1}^{\infty} s \int_0^{\infty} f_{2,2}(2\,\mathrm{ch}\,t)\, \mathrm{ch}\,t\,\mathrm{sh}^3 t\, e^{-6st}(1-\beta^2 e^{-4st})^{-1} dt \\ &= \sum_{p=0}^{\infty} M_p \beta^{2p} . \end{aligned}$$

In certain cases the coefficients M_p can be expressed by the logarithmic derivative of the Γ-function, ψ. For example, if the suspension is well-stirred, i.e. $f_{2,2} = 1$, then

$$(5.4)\quad \begin{aligned} M_p &= c_p^2 (2\psi(1 + c_p) - 2\psi(1 + 2c_p) - 0.5\, c_p (2\pi c_p / \sin 2\pi c_p - 1)), \\ c_p &= 1/(2p+3), \end{aligned}$$

a result given by Peterson & Hermans (1969). The first few of the coefficients (5.4) are

$$(5.5)\quad \begin{aligned} M_0 &= 1/6, \quad M_1 = 0.0139, \\ M_2 &= 0.0033, \quad M_3 = 0.0011, \end{aligned}$$

and, at $p \geq 4$, the asymptotic formula

$$(5.6)\quad M_p = 7.2122\, c_p^4$$

gives already very good results.

Upon combining (5.3) to (5.6), we can propose the following simple formula for the transverse conductivity of a well-stirred fibre-reinforced material

$$(5.7)\quad \begin{aligned} k^*/k_m = 1 + 2\beta c + 2\beta^2 c^2 (1 + \beta/3 &+ 0.0248\beta^3 + 0.0066\beta^5 \\ &+ 0.0022\beta^7 + \Delta(\beta)) + o(c^2) , \end{aligned}$$

where $\Delta(\beta) = O(\beta^9)$. More precisely,

$$(5.8)\quad |\Delta(\beta)| \leq 0.0023\beta^9 .$$

Note also that in the limiting case $\beta = 1$, i.e. $k_f/k_m = \infty$,

$$(5.9)\quad k^*/k_m = 1 + 2\beta c + 2.7400 c^2 + o(c^2)$$

Explicit results by means of the function ψ can be also obtained for the more general two-point function

$$(5.10)\quad f_{2,2}(z) = 1 + aA_1/|z| .$$

In this case, due to (5.3), we have

$$(5.11)\quad M_p = M_p^o + M_p^1 A_1 ,$$

where M_p^o are given in (5.4), and

$$(5.12)\quad \begin{aligned} M_p^1 = 1.5\, c_p^2 (\psi(1 + 0.5c_p) &- \psi(1 + 1.5c_p) + 1.5c_p \\ &(\pi c_p \cos(0.5\pi c_p)/\sin(1.5\pi c_p) - 2/3)) . \end{aligned}$$

Since A_1 should only exceeds -2 and can take arbitrary big values, eqn (5.11) suggests that while the statistics of the suspension does not influence the c-term in the virial expansion of the bulk conductivity, it can very strongly affect the c^2-term. (Note that increasing the parameter A_1 in (5.10) corresponds to an increasing tendency for the spheres to form groups at higher values of their volume fraction c.)

6 CONCLUDING REMARKS

In this lecture we have proceeded with the analysis of the applications of functional series in the theory of random heterogeneous materials. This approach, as already demonstrated by Christov & Markov (1985a,b) and Markov (1984), is highly advantageous in this field; it offers unique possibilities when relating micro and macro properties of the materials. This is well illustrated also in the foregoing analysis for a random suspension of spheres - a medium which has enjoyed considerable interest in the literature where mostly the problem of evaluating the bulk properties has been treated. Our aim here was much wider, namely, the construction of a full stochastic solution for the random temperature field under a prescribed constant average temperature gradient. The success is to be attributed to the idea of virial-orthogonality accomplished by the introduced modification of the Charlier polynomials. The obtained solution (3.10) appeared eventually to resemble a group expansion in the sense that it is broken up in a sum of consecutive terms which result from interaction within successively larger groups of spheres. A similar expansion was considered by Jeffrey (1974), again for the bulk conductivity only, who postulated its existence on the base of heuristic arguments. In our analysis such a group interpretation emerges in a natural way, when solving the basic stochastic problem (1.1) for the suspension in a virial form for the class of statistics satisfying the assumption (2.3). Moreover, our analysis resolves in passing the long disputed problem how to treat the conditionally convergent integrals which kept appearing in the theory of suspensions. Namely, it showed that they are inherent in this theory and prescribed also the mode of their integration.

REFERENCES

Batchelor, G.K. 1972. Sedimentation in a dilute suspension of spheres. J.Fluid Mech.52: 245-268.

Batchelor, G.K. & J.T.Green 1972. The determination of the bulk stress in a suspension of spherical particles to order c^2.J.Fluid Mech.56: 401-427.

Beran, M.J. 1968. Statistical continuum theories.New York:John Wiley.

Christov, C.I. 1981. Poisson-Wiener expansion in nonlinear stochastic systems. Ann.Univ.Sofia,Fac.Math. Méc.,L.2(Mécanique) 75: 143-164.

Christov, C.I. 1985. A further development of the concept of random density function with application to Volterra-Wiener expansion. C. R.Acad.bulg.Sci.38: 35-38.

Christov, C.I. & K.Z.Markov 1985a. Stochastic functional expansion for random media with perfectly disordered constitution.SIAM J. Appl.Math. 45: 289-312.

Christov, C.I. & K.Z.Markov 1985b. Stochastic functional expansion in elasticity of heterogeneous solids.Int.J.Solids Structures 21:in press.

Einstein, A. 1906. Eine neue Bestimmung der Moleküldimensionen. Ann. Phys.19:206.

Jeffrey, D.J. 1973. Conduction through a random suspension of spheres. Proc.R.Soc. A335: 355-367.

Jeffrey, D.J. 1974. Group expansions for the bulk properties of a statistically homogeneous, random suspension. Proc.R.Soc. A338:503 -516.

Markov, K.Z. 1984. An application of Volterra-Wiener series in mechanics of composite materials. Bulgar.Acad.Sci.,Theor.Appl.Mech. 15(1): 41-50.

Markov, K.Z. 1985. Application of Volterra-Wiener series for bounding the overall conductivity of heterogeneous media. SIAM J.Appl. Math. (submitted)

Ogura, H. 1972. Orthogonal functionals of the Poisson process. IEEE Trans.Inf.Theory 18: 473-481.

Peterson, J.M. & J.J.Hermans 1969. The dielectric constants of nonconducting suspensions. J.Composite Materials 3: 338-354.

Schetzen, M. 1980. The Volterra and Wiener theories of nonlinear systems. New York: John Wiley.

Stratonovich, R.L. 1963. Topics in the theory of random noises. Vol.1. New York: Gordon and Breach.

Wiener, N. 1958. Nonlinear problems in random theory. New York: Techn. Press MIT and John Wiley.

5th International Symposium on Continuum Models of Discrete Systems / Nottingham / 14-20 July 1985

Hydrostatics of rigid magnetic suspensions

R.K.T.Hsieh
Royal Institute of Technology, Stockholm, Sweden

ABSTRACT: Using a model of dilute suspension of rigid spherical magnetic particles, the hydrostatic behaviour of a magnetic fluid is given as well as the principles of a hydrostatic magnetic separator.

1 INTRODUCTION AND SUMMARY

Due to economic reasons among others, the process recovery of resources from waste materials and air pollution control gain importance. A central interest there is given to large scale density separations, sometimes also called sink-float principle. This paper is concerned with magnetohydrostatic density separations. Such concept is recent and has now been competitively industrially applied.

The hydrostatic equations for magnetic fluids are derived as the continuum, macroscopic description of a dilute suspension of rigid spherical particles, the magnetic particles at the microscopic scale having subdomain sizes. The principles of magnetohydrostatic separators are laid therefrom.

Such system which usually consists of an electromagnet, a separation cell and an equipment for recovery of magnetic fluid, has the advantage that the density of the liquid-medium can vary and cover a broad density range of the particles to be separated. It can also be continuously computer controlled through electromagnetic devices. Suspended particles with strong magnetization such as Co particles can be chosen. Magnetization and magnetic field can be not colinear, facilitating the design of the system and allowing for a weaker applied magnetic field. Compared to conventional magnetic separators, the medium here is itself magnetized rather than depending on the inherent magnetism of the particles to be separated.

2 MICROSTRUCTURE OF MAGNETIC FLUID

The engineering model of magnetic fluid is a monophase liquid which in the absence of an applied magnetic field reacts like a Newtonian fluid while in its presence augments the hydrostatic pressure by a magnetic contribution and exerts a magnetic body force on each volume of the magnetic fluid (Rosensweig, 1979). Moreover monodispersivity of the medium can be assumed due to the limitations imposed by the stability of the colloidal system on the particle sizes. This model has been proved to be successful in operating liquid seals, hydrostatic bearings... (see e.g. Perry, 1978).

In many respects the model is based on an analogy of the properties of a monodomain ferromagnetic fluid in the field with the properties of a homogeneously isotropic paramagnetic fluid, particularly with the tendency of both individual constant moments to align with magnetic field and the Brownian motion opposition encountered in this process. At thermodynamic equilibrium, the fluid magnetization has therefore the direction of the applied field and

$$\underset{\sim}{M} = nmL(\alpha)\frac{\underset{\sim}{H}}{H} \qquad (1)$$

with $L(\alpha) = \coth\alpha - \frac{1}{\alpha}$ and $\alpha = \frac{\mu_o mH}{kT}$.

This expression has been derived by neglecting particle to particle magnetic interaction. n is the number of particles per unit volume of magnetic fluid V. $m = m^d V^m$ is a constant magnetic moment of a magnetic particle of volume V^m, m^d is the magnetization of the domain. $mL(\alpha)$ is the average value of m. H_i is the field intensity. μ_o is the permeability of free space. k is the Boltzman constant and T is the absolute temperature. A basic difference between these two types of fluids is however that the electrodynamic action of the field on the magnetic fluid is resisted by fluid dynamic forces and thus the process is dissipative. These forces are large compared to the intertial forces. The particles therefore nearly always have a steady-state migration caused by a force balance between the magnetic force and the fluid-dynamic drag (see Freeman, 1982). The translational velocity estimated from such balance assuming Stoke's viscous drag force leads to a velocity of the order of 10^{-7} m/s. Thus a contribution of the translational motion of the particles to the balance energy and momentum can be neglected in many cases. Rotation of the particles relative to the liquid is due to the magnetization anisotropy possessed by the particle. For "easy-axis" type anisotropy, such effect can be accounted by the anisotropy energy

$$E^a = -(KV^m/m^2)(\underset{\sim}{m}\cdot\underset{\sim}{e})^2 \tag{2}$$

Here K is the anisotropy constant, e_i is the unit vector in the direction of the easiest axis frozen into the particle. The total energy of a particle has therefore a minimum when the vectors H_i, m_i and e_i are colinear. The magnetization of magnetic fluid is thus due to the alignment of the moment and of the particles. Whereas the former can be described through the parameter $\alpha = \mu_o mH/kT$, the latter is defined in terms of a parameter $\beta = KV^m/kT$. For $\beta<1$, the correlation of the direction of the moment and the particle may be neglected. For suspended particles with strong magnetization, this is not the case, e.g. for Co particles, β is of the order 60.

Actually magnetic fluids are multiphase and multicomponent media. The particles being under constant thermal motion are prevented from sedimentation and a polymeric coating prevents the particles to agglomerate. Several models have been proposed to describe such systems (see Rosensweig, 1979; Hsieh, 1980). In what follows, the aim is in first hand to extend the validity of the engineering model. It is intuitively clear that the different fluid properties are depending on the magnetization and the applied magnetic field. We shall therefore examine situations in which these two factors can be as large as possible, i.e. situations with the parameter β greater than unity.

The suspended particles are still rigid, of spherical shape, the magnetization is directed along the easy axis of magnetization and the suspension is diluted.

3 EQUATIONS OF MOTION

If the density ρ of the magnetic fluid is defined by the relation

$$\rho = n\mu + n^{(Nm)}\mu^{(Nm)} = n(\mu + \frac{n^{(Nm)}}{n}\mu^{(Nm)}) \tag{3}$$

where μ and $\mu^{(Nm)}$ are respectively the masses of magnetic and nonmagnetic particles, $n^{(Nm)}$ is the number of non-magnetic particles in unit volume V of the magnetic fluid. Because of a non- magnetic surfactant coating, this volume V is larger than V^m. For spherical Co particles, the diameter d^m is about 60 Å and the diameter d about 600 Å. For $n^{(Nm)}/n$ being a constant, there is no diffusion particle flow. This is the case of a dilute non-interacting suspension. Such assumption implies low volume concentration, up to 1%. The macroscopic continuum balance equations for the mass conservation, linear momentum and angular momentum can now be written as

$$\frac{d}{dt}\int_{V(t)} \rho dV = 0 \tag{4}$$

$$\frac{d}{dt}\int_{V(t)} \rho\underset{\sim}{v}dV = \int \underset{\sim}{t}dS + \int \rho\underset{\sim}{f}dV \tag{5}$$

$$\frac{d}{dt}\int_{V(t)} (I\underset{\sim}{\Omega} + \frac{\underset{\sim}{M}}{\gamma})dV = \int \underset{\sim}{N}dS + \int \rho\underset{\sim}{\ell}dV + \int \underset{\sim}{x}X\underset{\sim}{t}dS \tag{6}$$

where $I\underset{\sim}{\Omega}$ is the angular moment of particles. $I = ni$ is the particle inertia per unit volume V and i is the inertia moment of a spherical particle; $\frac{\underset{\sim}{M}}{\gamma}$ is the spin moment and γ is the gyromagnetic ratio. V(t) is a material volume.

In particular, equation (6) is the macroscopic continuum expression of the microscopic behaviour of the magnetization in which spin and mechanical moments have been simultaneously considered (Bashtovoi and Kashevsky, 1976)

$$\frac{d}{dt}(\frac{\underset{\sim}{m}}{\gamma} + i\underset{\sim}{\omega}) = \mu_o(\underset{\sim}{m}x\underset{\sim}{H}) - 6V\xi^o\underset{\sim}{\omega} \tag{7}$$

where ξ^o is the friction factor due to rotation of a spherical sphere.

For isotropic homogeneous magnetic fluids, the stress $t_i = t_{ij} n_j$ and the couple stress $N_i = N_{ij} n_j$ where n_i is the unit, are given by (Hsieh, 1980)

$$t_{ij} = -[p^* + (\lambda - \frac{2}{3}\mu) v_{k,k}]\delta_{ij} + \mu(v_{i,j} + v_{j,i}) + 2\xi\epsilon_{ijk}[(\nabla x \underset{\sim}{v})_k - \Omega_k] \quad (8)$$

$$N_{ij} = \gamma^1 \Omega_{k,k}\delta_{ij} + \gamma^2(\Omega_{i,j} + \Omega_{j,i}) + \gamma^3(\Omega_{i,j} - \Omega_{j,i}) \quad (9)$$

and the body force and body couple by

$$\rho \underset{\sim}{f} = \mu_o(\underset{\sim}{M} \cdot \nabla)\underset{\sim}{H} \text{ and } \rho \underset{\sim}{\ell} = \mu_o \underset{\sim}{M} x \underset{\sim}{H} \quad (10)$$

In quasi-steady approximation and in the absence of electric displacement and electric current, the electromagnetic field equations write

$$\int \underset{\sim}{B} \cdot d\underset{\sim}{S} = 0 \quad (11)$$

$$\int \underset{\sim}{E} \cdot d\underset{\sim}{\ell} = -\frac{d}{dt}\int \underset{\sim}{B} \cdot d\underset{\sim}{S} \quad (12)$$

$$\int \underset{\sim}{H} \cdot d\underset{\sim}{\ell} = 0 \quad (13)$$

and

$$\underset{\sim}{B} = \mu_o(\underset{\sim}{M} + \underset{\sim}{H}) \quad (14)$$

Equations (4) - (6), the Maxwell eqs. (11) - (13) together with the constitutive relations (8), (9), and (1) describe the behaviour of magnetic fluids. In general the magnetization (1) should be augmented with a relaxation term which clearly arises in the microscopic description. However in macroscopic experiments, the condition $\Omega\tau \ll 1$ where τ is the relaxation time is usually fullfilled and the macroscopic relaxation effect can be therefore neglected.

Several works in the literature aimed on the improvement of the expression of the magnetization (1). Jansson (1983) gave from statistical considerations general lines on how the Langevin function $\mathbb{L}(\alpha)$ can be modified to account for the interaction between magnetic particles. Hsieh (1981) introduce a continuum gradient of magnetization,... It seems however that for large scale uses of magnetic fluids, the expression of the magnetization (1) is accurate enough.

The above formulated system of equations should be supplemented with the boundary conditions. They are not listed here but their derivations present no difficulty.

If we disregard the magnetic effects but assume that the equations of balance (4)-(6) are still true with the couple stress M_{ij} now of purely mechanical origin, we recover the results obtained by Eringen (1984) for rigid suspensions. His model successfully predicts the behaviour of the effective viscosity of fluids even when all rigid fiber suspensions are parallelly aligned. If, in addition, the couple stress and rotary inertia are neglected and the particles are assumed being rigid fiber suspensions, the continuum model for the suspension rheology of slender fibers is recovered (Batchelor, 1970).

4 MAGNETOHYDROSTATIC SEPARATIONS

The characteristic time of the present problem is often much greater than 10^{-9}s, a calculation of the order of magnitude of the process of spin and mechanical moment shows that both of them are negligible. Also, using a simple argument, it has been shown that couple stresses are insignificant (Tsebers, 1974). Hsieh and Vörös (1981) using a structural model reached similar results for the solid case. Under these conditions and for negligible inertial forces, the equations of linear and angular momentum, (5) and (6), reduce to

$$0 = \int \underset{\sim}{t} dS + \int \rho \underset{\sim}{f} dV \quad (15)$$

$$0 = \int \rho \underset{\sim}{\ell} dV + \int \underset{\sim}{x} X \underset{\sim}{t} dS \quad (16)$$

In local form we shall have

$$0 = -\nabla p^* + (\lambda - \frac{\mu}{3})\nabla\nabla \cdot \underset{\sim}{v} + 2\xi\nabla x(\nabla x \underset{\sim}{v} - \underset{\sim}{\Omega}) + \mu_o(\underset{\sim}{M} \cdot \nabla)\underset{\sim}{H} \quad (17)$$

$$4\xi(\nabla x \underset{\sim}{v} - \underset{\sim}{\Omega}) = \mu_o(\underset{\sim}{M} x \underset{\sim}{H}) \quad (18)$$

Here the friction of rotation coefficient ξ can be estimated

$$\xi = \frac{3}{2}\Phi\xi^o \quad (19)$$

where $\Phi = nV$ is the hydrodynamic concentration of the particles.

Substituting equation (18) into (17) we will have

$$0 = -\nabla p^* + (\lambda - \frac{\mu}{3})\nabla\nabla \cdot \underset{\sim}{v} + \frac{1}{2}\mu_o \nabla x(\underset{\sim}{M} x \underset{\sim}{H}) + \mu_o(\underset{\sim}{M} \cdot \nabla)\underset{\sim}{H} \quad (20)$$

If a nonmagnetic body B of volume V^B and surface S^B is immersed in the here proposed magnetic fluid, the net forces acting on the body will be

$$\underset{\sim}{F}^B = \oint_{S_B} t^{MF} dS + \int_{V_B} \rho^B \underset{\sim}{g} dV \quad (21)$$

where ρ^B is the density of the body, and the inviscid magnetic fluid stress t^{MF} is given by

$$t_{ij}^{MF} = -[p^* - \frac{1}{2}\mu_o H^2]\delta_{ij} + H_i B_j \tag{22}$$

The force $\underset{\sim}{F}^B$ can then be written as

$$\underset{\sim}{F}^B = \int_{V^B}(\rho^B - \rho)\underset{\sim}{g}dV + \int_{V^B}[\mu_o(\underset{\sim}{M}\cdot\nabla)\underset{\sim}{H} + \frac{\mu_o}{2}\nabla x(\underset{\sim}{M}x\underset{\sim}{H})]dV \tag{23}$$

Defining an apparent density ρ^a of the magnetic liquid by the expression

$$\rho^a = \rho - \frac{\mu_o}{g}[(\underset{\sim}{M}\cdot\nabla)\underset{\sim}{H} + \frac{1}{2}\nabla x(\underset{\sim}{M}x\underset{\sim}{H})]_z \tag{24}$$

in which ρ is the true density, g is the gravitation constant, z is vertical in upward direction, we see that with H decreasing in the direction of positive z, the sign of ∇H is negative, so the apparent density exceeds the true liquid density.

The magnetohydrostatic levitation force on the body B can then be written as

$$\underset{\sim}{F}^B = \int_{V^B}(\rho^B - \rho^a)\underset{\sim}{g}dV \tag{25}$$

The expression (25) is the basic relationship used in magnetohydrostatic separators. The body B will sink for $\rho^B < \rho^a$ and float for $\rho^B > \rho^a$.

For $\underset{\sim}{M}$ and $\underset{\sim}{H}$ colinear, we recover the result of Rosensweig (1979) and Shimoiizaka, Nakatsuka, Fujita, Kounosu (1980). For a magnetic body B with magnetization $\underset{\sim}{M}^B$, in the presence of an unhomogeneous field H, expression (23) should be augmented with the force

$$\mu_o(\underset{\sim}{M}^B\cdot\nabla)\underset{\sim}{H} + \frac{\mu_o}{2}\nabla x(\underset{\sim}{M}^B x\underset{\sim}{H}).$$

In such case, it is therefore sufficient to replace $\underset{\sim}{M}$ with $\underset{\sim}{M} - \underset{\sim}{M}^B$ in the expressions (21)-(24).

Based on these principles, sink-float separators in which density of the liquid medium can be controlled through applied magnetic fields have been designed and industrially operated. For water based magnetic fluids, particles of density up to 9000 kg/m^3 can be separated.

REFERENCES

Bashtoroi, V.G., Kashevsky, B.E. 1976. An asymmetric model of magnetic fluid with finite anisotropy of ferromagnetic particles. Magn.Gidrodin. 4:24-32.

Batchelor, G.K. 1970. The stress system in a suspension of force-free particles. J.Fl.Mech. 41:545-570.

Freeman, M.P. 1982. Separation mechanisms for liquid suspensions in Handbook of multiphase systems, G. Hetsroni Ed., Hemisphere, 9.98-9.146.

Eringen, A.C. 1984. A continuum theory of rigid suspensions. Lett.Appl.Engng.Sci. 22:1373-1388.

Hsieh, R.K.T. 1980. Continuum mechanics of a magetically saturated fluid. IEEE Trans.Magnetics 16:207-210.

Hsieh, R.K.T. and Vörös, G. 1981. Microanelasticity of a vibrating rod containing particles with magnetic moment. Int.J.Engng.Sci. 19:1369-1376.

Jansons, K.M. 1983. Determination of the constitutive equations for a magnetic fluid. J.Fluid Mech. 137:187-216.

Perry, M.P. 1978. A survey of magnetic fluid applications in Thermomechanics of magnetic fluids, B. Berkovsky Ed., Hemisphere, 219-230.

Rosensweig, R.E. 1979. Fluid dynamics and science of magnetic liquids. Advances in Electronics an Electron Physics 48:103-199.

Shimoiizaka, I., Nakatsuka, K., Fujita, T. and Kounosu, A. 1980. Sink-float separators using permanent magnets and water based magnetic fluid. IEEE Trans.Magnetics 16:368-371.

Tsebers, A.O. 1974. On the rotating magnetic effect on ferrofluid. Magnit.Gidrodin. 3:151-155.

5th International Symposium on Continuum Models of Discrete Systems / Nottingham / 14-20 July 1985

The maximum entropy principle and the moment truncation procedure

Z.Banach
Theoretische Physik, Universität-Gesamthochschule-Paderborn, FR Germany

ABSTRACT: The moment representation for the one-point distribution function f_1 (see (1)) describing nonequilibrium processes which are quadratic in the heat flux q_1 and in the friction pressure tensor ω_{1j} is now independently derived from the maximum entropy principle.

1 FORMULATION OF THE BASIC PROBLEM

The finite set of Maxwell's equations of transfer (2) for the mass density ρ, the momentum density ρv_1, the internal energy density ε, the symmetric traceless part ω_{1j} of the friction pressure and the heat flux q_1 never closes on itself. Indeed, the macroscopic moment which plays a role of the flux in the nth equation acts, roughly speaking, as the density in the $(n+1)^{th}$ equation. At the same time, the production dissipative terms appearing on the r.h.s. of the deduced equations of transfer depend in a very complicated way on moments of all orders. As a consequence, we are faced with a problem of expressing the higher order moments and collisional productions in terms of moments of lower order.

2 THE SYSTEM OF BALANCE EQUATIONS

Let us consider the simplest case of a system of identical monatomic molecules with the molecular mass μ which is described by the Boltzmann kinetic equation. Following the same line of arguments as that in (1,3), we arrive at the system of balance equations for ρ, ρv_1, ε, ω_{1j} and q_1:

$$\frac{\partial \rho}{\partial t} + \frac{\partial \rho v_k}{\partial x_k} = 0, \qquad (2.1)$$

$$\frac{\partial \rho v_1}{\partial t} + \frac{\partial(\rho v_1 v_k + \frac{2}{3}\varepsilon\delta_{1k} + \omega_{1k})}{\partial x_k} = 0, \qquad (2.2)$$

$$\frac{\partial \varepsilon}{\partial t} + \frac{\partial(\varepsilon v_k + q_k)}{\partial x_k} + (\tfrac{2}{3}\varepsilon\delta_{1k} + \omega_{1k})L_{k1} = 0, \qquad (2.3)$$

$$\frac{\partial \omega_{1j}}{\partial t} + \frac{\partial(\omega_{1j}v_k + \frac{4}{5}q_{<1}\delta_{j>k} + \omega_{1jk})}{\partial x_k} + 2L_{k<1}(\tfrac{2}{3}\varepsilon\delta_{j>k} + \omega_{j>k}) = P_{<1j>}, \qquad (2.4)$$

$$\frac{\partial q_1}{\partial t} + \frac{\partial(q_1 v_k + \frac{1}{3}q\delta_{1k} + s_{1k})}{\partial x_k} + (\omega_{1sk}$$

$$+ \frac{6}{5}q_{(l}\delta_{sk)} + q_k\delta_{ls})L_{ks} - \frac{1}{\rho}(\frac{5}{3}\epsilon\delta_{ls}$$

$$+ \omega_{ls})\frac{\partial(\frac{2}{3}\epsilon\delta_{sk} + \omega_{sk})}{\partial x_k} = \frac{1}{2}P_{ppl}, \qquad (2.5)$$

where

$$L_{lj} = \frac{\partial v_j}{\partial x_l}$$

and angle brackets $A_{<lj...k>}$ enclosing n indices represent the traceless part of a tensor (see, however, (1,3) for more details). In addition, to shorten the writing we denote by parentheses $A_{(lj...k)}$ the sum over the n! permutations of the indices, divided by n!.

In neither case can the set (2.1)-(2.5) of balance equations serve as field equations for the fields $\{\rho, v_l, \epsilon, \omega_{lj}, q_l\}$ because additional quantities have appeared, namely $\{q, s_{lk}, \omega_{lsk}, P_{<lj>}, P_{ppl}\}$. In (1,3) we are concerned with the explicit functional expressions for the higher order moments $\{q, s_{lk}, \omega_{lsk}\}$ and for the collisional productions $\{P_{<lj>}, P_{ppl}\}$ in terms of the one-point distribution function f_1 and of the interaction potential Φ. The main problem of this letter is to specify the dependence of f_1 on the basic fields $\{\rho, v_l, \epsilon, \omega_{lj}, q_l\}$ we wish to determine.

3 THE MAXIMUM ENTROPY PRINCIPLE

According to the maximum entropy principle (4), the probability distribution function f_1, having prescribed the moments

$$\rho = \mu\int d\underset{\sim}{\lambda} f_1, \quad \rho v_l = \int d\underset{\sim}{\lambda}\lambda_l f_1, \qquad (3.1)$$

$$\epsilon = \frac{1}{2\mu}\int d\underset{\sim}{\lambda}\bar{\lambda}^2 f_1, \quad \omega_{lj} = \frac{1}{\mu}\int d\underset{\sim}{\lambda}\bar{\lambda}_{<l}\bar{\lambda}_{j>}f_1, \qquad (3.2)$$

$$q_l = \frac{1}{2\mu^2}\int d\underset{\sim}{\lambda}\bar{\lambda}^2\bar{\lambda}_l f_1, \qquad (3.3)$$

where

$$\bar{\lambda}_l = \lambda_l - \mu v_l$$

is the microscopic momentum of a particle measured in a frame moving with the gas, may be found by maximisation of the entropy density

$$h = -\int d\underset{\sim}{\lambda} f_1 \mathrm{Ln} f_1$$

under the constraints (3.1)-(3.3). A formal derivation of the entropy-maximising distribution can be carried out by means of the Lagrange multipliers method, which yields

$$f_1 = f_1^M \exp(x), \qquad (3.4)$$

where

$$f_1^M = \frac{\rho}{\mu(2\pi\mu T)^{3/2}}\exp(-\frac{\bar{\lambda}^2}{2\mu T}), \qquad (3.5)$$

$$T = \frac{2\mu}{3\rho}\epsilon, \qquad (3.6)$$

$$x = \mu\Delta^\rho + \Delta_l^v\bar{\lambda}_l + \frac{1}{2\mu}\Delta^\epsilon\bar{\lambda}^2$$

$$+ \frac{1}{\mu}\Delta_{lj}^\omega\bar{\lambda}_l\bar{\lambda}_j + \frac{1}{2\mu^2}\Delta_l^q\bar{\lambda}^2\bar{\lambda}_l. \qquad (3.7)$$

In the expression (3.7) the nonequilibrium Lagrange multipliers $\{\Delta^\rho, \Delta_l^v, \Delta^\epsilon, \Delta_{lj}^\omega, \Delta_l^q\}$ remain still to be determined.

Here we shall be interested in relatively rapid nonequilibrium phenomena that are at most quadratic in the heat flux q_l and in the symmetric traceless part ω_{lj} of the friction pressure; we propose

$$\Delta^\rho = A_2\omega_{pq}\omega_{pq} + A_3 q_p q_p, \qquad (3.8)$$

$$\Delta_l^v = B_1 q_l + B_2\omega_{lp}q_p, \qquad (3.9)$$

$$\Delta^{\varepsilon} = C_2\omega_{pq}\omega_{pq} + C_3 q_p q_p, \tag{3.10}$$

$$\Delta^{\omega}_{1j} = D_1\omega_{1j} + D_2\omega_{p<1}\omega_{j>p} + D_3 q_{<1}q_{j>}, \tag{3.11}$$

$$\Delta^{q}_{1} = E_1 q_1 + E_2\omega_{1p}q_p. \tag{3.12}$$

The assumtions (3.8)-(3.12) permit an expansion of (3.4) which may be written as

$$f_1 = f_1^M(1 + x + \tfrac{1}{2}x^2). \tag{3.13}$$

Substituting (3.8)-(3.12) into (3.7), retaining in the formula (3.13) those resulting terms which are at most quadratic in the two basic nonequilibrium characteristics $\{\omega_{1j}, q_1\}$, using finally the constraints (3.1)-(3.3), we arrive at the elementary equations for the expansion coefficients in (3.8)-(3.12)

$$\mu B_1 + \tfrac{5}{2}TE_1 = 0, \tag{3.14}$$

$$\mu B_1 + \tfrac{7}{2}TE_1 - \frac{2\mu^2}{5\rho T^2} = 0, \tag{3.15}$$

$$D_1 - \frac{\mu}{2\rho T^2} = 0, \quad D_2 + 2TD_1^2 = 0, \tag{3.16}$$

$$\mu A_2 + \tfrac{3}{2}TC_2 + T^2D_1^2 = 0, \tag{3.17}$$

$$\mu A_2 + \tfrac{5}{2}TC_2 + \tfrac{7}{3}T^2D_1^2 = 0, \tag{3.18}$$

$$\mu B_2 + \tfrac{5}{2}TE_2 + 2\mu TD_1B_1 + 7T^2D_1E_1 = 0, \tag{3.19}$$

$$\mu B_2 + \tfrac{7}{2}TE_2 + \tfrac{14}{5}\mu TD_1B_1 + \tfrac{63}{5}T^2D_1E_1 = 0, \tag{3.20}$$

$$D_3 + \tfrac{7}{2}TB_1E_1 + \tfrac{1}{2}\mu B_1^2 + \frac{63T^2}{8\mu}E_1^2 = 0, \tag{3.21}$$

$$\mu A_3 + \tfrac{3}{2}TC_3 + \tfrac{5}{2}T^2B_1E_1 + \tfrac{1}{2}\mu TB_1^2 + \frac{35T^3}{8\mu}E_1^2 = 0, \tag{3.22}$$

$$\mu A_3 + \tfrac{5}{2}TC_3 + \tfrac{35}{6}T^2B_1E_1 + \tfrac{5}{6}\mu TB_1^2 + \frac{105T^3}{8\mu}E_1^2 = 0 \tag{3.23}$$

with the simple solution as follows:

$$A_2 = \frac{\mu}{4\rho^2T^2}, \quad A_3 = \frac{2\mu^2}{5\rho^2T^3}, \tag{3.24}$$

$$B_1 = -\frac{\mu}{\rho T^2}, \quad B_2 = \frac{7\mu^2}{5\rho^2T^3}, \tag{3.25}$$

$$C_2 = -\frac{\mu^2}{3\rho^2T^3}, \quad C_3 = -\frac{2\mu^3}{5\rho^2T^4}, \tag{3.26}$$

$$D_1 = \frac{\mu}{2\rho T^2}, \quad D_2 = -\frac{\mu^2}{2\rho^2T^3}, \tag{3.27}$$

$$D_3 = -\frac{9\mu^3}{25\rho^2T^4}. \tag{3.28}$$

4 THE MOMENT STRUCTURE OF THE ONE-POINT DISTRIBUTION FUNCTION

By substituting the result (3.24)-(3.28) into (3.13) we obtain the moment formula for f_1 in its final form; the basic result is:

$$\begin{aligned} f_1 = f_1^M\{1 &+ \tfrac{1}{2}\tilde{\omega}_{1j}\tilde{\lambda}_1\tilde{\lambda}_j - \tilde{q}_1\tilde{\lambda}_1(1 - \tfrac{1}{5}\tilde{\lambda}^2) \\ &+ \tfrac{1}{10}\tilde{\omega}_{pq}\tilde{\omega}_{pq}(\tfrac{5}{2} - \tfrac{5}{3}\tilde{\lambda}^2 + \tfrac{1}{6}\tilde{\lambda}^4) \\ &- \tfrac{1}{14}\tilde{\omega}_{p<1}\tilde{\omega}_{j>p}\tilde{\lambda}_1\tilde{\lambda}_j(7 - \tilde{\lambda}^2) \\ &+ \tfrac{1}{8}\tilde{\omega}_{<1j}\tilde{\omega}_{ir>}\tilde{\lambda}_1\tilde{\lambda}_j\tilde{\lambda}_i\tilde{\lambda}_r \\ &+ \tfrac{1}{25}\tilde{\omega}_{1p}\tilde{q}_p\tilde{\lambda}_1(35 - 14\tilde{\lambda}^2 + \tilde{\lambda}^4) \\ &- \tfrac{1}{2}\tilde{\omega}_{<1j}\tilde{q}_{i>}\tilde{\lambda}_1\tilde{\lambda}_j\tilde{\lambda}_i(1 - \tfrac{1}{5}\tilde{\lambda}^2) \\ &+ \tfrac{1}{5}\tilde{q}_p\tilde{q}_p(2 - \tfrac{1}{6}\tilde{\lambda}^2 - \tfrac{1}{3}\tilde{\lambda}^4 + \tfrac{1}{30}\tilde{\lambda}^6) \\ &+ \tfrac{1}{50}\tilde{q}_{<1}\tilde{q}_{j>}\tilde{\lambda}_1\tilde{\lambda}_j(7 - 10\tilde{\lambda}^2 + \tilde{\lambda}^4)\}, \end{aligned} \tag{4.1}$$

where

$$\tilde{\omega}_{1j} = \frac{\mu}{\rho T}\omega_{1j}, \quad \tilde{q}_1 = \frac{1}{\rho}\left(\frac{\mu}{T}\right)^{3/2} q_1, \tag{4.2}$$

$$\tilde{\lambda}_1 = \frac{\bar{\lambda}_1}{(\mu T)^{1/2}} \ . \tag{4.3}$$

5 CONCLUSIONS

The often used 13 moment expression for f_1 (see (2))

$$f_1 = f_1^M\{1 + \frac{1}{2}\tilde{\omega}_{1j}\tilde{\lambda}_1\tilde{\lambda}_j - \tilde{q}_1\tilde{\lambda}_1(1 - \frac{1}{5}\tilde{\lambda}^2)\} \tag{5.1}$$

gives rise to constitutive relations for excessive unknowns in (2.1)-(2.5) that are linear in ω_{1j} and in q_1. To within questions concerning applications, (4.1) reproduces correctly more general nonlinear constitutive relations, the latter being quadratic in ω_{1j} and in q_1 (1). All our discussion so far has been elementary and simple in the sense that the compatibility of (4.1) with the original Grad moment representation for f_1 (2) on the one hand and with the second axiom of thermodynamics on the other hand have not yet been studied.

So, it is now interesting to report that Eq.(4.1) results from the Grad moment representation as well (1) and, in addition, that the maximum entropy principle is entirely consistent with the Clausius-Duhem inequality. However, for complementary understanding of Grad's truncation procedure from the thermodynamic view point, at least for the Boltzmann kinetic equation, the reader should consult our recent work in submission (1).

ACKNOWLEDGEMENT

The result of this comment is attained with the assistance of the Alexander von Humboldt Foundation (AvH).

REFERENCES

1.Z.Banach, "Explanation and Generalization of Grad's Moment Truncation Procedure for a Gas of Non-Maxwellian Molecules" (submitted to Ann. Phys. (N.Y)).
2.H.Grad, Comm. Pure and Appl. Math. 2 (1949), 331-407.
3.Z.Banach, Physica 129A (1984), 95-124.
4.I.Müller, Thermodynamics, Pitman Publ., London (1985).

ERRATUM

The expressions (3.24)-(3.28) should be supplemented by

$$E_1 = \frac{2\mu^2}{5\rho T^3} \ , \quad E_2 = -\frac{18\mu^3}{25\rho^2 T^4} \ .$$

5th International Symposium on Continuum Models of Discrete Systems / Nottingham / 14-20 July 1985

Solitons in microstructured elastic media: Physical and mechanical aspects

G.A.Maugin & J.Pouget
Université Pierre-et-Marie Curie, Paris, France

ABSTRACT: Elastic crystals with a microstructure present all the ingredients needed to allow for the propagation of nonlinear waves of the soliton type: the nonlinearity in some parameter which is often related to phase-transition phenomena and the long-range interactions giving rise to dispersion. These two ingredients, when compensated, favor the existence of solitary waves. To illustrate our talk we consider three cases whose starting equations are quite dissimilar. They nonetheless yield dispersive systems of nonlinear hyperbolic equations of a unique type. More precisely, in the first case we consider ferroelectric crystals of the molecular-group type with a continuum description obtained by passing to the long-wave limit in a lattice-dynamics model. The second case concerns the nonlinear continuum theory of elastic ferromagnets. The third case is provided by nonlinear micropolar elastic solids either in the formulation of Kafadar and Eringen or in the director formulation. The notion of structure in "domains and walls" is introduced for the three classes of materials, examples of which are provided by sodium nitrite, nickel or yttrium-iron garnets and potassium nitrate, respectively. Walls of the Bloch and Néel types may be envisaged. The common mathematical problem obtained concerns a sine-Gordon equation which is nonlinearly coupled to one or two wave equations that govern two elastic displacements. Stable exact solitary-wave solutions can be obtained and this provides a unifying view of moving walls generating acoustic effects in a variety of crystals. By means of perturbation theory, an almost soliton-like behaviour is also exhibited by these solutions.

> "and some are very naughty crystals indeed"
>
> John Ruskin(The Ethics of the Dust:Ten Lectures to Little Housewives on the elements of Crystallisation, 1866).

1 PREREQUISITES

The present work combines three ingredients that we shall review successively.

1.1 Domains and walls

Many materials of which the study concerns both condensed-matter physics and engineering sciences present an ordered structure below a certain critical temperature. This structure may be of various types:the crystalline structure of usual (solid)crystals (regular arrangement of "atoms" building up the solid), the ordered arrangement of "particles" constituting a liquid crystal of the nematic type, the ordered parallel structure of magnetic spins in ferromagnetic crystals, the alternatively parallel and antiparallel structure of magnetic spins in antiferromagnetic crystals, the parallel ordered structure of permanent microscopic dipoles in a ferroelectric crystal (Kittel,1971). In practice, however, this perfect ordered structure seldom exists for a bulky specimen because of the existence of defects. In particular, Nature may found it energetically more advantageous that this order exists, but only locally in so-called domains. Between domains, there exist therefore transition layers of small "thickness" as compared to the overall sizes of domains, called walls, which then provide a reasonably smooth transition between one ordered configuration in a domain to that one in the adjacent domain.

This situation is best illustrated by ferromagnetism which, historically, has provided the first example of "structure in domains and walls", a structure made

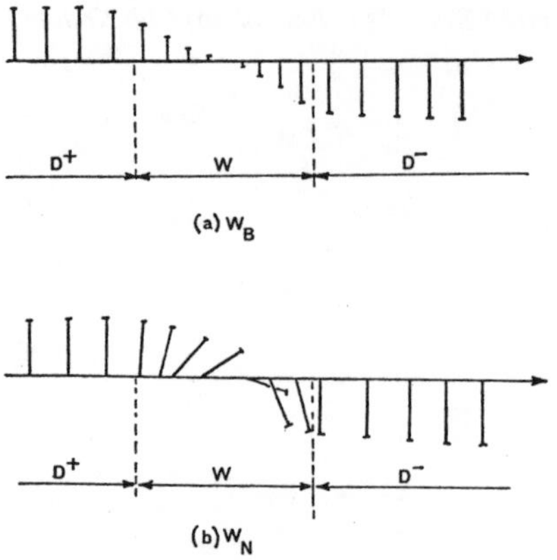

Fig. 1 Bloch and Néel walls

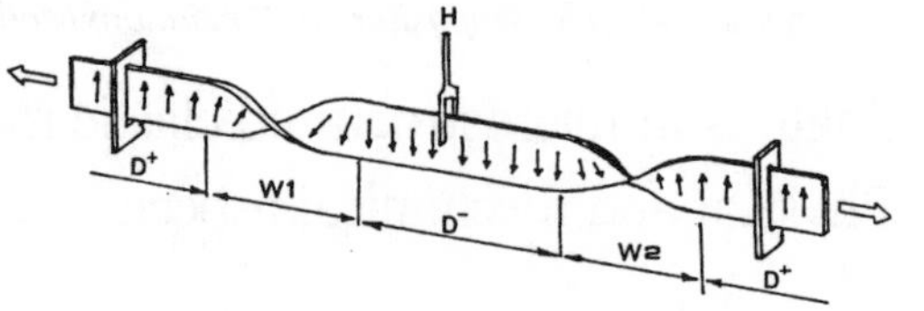

Fig. 2 The twisted elastic flat band. On removing the diapason holder H, walls W1 and W2 (soliton and anti-soliton) move toward one another and collide, etc)

readily visible by Bitter's technique. Two examples are shown in Figure 1 of so-called 180° walls between a ferromagnetic domain ⊕ [all magnetic spins aligned parallel up in the left part of Figures 1a and 1b] and a ferromagnetic domain ⊖ [all magnetic spins aligned parallel down in the right part of Figures 1a and 1b]. However, whereas in Figure 1a magnetic spins in the transition zone rotate by 180° effecting a rotation about the x-axis (so-called "Bloch wall" after F.Bloch), hence an "out-of plane rotation, in Figure 1b the axis of rotation is perpendicular to the plane of the figure, hence an "in-plane rotation". We have then a so-called "Néel wall" (after the French physicist L.Néel). These notions of Bloch and Néel walls will recur in the forthcoming development. In the process of magnetization of a ferromagnetic sample or under the action of an applied stress, ferromagnetic walls may move. In fact, the magnetization of a sample and a large part of the well-known hysteresis cycle result from more or less successive jerky displacements (so-called Barkhausen jumps) of walls already present in the sample, so that domains with favorably oriented magnetization-i.e., more or less aligned with the magnetizing field-grow at the expenses of others. Looked upon at a gross scale, walls appear as defects in the magnetic organization of the crystals, and these are to be studied in a general framework along with dislocations and other defects (Kléman,1983).

An interesting analogy helping one visualize the notion of Bloch wall relates to the mechanical behaviour , in torsion and under an applied tension, of a metallic flat band (thus with a nonsymmetric rectangular cross section). The small arrows pictured on the wide faces of the band (Figure 2) give an image of the arrangement of magnetic spins. The mechanical equation governing the elasti band happens to be identical to the equation governing the change in orientation of the magnetic spins within a Bloch wall [equations (1.1) below; Wesołowski(1983)].

The first comprehensive phenomenological theory of stationary Bloch walls in rigid ferromagnets was given in a celebrated paper of Landau and Lifshitz (1935). The distribution of spin orientation ($\vartheta = \phi/2$) on the infinite interval $(-\infty,+\infty)$ is governed by the one-dimensional (adimensionalized) problem

$$(1.1)\qquad \frac{d^2\phi}{dX^2} = \sin\phi \quad , \; X \in (-\infty,+\infty)$$

$$\phi \rightarrow 0, \; \frac{d\phi}{dX} \rightarrow 0 \; , \; |X| \rightarrow \infty \, ,$$

which admits the front-like solution (see Figure 3b below for a sketch)

$$(1.2)\qquad \cos\vartheta = -\tanh X$$

The drastic change in ϑ occurs over a distance of order unity, hence the wall is said to have thickness unity. Within the wall the out-of-plane rotation of magnetic spins has a positive helicity (direct screw) with increasing X. Notice that eqn.$(1.1)_a$ is the exact equation of a pendulum (but the motion of a pendulum requires boundary conditions at the ends of a finite interval so that its solution involves elliptic integrals). More on magnetic domain walls in rigid ferromagnets may be found in Soohoo(1965) and Wadas (1974). There also exist 90° walls as well as walls with rather complex structure involving ,for instance, ondulant structures for which one needs a full three-dimensional description. The notions of domains, walls and defects apply to all cases listed at the beginning of the paragraph.

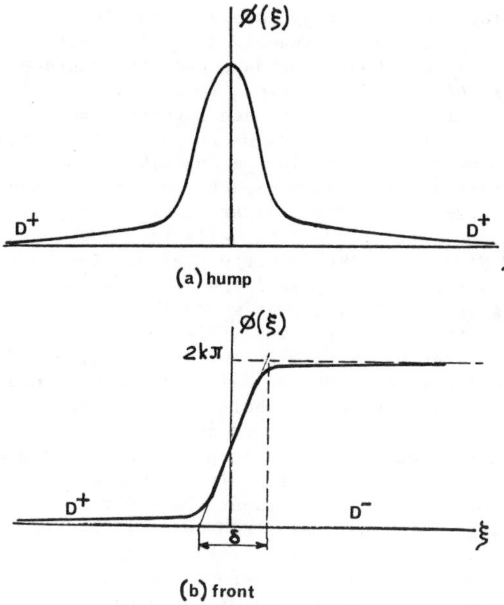

Fig. 3 Solitary waves:(a)hump type;
(b) front type

1.2 Microstructured media

Simultaneously, from the point of view of continuum mechanics, all the above-mentioned ordered structures are real physical examples of what is commonly known as continua endowed with a microstructure. Indeed, these structures may be considered as continua at a macroscopic scale of observation but their specific ordering properties of quantum origin requires a finer description than the one usually given, for instance, by elasticity, even on account of anisotropy and nonlinearity. This is induced by the fact that the microstructure introduces in the picture additional degrees of freedom. In a general manner, a microstructured continuum is a material continuum of which the usual constituents, so-called material points, present, in addition to their usual motion giving rise to the classical notions of velocity and strains, internal degrees of freedom. The latter may be in a finite number (case of a rigid microstructure) or they may be infinitely numerous (deformable microstructure). Such a microstructure is kinematically described in some abstract manner by means of a field of three vectors, called directors, which are attached to each material point, the latter being equipped with a tensor of inertia in addition to the usual mass density . Whenever the directors form at all times a rigid triad (Kafadar, Eringen, 1971), we say that the medium is "micropolar" or, else, it is a "Cosserat continuum "(to honour the Cosserat brothers who apparently first introduced such a microstructure in 1909). Otherwise (deformable microstructure), the medium is said to be "micromorphic" in agreement with the term coined by A.C.Eringen (1964).

It is clear that the analytical study of micromorphic continua is a very difficult task. Two physical examples of such continua are provided by solutions of deformable macromolecules (Maugin, Drouot, 1983) and piezoelectric powders where each grain deforms through electromechanical couplings under the action of an electric field (Pouget, Maugin, 1982). Insofar as micropolar continua are concerned, liquid crystals of the nematic type and suspensions of rigid fibers (microstructure with one rigid director) provide two wellknown examples (Stokes, 1984).

Certain deformable crystals of the "fully solid" type also have long been recognized as elastic continua with a microstructure (Laval, 1957; Nowacki, 1981; Kunin, 1982-3). More recently, it has also been shown that elastic ferromagnetic crystals at low temperature (e.g., nickel, cobalt, yttrium-iron garnets at room temperature) were also examples of micropolar continua with, however, peculiar inertia properties of the microstructure as a consequence of the original nature of magnetic spin (Maugin, Eringen, 1972) —this is itself proportional to an angular momentum via the gyromagnetic effect. This is also true of elastic antiferromagnets, in which case the microstructure is described by two co-existing magnetic moments which may compensate only in certain configurations of the material (Maugin, 1976). Classical crystals with a molecular group [e.g., potassium nitrate KNO_3, Askar, 1973] and ferroelectric crystals with a molecular group [e.g., sodium nitrite $NaNO_2$ (Pouget et al, 1985b)] also provide good examples of elastic micropolar continua. In the last case a density of electric dipole is rigidly attached to the microstructure (the central molecular group). As compared to classical continuum mechanics, the mechanics of microstructured continua differs essentially by the nonsymmetry of the Cauchy stress tensor and, therefore, the active role played by the moment of momentum equation or any equivalent equation. A general and safe approach to the governing equations of such complex continua is the use of the principle of virtual power in the form presented by the first author (Maugin, 1980a).

1.3 Solitary waves and solitons

The first known example of solitary wave is due to the Scottish scientist Scott-Russell in his 1834 observations on the wave propagation at the surface of a shallow-water canal. It is now known that a "solitary wave" is a progressive wave which is localized in phase space, which is an exact solution to a nonlinear hyperbolic equation where nonlinearity and a dispersive contribution compensate one another, and which, therefore, propagates without deformation in its profile with a velocity-dependent amplitude (see Figures 3). A solitary wave is said to be a "soliton" if and only if, after interaction (head-on collision, catching up) with another such "soliton", it recovers the same essential characteristics (profile and velocity), presenting thus in its dynamical behaviour an obvious analogy with a particle subjected to an elastic collision. Equations that yield such behaviours for wave-like solutions are ,among others, the celebrated Korteweg-de Vries equation, the nonlinear Schrödinger equation for certain potentials and, closer to our present concern, the sine-Gordon equation (an obviously nonlinear extension of the Klein-Gordon equation via a sine term hence the facetious naming;Barone et al, 1971):

$$\frac{\partial^2 \phi}{\partial t^2} - \frac{\partial^2 \phi}{\partial X^2} + \sin \phi = 0 \quad . \tag{1.3}$$

Such an equation was apparently written down in a physical context for the first time in another celebrated paper, but this one by Frenkel and Kontorova (1939), as a continuum approximation to the system of difference-differential equations formulated with a view to modelling certain defects in crystal lattices. The same trend was further developed by several authors , of whom Seeger (1955,1979) for modelling dislocations in crystals and kinks on dislocations. In statics, eqn.(1.3) reduces to eqn.$(1.1)_a$, i.e., the "stationary" Landau-Lifshitz Bloch-wall equation (alias the pendulum equation, alias the "elastica" equation in rod theory;Kafadar,1972). Using the characteristic variables $\xi = X + t$ and $\zeta = X - t$ of the linearized equation , (1.3) also yields the Enneper equation(Seeger,Wesolowski,1981)-with a change in the determination of ϕ :

$$\frac{\partial^2 \phi}{\partial \xi \partial \zeta} = \sin \phi \tag{1.4}$$

known in the geometry of pseudo-spheres (surfaces of constant negative curvature) since the last century (Eisenhart,1960). This led to the discovery of Bäcklund transformations (Bäcklund,1882;Goursat, 1925) which are nowadays used to generate multi-soliton solutions (Miura,1976).

The similarity between Figures 1 and 3 is sufficiently obvious to understand why many authors, following Enz (1964) in the case of rigid ferromagnetic crystals, have thought of using solitary waves to represent static or moving walls [see,e.g., Feldkeller (1965),Currie (1977)]. The thickness of the wall then is given by the spatial interval over which the amplitude of the solitary waves varies significantly between one spatially uniform solution and another one (the two being possibly identical; case of a hump soliton in Figure 3).

More recently, Pouget and Maugin (1984, 1985a) have gone one step further in complexity in showing that solitary waves and solitons could represent the evolving structure in domains and walls in a deformable ferroelectric crystal of the molecular-group type (hence a deformable solid with a practically rigid microstructure), the nonlinearity needed being related not to elasticity as is the case in Auld and Fester (1981), but to the large amplitude variations in the orientation, and the possible flip-flop motion of, the molecular group or the attached electric dipole. This new possibility arises from the fact that, under certain simplifying assumptions, the relevant coupled electromechanical system is governed by a sine-Gordon equation that is nonlinearly coupled to a wave equation for the transverse elastic displacement. In the case of a single wall, it is thus shown that the complete dynamical electromechanical problem is equivalent to solving a so-called"double sine-Gordon equation"

$$\frac{\partial^2 \phi}{\partial t^2} - \frac{\partial^2 \phi}{\partial X^2} - \sin \phi - \gamma \sin(2\phi) = 0 \tag{1.5}$$

under proper limit conditions at $\pm \infty$, a problem which also admits solutions of the soliton type (Bullough et al,1980). The interaction between such solutions (Pouget, Maugin,1985a) and the starting motion from rest of one such electromechanical solitary wave under sudden application of an external stimulus have also been studied by various analytical methods as also numerically (Pouget,Maugin,1985b,d).

Simultaneously, Motogi and Maugin(1984a,b) have studied the small-amplitude vibrations of Bloch and Néel walls [hence small harmonic perturbations superimposed on static,strongly disuniform solutions of the type of (1.2)- a "stationary" soliton] in elastic ferromagnetic crystals both in the absence and the presence of mechanical and magnetic dissipative processes [viscosity and spin-lattice relaxation (Maugin,1975)], a problem closely related to the stimulation of ultrasound in ferromagnets (Baldokhin et al,1972).

The three ingredients that have just been presented at some length help one explain and justify the contents of the present work which is confined to the most representative physical and mechanical aspects of the problems. The more mathematical aspects aspects are dealt with in a unified manner,for all presently evoked cases, in a companion lecture (Maugin,1985).

1.4 The contents of this work

First, we shall review, but in a wider context, the approach of Pouget and Maugin to the case of elastic ferroelectric crystals (Section 2).One of the strong point here is the relationship exhibited by these authors between the domain-wall structure represented by solitons and incommensurate-commensurate phase transitions that occur in certain ferroelectric crystals. Then , on the basis of recent works by Maugin and Miled (1985a), we consider the cases of walls in elastic ferromagnets, more specifically, Bloch walls in an infinite crystal and Néel walls in a thin elastic film (Section 3) and micropolar elastic continua (Maugin,Miled,1985b) according to the nonlinear kinematical description of Kafadar and Eringen (1971) who use orthogonal geometrical objects to represent the microstructure (Section 4;Bloch and Néel types of walls are introduced through the notion of solitary waves). Finally, another kinematical description of elastic micropolar continua is also introduced (Section 5) which is closer to the one often used in liquid crystals, the so-called Ericksen-Leslie approach in which a single director is taken into account. For the case of elastic solids (Pouget,Maugin,1985e), the corresponding model is extracted from a more general work of these authors (1982). In all, the case yet to be studied is that of elastic anti-ferromagnets. Certain works of A.E.Turov and co-workers, however, come very close to the solution of this problem (Turov,Tchvrov, 1983; Turov,1984;Lugovoĭ et al,1983; Lebedev et al,1983).

2 ELASTIC FERROELECTRIC CRYSTALS

2.1 The model

Following Pouget et al (1985a) -also Pouget (1984) - in the lattice dynamics of ferroelectric crystals with a molecular group,we consider the exemplary case of sodium nitrite $NaNO_2$ (Figure 4a)in which the central molecular group NO_2 equipped with a permanent electric dipole $\vec{e}$, may suffer large-amplitude rigid-body rotations about the $\vec{c}$-axis and rather small-amplitude rotations about the $\vec{a}$-xis . In the lattice-dynamics model that we do not detail here (see Figure 4b when only the rotation about the $\vec{c}$-axis is accounted for), the forces acting on the lattice result from :(i) short and long-range interactions between neighboring ions,(ii) mutual interactions between microscopic electric dipoles and, possibly,(iii) electrostatic interactions of these dipoles with an external electric field. In the range of frequencies of interest it is supposed that all ions vibrate in phase (optical modes are discarded) so that only acoustic vibrations and rotational oscillations of the molecular groups are excited. Passing to the long-wavelength limit in the case where the two above-mentioned rotations are taken into account (see Figure 5 for notation),one obtains the following continuous system that governs the longitudinal elastic displacement u, the transverse elastic displacement v and twice the real angles of rotations ($\phi = 2\vartheta$, $\alpha = 2\psi$) in a so-called one-dimensional nonlinear motion (X:nondimensional spatial coordinate,t:nondimensional time) and in the absence of forcing (Pouget,unpublished,1985):

$$(2.1)_a \quad \frac{\partial^2 u}{\partial t^2} - V_L^2 \frac{\partial^2 u}{\partial X^2} = \frac{1}{2} \nu \frac{\partial}{\partial X}[(1+\cos\alpha)\times (1+\cos\phi)] \quad ,$$

$$(2.1)_b \quad \frac{\partial^2 v}{\partial t^2} - V_T^2 \frac{\partial^2 v}{\partial X^2} = -\eta\frac{\partial}{\partial X}[\sin\phi \cos(\alpha/2)],$$

$$(2.1)_c \quad \frac{\partial^2 \phi}{\partial t^2} - \frac{1}{2}(3\cos\alpha - 1)\frac{\partial^2\phi}{\partial X^2} = \frac{1}{2}(1+\cos\alpha)\sin\phi + \frac{1}{2}\nu(1+\cos\alpha)(\sin\phi)\frac{\partial u}{\partial X} + \eta\cos\phi\left(\cos\frac{\alpha}{2}\right)\frac{\partial v}{\partial X}$$

$$(2.1)_d \quad \frac{1}{\kappa}\frac{\partial^2\alpha}{\partial t^2} - \frac{1}{2}V_S^2(1+\cos\phi)\frac{\partial^2\alpha}{\partial X^2} = \frac{1}{2}(1+\cos\phi)\sin\alpha + \frac{1}{2}\nu(1+\cos\phi)(\sin\alpha)\frac{\partial u}{\partial X} - \frac{1}{2}\eta\sin\phi\left(\sin\frac{\alpha}{2}\right)\frac{\partial v}{\partial X}$$

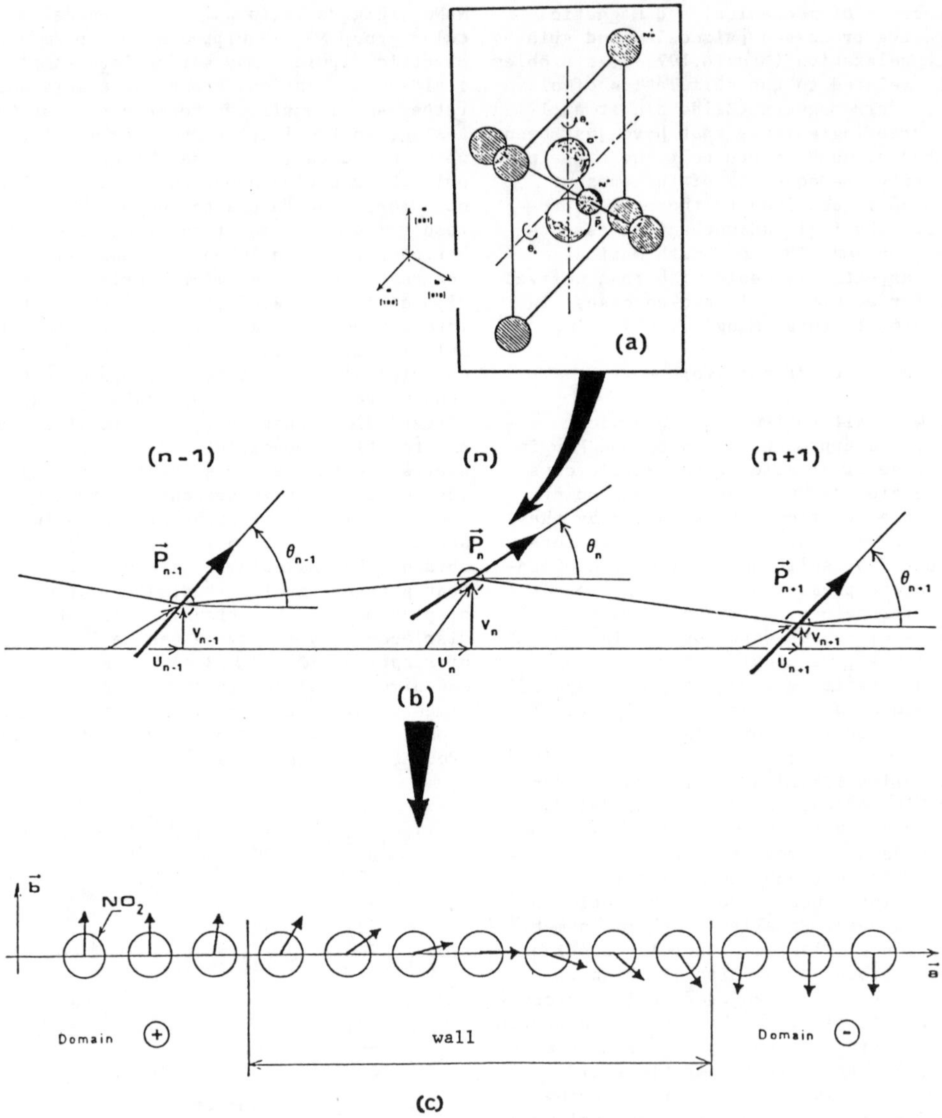

Fig. 4 Structure of $NaNO_2$, simplified lattice model and wall structure

where all quantities are nondimensional. Speeds are normalized with respect to the characteristic speed of ϕ. The coefficients γ and η represent electromechanical couplings and these may be assumed small as compared to unity if necessary; $\varkappa$ is a ratio of inertia moments. In agreement with physical observation Ψ, hence α, may always be considered small (rotation about the $\vec{a}$-axis) so that the above-given system reduces to the following one:

$$(2.2)_a \quad \frac{\partial^2 u}{\partial t^2} - V_L^2 \frac{\partial^2 u}{\partial X^2} = \gamma \frac{\partial}{\partial X}(1+ \cos \phi) \quad ,$$

$$(2.2)_b \quad \frac{\partial^2 v}{\partial t^2} - V_T^2 \frac{\partial^2 v}{\partial X^2} = -\eta \frac{\partial}{\partial X} (\sin \phi) \quad ,$$

$$(2.2)_c \quad \frac{\partial^2 \phi}{\partial t^2} - \frac{\partial^2 \phi}{\partial X^2} - \sin \phi = \gamma \sin \phi \frac{\partial u}{\partial X} + \eta \cos \phi \frac{\partial v}{\partial X} \quad ,$$

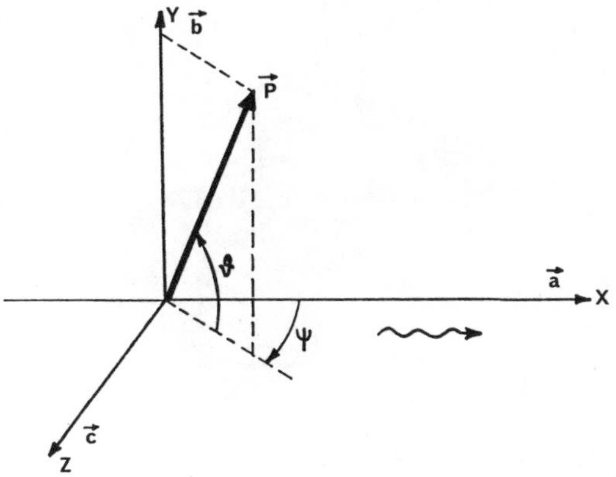

Fig. 5 Notation of Section 2

$$(2.2)_d \quad \frac{1}{\chi}\frac{\partial^2\alpha}{\partial t^2} - \frac{1}{2}V_S^2\frac{\partial^2\alpha}{\partial X^2} = \frac{1}{2}[(1+\cos\phi)\times$$

$$\times(1+\nu\frac{\partial u}{\partial X}) - \frac{1}{2}\eta(\sin\phi)\frac{\partial v}{\partial X}$$

$$+(\cos\phi)V_S^2\frac{\partial^2}{\partial X^2}]\alpha$$

The structure of this system of dispersive nonlinear hyperbolic equations is of special interest because it already gives the general features of the systems that we shall meet in the forthcoming cases. In particular, the first system of three equations for the three unknowns (u,v,ϕ) is made of two wave equations for two elastic displacements in the plane of the figure (like in a Rayleigh surface wave where the elastic displacement is confined to the sagittal plane) and a sine-Gordon equation for ϕ, each elastic displacement being nonlinearly coupled to ϕ through one of the electromechanical coupling coefficients ν or η. As to the remaining angle α, it is governed by a Klein-Gordon equation which is nonlinearly perturbed by the coupled solution of the first system. The interesting system is the one formed by eqn. $(2.2)_{a,b,c}$. Whenever α remains practically zero at all times and places, we can be satisfied with this latter system. It is further noticed in the case of $NaNO_2$ (Pouget, 1984) that $\nu \ll \eta$, so that we may in fact study the following system of two partial differential equations:

$$(2.3)_a \quad \frac{\partial^2 v}{\partial t^2} - V_T^2\frac{\partial^2 v}{\partial X^2} = -\eta\frac{\partial}{\partial X}(\sin\phi) \quad ,$$

$$(2.3)_b \quad \frac{\partial^2\phi}{\partial t^2} - \frac{\partial^2\phi}{\partial X^2} = \sin\phi + \eta\cos\phi\frac{\partial v}{\partial X} \quad ,$$

which uncouples from the longitudinal displacement and where ϕ is an in-plane rotation, hence the notion of Néel wall.

2.2 Solitary waves and solitons, solutions of the system (2.3)

A. Single soliton

The system (2.3) consists of a sine-Gordon equation which is nonlinearly coupled to a d'Alembert equation. One could first think of using a perturbation scheme since η is an infinitesimally small parameter in many cases. However, the system (2.3) as a whole happens to admit exact nonlinear propagative solutions (Pouget, Maugin, 1984). Indeed, seeking a solution that depends on a single phase variable $\xi = QX - \Omega t + \xi_o$, with the following behaviour at infinity:

$$\phi \to \pm\pi \quad \text{as} \quad \xi \to \mp\infty \quad ,$$

$$(2.4) \quad \frac{d\phi}{d\xi} \to 0 \quad \text{as} \quad |\xi| \to \infty \quad ,$$

$$\frac{dv}{d\xi} \to 0 \quad \text{as} \quad |\xi| \to \infty \quad ,$$

one finds that the system (2.3) is equivalent to a single ordinary differential equation which admits a first (energy) integral

$$(\Omega^2 - Q^2)\left(\frac{d\phi}{d\xi}\right)^2 + V(\phi) = E_o \quad ,$$

$$(2.5) \quad V(\phi) = 2\cos\phi - \gamma\cos(2\phi) \quad ,$$

$$\gamma = \frac{\eta^2}{2}\frac{Q^2}{\Omega^2 - \Omega_T^2} \; , \; \Omega_T = V_T Q \quad ,$$

which integrates immediately to give (with $E_o = 2 - \gamma$)

$$(2.6) \quad \phi = -2\tan^{-1}\left[\frac{\sinh\xi}{\sqrt{1+2\gamma}}\right]$$

and where Ω and Q are necessarily related by the pseudo "dispersion relation"

$$(2.7) \quad \Omega^2 - Q^2 = -(1+2\gamma) \quad .$$

The stable solution ($\Omega < Q$, $d\phi/d\xi \leq 0$) corresponds to points (Ω, Q) belonging to the branch (c) on the dispersion diagram of Figure 6, where Ω_T and Ω_F would be the uncoupled transverse elastic and ferroelectric harmonic modes while (i) and (j) would be the coupled modes for harmonic linear waves [see Pouget, Maugin (1980) for the continuum approach and Pouget et al (1985a) for the discrete one], and (e), (f) and (g)

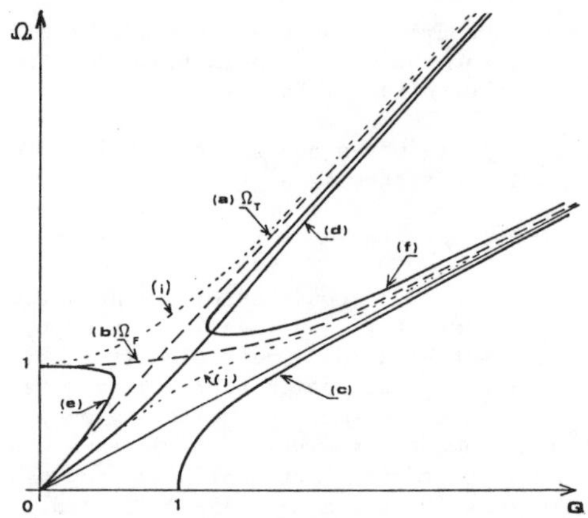

Fig. 6 Dispersion relation for solitary waves in elastic ferroelectric crystals of the molecular-group type

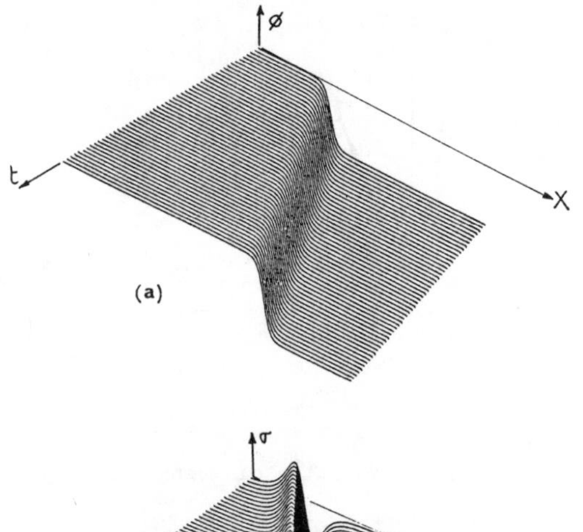

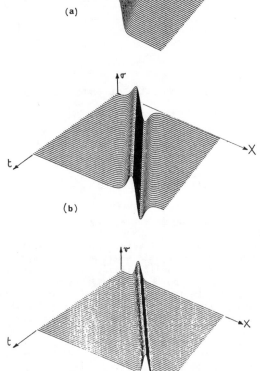

Fig. 7 Solitary wave in the electroacoustic of elastic ferroelectric crystals of the molecular-group type

are other branches corresponding to unstable solitary-wave solutions. It can also be noticed on account of (2.4) and (2.5) that (2.6) is also a solution of the double sine-Gordon equation (1.5) which admits a reduction to the same nonlinear ordinary differential equation for solution depending on a single phase variable, with limits (2.4) and where $\gamma = \gamma(\Omega, Q)$ is supposed to be known according to $(2.5)_3$. A numerical simulation of $\emptyset$, the accompanying stress field σ , or deformation $\partial v/\partial X$, and elastic displacement v (which can be shown to be integrable since the strain compatibility conditions are fulfilled) is reproduced in Figures 7. The displacement corresponding to (2.6) is given by

$$(2.8)\quad v-v_o = \frac{2\eta Q}{\Omega^2 - \Omega_T^2}\left(\frac{1+2\gamma}{-2\gamma}\right)^{1/2} \times$$

$$\times \tanh^{-1}[(-2\gamma)^{-1/2} \cosh \xi] .$$

The solution (2.6)-(2.8) represents a moving ferroelectric 180° Néel wall with the accompanying acoustic field. The speed of propagation (see below) of this wall is rather low (as compared to acoustic phenomena so that, here, $V_T \gg 1$), of the order of 1 mm/sec at 120°C under an applied field of 5×10^{-2} MVolt/m(Tran et al,1981). The thickness δ_w of the wall computed from the microscopic parameters contained in the macroscopic coefficients of eqns.(2.3) is obtained of the order of 0.6 μm, which compares very well with the estimates (0.3-1. μm) directly measurable on electron-microscopy pictures (Suzuki,Takagi,1971), and the wall energy Σ_w , computed by summing the Hamiltonian of the system (see E_o in eqn.$(2.5)_1$) over the whole real line $(-\infty, +\infty)$ is found of the order of 1 erg/cm^2 . It is found to vary according to the rather simple formula

$$(2.9)\quad \Sigma_w = \frac{\sqrt{2}}{4\pi}\ X'\ (1-\bar{e}^2)^{1/2}(P_o^2)\ \delta_w ,$$

where P_o is the macroscopic electric polarization at $\pm\infty$, $\bar{e}$ is the reduced coefficient of electromechanical coupling and X' is a reduced electric susceptibility.

It must be noticed that the model (2.3)

can be written at once and does not need at all the introduction of the angle α (compare the ferromagnetic case below).

B. Multiple-soliton solutions

The sine-Gordon equation of (2.3) for $\eta=0$ admits multiple-soliton solutions which can be generated by means of Bäcklund transformations (see Barone et al,1971). What about the whole system (2.3)? That is, what about solutions depending on several phase variables ? Do the constituents of such solutions behave like solitons ? The mathematical answer is not obvious and requires the use of sophisticated techniques (Pouget,Maugin, 1985a;Maugin,1985). Here we simply give a flavor of the results and methodology. The system (2.3) can be cast in the form of a system of evolution for a four-vector $\vec{U}=(v, \frac{\partial v}{\partial t}, \phi, \frac{\partial \phi}{\partial t})^T$ as (Pouget,Maugin,1985a)

$$\frac{\partial \vec{U}}{\partial t} + N(\vec{U}) = \eta \vec{F}(\vec{U}) \quad , \tag{2.10}$$

where N is a nonlinear operator and η is an infinitesimally small parameter. Then a singularly perturbed solution of (2.10) is sought in the form of the asymptotic expansion

$$\vec{U} = \vec{U}_o + \eta \vec{U}_1(X,t)+\cdots \quad , \tag{2.11}$$

where $(v_o, \frac{\partial v_o}{\partial t})$ corresponds to transverse elastic waves and $(\phi_o, \partial\phi_o/\partial t)$ is a multiple-soliton solution (generated by Bäcklund transformation) containing free parameters for the uncoupled sine-Gordon equation. Then $\vec{U}_1$ is found by using the Green function associated with the zeroth-order solution $\vec{U}_o$ and the secularity condition

$$\lim_{\eta \to 0} \bar{U}_1(X,t/\eta) = 0 \quad , \tag{2.12}$$

is shown to impose a modulation, on the time scale ηt, on the free parameters of the zeroth-order solution. Then $\vec{U}_1$ is fully constructed with accompanying radiations in ϕ and v, and a pure soliton behaviour cannot be obtained in reason of the very structure of the system (2.10). For numerical simulations (of which one example is given in Figure 8 for a "soliton-antisoliton" head-on collision) the system (2.10) is symbolically rewritten in the following nonlinear hyperbolic form (the left-hand side has the form of a conservative contribution)

$$\frac{\partial \vec{U}}{\partial t} + \frac{\partial}{\partial X} \vec{\mathfrak{F}}(\vec{U}) = \vec{G}(\vec{U}) \tag{2.13}$$

and use is made of a leap-frog Lax-Wendroff scheme on a finite spatial interval with the

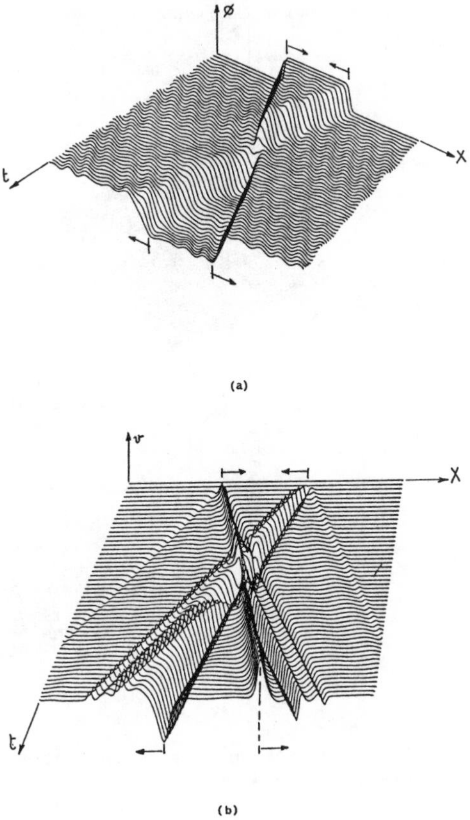

Fig. 8 "Soliton-antisoliton"collision in elastic ferroelectric crystals of the molecular-group type.

asymptotic values of the analytical scheme (2.10)-(2.12) as end conditions. The graphs of Figures 8 exhibit superimposed oscillations related to radiation phenomena. Note that the numerical solutions obtained cannot be extended over too long intervals of time.

C. Transient motion -Action of an external stimulus

Consider an electric field $\vec{E}_o$ in the y-direction. In the wall region where electric dipoles are no longer aligned with $\vec{E}_o$ this creates a torque and summing up the action of these elementary torques over the wall thickness will give rise to a motive force that will move the wall so that the favorably oriented domain should grow at the expense of the other one (if we have a 180°-wall and domain structure). Mathematically, the action of such an electric field yields a forcing term in eqns.(2.3) that are now written as

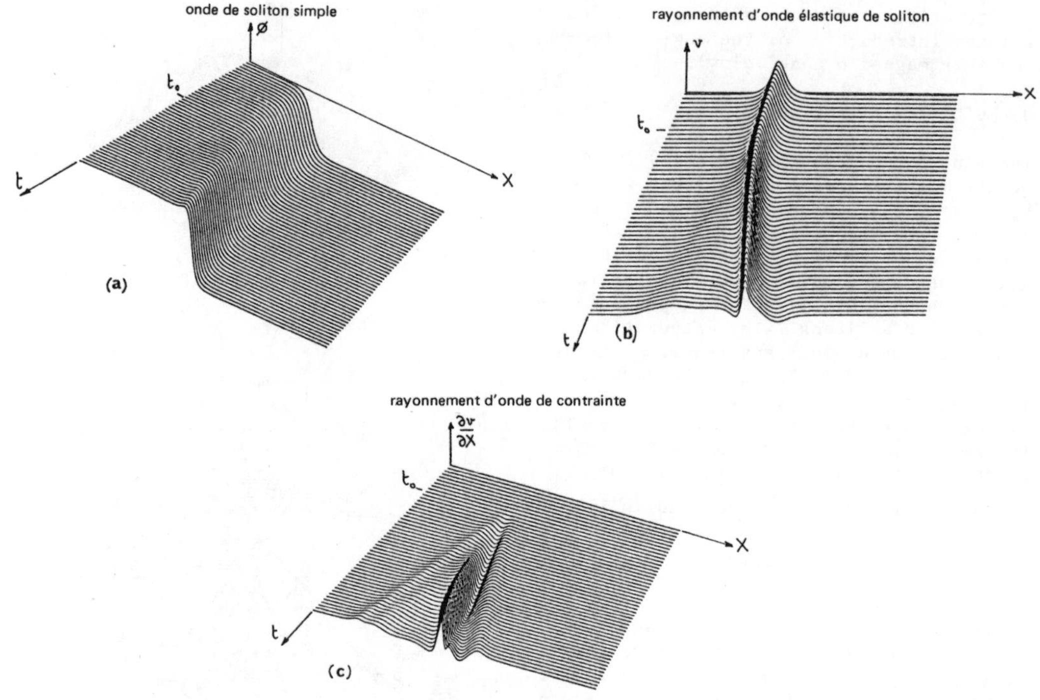

Fig. 9 Transient motion of a ferroelectric wall under the action of a suddenly applied electric field

Table I. The smoothing out of disorder-order transition in orthorhombic ferroelectric crystals ($NaNO_2$; T_I = 164°C , T_{II}=162.8°C), after Pouget, Maugin(1984)

	First-order phase transition		2nd-order phase transition	
	T_{II}	T_o — piezo — T'_o	T_I	T
phase :	FERROELECTRIC PHASE	FORMATION OF DOMAINS	INCOMMENSURATE PHASE	PARAELECTRIC PHASE
symmetry class :	Im2m	governed by anharmonicity	I2mm	Immm
order parameters:	P_X=0 , P_Y = P_S	$P_X = \frac{Q_o}{\cosh qX}$, $P_Y = P_o \tanh qX$ $q^{-1} \simeq$ thickness of a wall	$P_X = P_1 \cos kX$, $P_Y = P_2 \sin kX$ $k \downarrow 0$ as $T \downarrow$, hence $\lambda \uparrow$	$P_X = P_Y = 0$, $e_{ij} = 0$
energy :	$\Phi_C(T)$	$\Phi_{wall}(T)$	$\Phi_{inc}(T)$	
electromechanical couplings :	piezo + electrostriction	only electrostriction plays a role (→ internal strains which are not integrable, hence quasi-dislocations)	electrostriction plays no role	

$\Phi_C(T_{II}) = \Phi_{inc}(T_{II})$

$$(2.14)\quad \frac{\partial^2 v}{\partial t^2} - V_T^2 \frac{\partial^2 v}{\partial X^2} = -\eta \frac{\partial}{\partial X}(\sin \phi) \ ,$$
$$\frac{\partial^2 \phi}{\partial t^2} - \frac{\partial^2 \phi}{\partial X^2} - \sin\phi = \eta \frac{\partial v}{\partial X}\cos\phi + F\cos(\phi/2).$$

The new forcing term, where F is the non-dimensional magnitude of the applied field, may be considered as a perturbation for sufficiently small F's and the scheme devised in Paragraph B for multiple-soliton solutions can be used to treat this perturbation. However, much simpler approaches can already provide some information on the transient motion from rest when F is a Heaviside step function. One easy way is to consider that for $t < t_o$ we have a "static" solution of the type of (2.6). The applied field F at $t = t_o$ perturbs this basic solution by putting it into motion and the asymptotic velocity reached by the wall can be deduced through a simple energy argument (Pouget,Maugin,1985b). The total Hamiltonian of the system is conserved after switching on the electric field,i.e.,

$$(2.15)\quad \frac{dH}{dt} = 0 \ ,$$

where

$$(2.16a)\quad H = H_o + H_F \ ,$$

$$(2.16b)\quad H_o = \int_{-\infty}^{+\infty}\Big[\frac{1}{2}\Big\{\Big(\frac{\partial v}{\partial t}\Big)^2 + \Big(\frac{\partial \phi}{\partial t}\Big)^2 + V_T^2\Big(\frac{\partial v}{\partial X}\Big)^2 + \Big(\frac{\partial \phi}{\partial X}\Big)^2\Big\} + (1+\cos\phi) - \eta\frac{\partial v}{\partial X}\sin\phi\Big]\, dX \ ,$$

$$(2.16c)\quad H_F = 2\int_{-\infty}^{+\infty} F\sin(\phi/2)\, dX \ .$$

The predominant effect of the electric field is sought in the form of a modulation of the velocity of the solitary wave $V = \Omega/Q = V(t)$. Thus a solution (2.6) with modulated velocity is carried in (2.15) and the latter, after tedious calculations in which η^2 is considered as a small parameter, yields the equation

$$(2.17)\quad \frac{d}{dt}\Big[Q + \frac{1}{6}\eta^2 Q G(Q)\Big] = \frac{1}{2}\frac{F}{Q}(Q^2-1)^{1/2}, Q \geq 1,$$
$$G(Q) = \frac{1}{\Omega^2 - \Omega_T^2}\Big[Q^2 + \frac{\Omega^2}{(\Omega^2 - \Omega_T^2)}\Big] \ ,$$

along with the "dispersion relation" (2.7). With the initial condition $V(t_o)=0$ and for large Q's, we deduce from (2.17) the following asymptotic expression

$$(2.18)\quad V(t) = \bar{V}(t-t_o)\Big\{1 + \frac{\eta^2}{6(V_T^2-1)}\big[1-\bar{V}^2(t-t_o)\big]\Big\}$$
$$\bar{V}(t) := \frac{\bar{F}\, t}{(1+\bar{F}^2 t^2)^{1/2}} \ , \quad \bar{F} = F/2 \ .$$

For $\eta = 0$, the velocity solution (2.18) coincides with that of a relativistic particle of unit rest mass which is being uniformly accelerated by F. This follows from the very Lorentz invariance of $(2.14)_2$ for $\eta=0$.

The above-sketched out "direct energy method" provides only a very few informations (nothing on the change in profile). A more elaborate one consists in applying the averaged-Lagrangian method of Whitham(1974)-see Pouget,Maugin(1985d) and Maugin,Pouget (1985c) for its application to the present problem. In this method where the system (2.14) is written in the evolutionary form ,or Hamiltonian form, (2.10) and where F is a perturbation parameter, the Lagrangian of the unperturbed system is optimalized with wave parameters depending on time (so-called adiabatic perturbation) and this provides a variational formulation for the said parameters (speed and position or phase) of the solitary wave. This already gives more informations than the direct energy method . However, to obtain radiations, one would have to use the whole machinery of the inverse scattering method. The transient motion, treated numerically (thus with the accompanying change in profile), is represented in Figures 9. In reality, various obstacles (defects) hinder a smooth starting motion,or stop the progression,of walls.

2.3. Relation with the commensurate-incommensurate phase transition (Table I)

The above treatment clearly exhibits the relationship between electroacoustic solitary waves and the structure in domains and walls in certain ferroelectric crystals. If, now, the temperature parameter T is introduced in the picture, then it can be conjectured that the domain formation appears as the limit of an incommensurate phase (periodic solution) when T tends to T_c (transition temperature). An analysis by Pouget and Maugin(1984) briefly sketched out in Table I, shows that a Landau-Ginzburg type of approach can be formulated in which a so-called

Lifshitz invariant, exchange-dipole interactions, piezoelectricity and electrostriction are taken into account, allows one to exhibit the initiation of the ferroelectric phase within the incommensurate phase, locally in the crystal, with the formation of domains and the motion of walls. In the process the components of electric polarization on the X and Y axes (instead of the orientation angle $\emptyset$) are taken as two order parameters. In the incommensurate phase between T_I (second-order phase transition) and T'_o (piezoelectrically "stiffened" lock-in temperature), the wavelength of the modulation in P increases with decreasing temperature while domains form in the temperature range between T_{II} (first-order phase transition) and T'_o , which is governed by anharmonicity. The existence of the incommensurate phase [wellknown in $NaNO_2$ within a small temperature interval, Hatta et al, (1980)] has the effect of smoothing out the transition between the disordered paraelectric phase and the fully ordered low-temperature phase. We refer the reader to Pouget,Maugin(1984) for full details. A numerical simulation of periodic solutions of the double sine-Gordon equation (1.5) with variable period - through a temperature dependence of the electric susceptibility - corroborates this phenomenon (Figures 10 previously unpublished).

3 ELASTIC FERROMAGNETIC CRYSTALS

3.1 Continuum equations

For the sake of simplicity we consider the simplified theory of cubic elastic ferromagnetic crystals (such as nickel, iron, yttrium-iron garnets) as given, for instance, in Maugin (1979) — for the fully rotationally invariant theory, see Maugin, Eringen (1972). In a continuous region of the body where fields do not suffer jump discontinuities, we have the following field equations:

. Balance of linear momentum:

$$(3.1) \quad \rho \frac{d\underset{\sim}{v}}{dt} = \text{div}\, \underset{\sim}{t} + {}_M\underset{\sim}{f} \qquad ;$$

. Magnetic-spin precession equation (equivalent to the moment-of-momentum equation):

$$(3.2) \quad \frac{d\underset{\sim}{\mu}}{dt} = (-\gamma\, \underset{\sim}{H}^{eff}) \times \underset{\sim}{\mu} \;, \quad \underset{\sim}{\mu} = \underset{\sim}{M}/\rho \qquad ;$$

. Maxwell's magnetostatic equations:

$$(3.3) \quad \underset{\sim}{\nabla} \times \underset{\sim}{H} = \underset{\sim}{0} \;, \quad \underset{\sim}{\nabla} . (\underset{\sim}{H} + \underset{\sim}{M}) = 0 \qquad ,$$

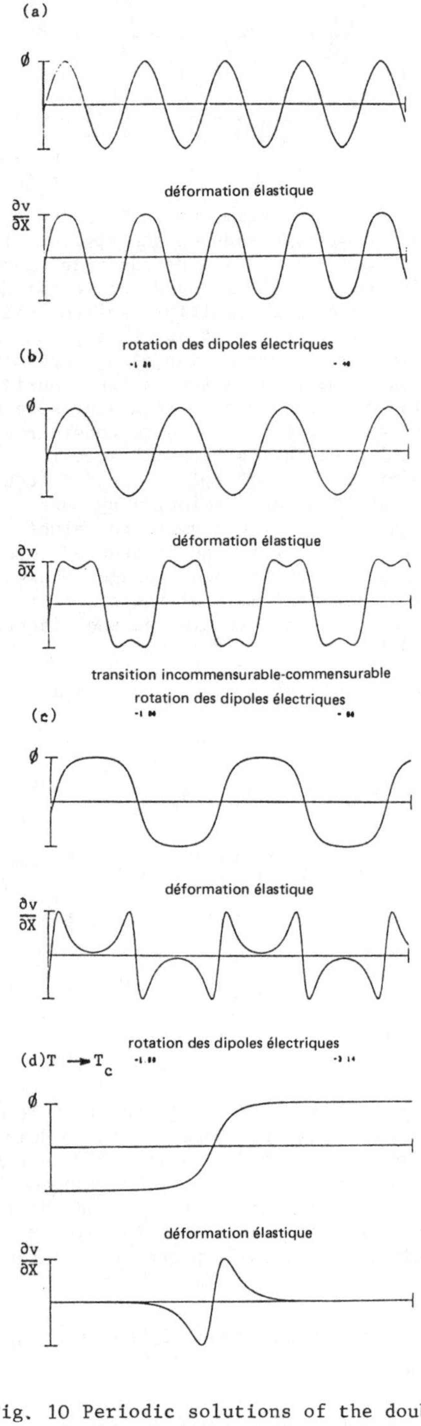

Fig. 10 Periodic solutions of the double sine-Gordon equation with decreasing temperature as $T \downarrow T_c$

with

$$\mathbf{M}^f = (\mathbf{M}.\nabla)\,\mathbf{H} \quad , \quad \mathbf{t} = \partial\Sigma/\partial\mathbf{e} \quad ,$$

(3.4)

$$\mathbf{H}^{eff} = \mathbf{H}^o + \mathbf{H}^d - \frac{1}{M_S}\left[\frac{\partial\Sigma}{\partial\boldsymbol{\alpha}} - \nabla.\left(\frac{\partial\Sigma}{\partial\nabla\boldsymbol{\alpha}}\right)\right]$$

and

$$(3.5)\quad \Sigma = \Sigma_{anis} + \Sigma_{ex} + \Sigma_{el} + \Sigma_{magel} \quad ,$$

$$\Sigma_{anis} = -\frac{1}{2} K M_S^2 (\boldsymbol{\alpha}.\mathbf{d})^2 \quad ,$$

$$\Sigma_{ex} = \frac{1}{2}\lambda M_S^2 \alpha_{i,j}\alpha_{i,j} \quad ,$$

(3.6)

$$\Sigma_{el} = \frac{1}{2}c_{11}(e_{xx}^2 + e_{yy}^2 + e_{zz}^2) + c_{44}(e_{xy}^2 + e_{yx}^2 + e_{zx}^2) + c_{12}(e_{xx}e_{yy} + e_{yy}e_{zz} + e_{zz}e_{xx}),$$

$$\Sigma_{magel} = [B_1(\alpha_x^2 e_{xx} + \alpha_y^2 e_{yy} + \alpha_z^2 e_{zz}) + 2B_2(\alpha_x\alpha_y e_{xy} + \alpha_y\alpha_z e_{yz} + \alpha_z\alpha_x e_{zx})]M_S^2$$

where

$$(3.7)\quad \boldsymbol{\alpha} = \boldsymbol{\mu}/\mu_S = \mathbf{M}/M_S \quad ,$$

$$\mathbf{v} = \partial\mathbf{u}/\partial t \quad , \quad \mathbf{e} = \frac{1}{2}[\nabla\mathbf{u} + (\nabla\mathbf{u})^T] \quad .$$

Here ρ is the matter density in the natural state, $\mathbf{u}$ is the elastic displacement, $\mathbf{v}$ is the matter velocity, $\mathbf{e}$ is the tensor of infinitesimal strains, $\mathbf{H}$ is the Maxwellian magnetic field, $\boldsymbol{\mu}$ is the magnetization per unit mass, γ is the gyromagnetic ratio, $\mathbf{t}$ is the stress tensor, $\mathbf{M}^f$ is the ponderomotive magnetic force and Σ is the internal energy per unit volume. The quantities Σ_{anis}, Σ_{ex}, Σ_{el} and Σ_{magel} are, respectively, the energy of magnetic anisotropy (or magnetocrystalline energy; $\mathbf{d}$ is a unit vector in the direction of so-called easy magnetization, later taken along the y-axis), the Heisenberg exchange energy in the continuum approximation, the elastic energy and the magnetoelastic (magnetostrictive) energy. Correspondingly, K is a constant of magnetic anisotropy, λ is the exchange constant, c_{11}, c_{44} and c_{12} are the elasticity coefficients and B_1 and B_2 are the magnetostriction coefficients. The magnetization has reached its saturation value M_S and further magnetization processes occur ,within a domain, only through

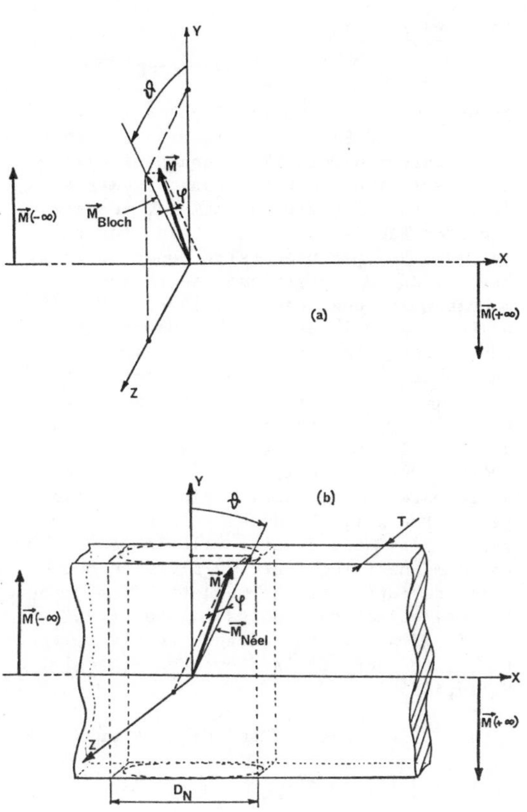

Fig.11 Walls in elastic ferromagnets:(a) Bloch wall in an infinite crystal; (b) Néel wall in a thin film

rotation of the magnetization. Finally, $\mathbf{H}^o$ is the externally applied field and $\mathbf{H}^d$ is the so-called demagnetizing field.

The essential nonlinearities and dispersion effects are contained in the precession equation (3.2), an extension of the Landau-Lifshitz equation, while Σ_{magel} and $\mathbf{M}^f$ provide a coupling between the equations of motion of (3.1) and the spin system. We shall study two cases of domain-wall structures in this magnetoelastic framework (Maugin, Miled, 1985a) in agreement with Figures 11. Let x be the direction of propagation. For a 180° Bloch wall we consider

$$\boldsymbol{\alpha} \simeq (\varphi, \cos\vartheta, \sin\vartheta); -\infty < x < +\infty ,$$

(3.8) Bloch:

$$\boldsymbol{\alpha} = (0, \pm 1, 0) \text{ as } x \longrightarrow \mp\infty \quad ,$$

while for a 180° Néel wall

$$\boldsymbol{\alpha} \simeq (\sin\vartheta, \cos\vartheta, \varphi); -\infty < x < +\infty ,$$

(3.9) Néel

$$\alpha = (0, \pm 1, 0) \text{ as } x \rightarrow \mp\infty$$

In writing down eqns.(3.8) and (3.9) we have followed Enz (1964) in assuming that there exist no pure Bloch and Néel walls. That is, in the Bloch case one assumes a slight out-of-plane deviation φ and the same for Néel's case. The reason for this is that the spin precession equation (3.2) has already μ proportional to an angular momentum and one needs eliminating an angle (i.e., φ) between the components of the vectorial equation (3.2) to arrive at the equation governing the "essential" angle ϑ. In the process we assume that the whole theory is linearized in φ, the latter varying slowly along the x-axis. That is, $|\varphi|$, $|\partial\varphi/\partial x|$ and $|\partial^2\varphi/\partial x^2|$ remain small as compared to characteristic quantities (e.g., the wall thickness for the second quantity). Thena crucial role is played by the demagnetizing field H^d since an out-of-plane deviation of the magnetization opposes the magnetization process (Winters, 1961). The field H^d is therefore directly related to φ. In fact, it is shown that (Maugin, Miled, 1985a):

. for a Bloch wall in an infinite crystal:

$$(3.10) \quad H^d(\text{Bloch}) \simeq (-M_S\varphi,\ 0,\ 0) \quad ;$$

. for a Néel wall in a thin film:

$$(3.11) \quad H^d(\text{Néel}) \simeq (-N_{11}M_S\alpha_x, 0, -N_{33}M_S\alpha_z)$$

with

$$(3.12) \quad N_{11} = \frac{T}{D_N + T}, \quad N_{33} = \frac{D_N}{D_N + T},$$

where (Néel, 1955; see also Soohoo, 1961) T is the thickness of the film and D_N is the "size" of the elliptical cylinder where most of the demagnetizing effect occurs.

3.2 Bloch wall in an infinite magnetoelastic crystal

In this case, in regions far from the wall, the ferromagnet may be considered as uniformly magnetized and the uniform magnetization creates internal strains e^o which are readily computed as [Maugin(1979), pp. 272-273]

$$(3.13) \quad e^o_{yy}(\pm\infty) = -\frac{B_1 M_S^2}{c_{11} - c_{12}}, \text{ other } e^o_{ij}(\pm\infty) = 0 .$$

Setting

$$(3.14) \quad 1 \gg H^o_y/M_S,\ H^o_z/M_S \text{ and } \hat{K}$$

with

$$\hat{K} := K - 2B_1 e^o_{yy} = K + \frac{2B_1^2 M_S^2}{(c_{11} - c_{12})},$$

it is shown that the motion equation (3.1) fully uncouples from the spin precession equation to a zeroth-order approximation in φ, while eliminating φ between remaining components of eqn.(3.2) leads to the non-dimensional equation

$$(3.15) \quad \frac{\partial^2\phi}{\partial t^2} - \frac{\partial^2\phi}{\partial X^2} + \sin\phi - 2\frac{\partial}{\partial t}(\omega_{Hx}/\hat{\omega}_M) - \frac{2}{\sqrt{\hat{K}}}\left(\frac{\omega_{Hz}}{\hat{\omega}_M}\cos\frac{\phi}{2} - \frac{\omega_{Hy}}{\hat{\omega}_M}\sin\frac{\phi}{2}\right) = 0,$$

where we have set (τ: dimensional time)

$$\phi = 2\vartheta, \quad X = x/\delta_B, \quad t = \tau\hat{\omega}_M,$$

$$(3.16) \quad \omega_M = \gamma M_S, \quad \omega_{Hx,y,z} = H^o_{x,y,z},$$

$$\hat{\omega}_M^2 = \hat{K}\omega_M^2,$$

and

$$(3.17) \quad \delta_B = \sqrt{\frac{\lambda}{\hat{K}}} = \left[\frac{\lambda}{K + \dfrac{2B_1^2 M_S^2}{(c_{11} - c_{12})}}\right]^{1/2}$$

is the "magnetostrictively" reduced wall thickness found by Motogi and Maugin (1984a) while studying small-amplitude vibrations of "magnetoelastic" Bloch walls. This is the only effect of magnetoelastic couplings in this case. It follows from internal strains generated at $\pm\infty$ and it results in a dilatation of the spatial scale in eqn.(3.15). For $H^o = 0$, eqn.(3.15) reduces to the sine-Gordon equation (1.3). For $H^o \neq 0$, a perturbation scheme of that equation must be envisaged.

3.3 Néel wall in a thin magnetoelastic film

This case is more involved than the preceding one in that the elasticity equations remain coupled to the spin-precession equation. Indeed, while the transverse z-component of the elastic displacement uncouples, one finds after some lengthy algebra in which the out-of-plane deviation φ is eliminated, the following nondimensional system of equations for the longitudinal elastic displacement u, the transverse elastic displacement v (in the plane of the film) and twice the Néel precession angle (Maugin, Miled,1985a):

$$(3.18)_a \quad \frac{\partial^2 u}{\partial t^2} - V_L^2 \frac{\partial^2 u}{\partial X^2} = \alpha \frac{\partial}{\partial X}(1-\cos\phi) \quad ,$$

$$(3.18)_b \quad \frac{\partial^2 v}{\partial t^2} - V_T^2 \frac{\partial^2 v}{\partial X^2} = \beta \frac{\partial}{\partial X}(\sin\phi) \quad ,$$

$$(3.18)_c \quad \frac{\partial^2 \phi}{\partial t^2} - \frac{\partial^2 \phi}{\partial X^2} + \sin\phi$$

$$= -(\alpha \frac{\partial u}{\partial X}\sin\phi + \beta \frac{\partial v}{\partial X}\cos\phi)$$

$$- 2\frac{\partial}{\partial t}(\omega_{Hz}/\hat{\omega}_M)$$

$$+2\hat{a}(\frac{\omega_{Hx}}{\hat{\omega}_M}\cos\frac{\phi}{2} - \frac{\omega_{Hy}}{\hat{\omega}_M}\sin\frac{\phi}{2}) \quad ,$$

where α and β are directly proportional to B_1 and B_2 , respectively, and

$$(3.19) \quad X=x/\hat{\delta}_N, \quad t=\tau\hat{\omega}_M, \quad \hat{\omega}_M^2 = N_{33}\hat{K}\,\omega_M^2 \quad ,$$

$$\hat{K}=K+N_{11}-2B_1e^o_{yy}, \quad \hat{a}=(N_{33}/\hat{K})^{1/2} \quad ,$$

and

$$(3.20) \quad \hat{\delta}_N = \sqrt{\frac{\lambda}{\hat{K}}} = \left[\frac{\lambda}{K + \frac{2B_1^2M_S^2}{(c_{11}-c_{12})} + \frac{T}{D_N+T}}\right]^{1/2}$$

is the"wall thickness "obtained by Motogi and Maugin (1984a) in their study of small-amplitude vibrations of Néel walls in magnetostrictive thin films. Both the film "aspect-ratio" factor $T/(D_N+T)$ and the internal strains created by the state of uniform magnetization at $\pm\infty$ reduce the thickness of the wall as compared to the Landau-Lifshitz value $\delta=\sqrt{\lambda/K}$.

For $\underset{\sim}{H}^o = \underset{\sim}{0}$, the system (3.18) takes the same form as the system $(2.2)_{a,b,c}$. In particular, for dynamical magnetoelastic solutions depending on a single phase variable $\xi = QX - \Omega t + \xi_o$ only, one can show that the system (3.18) with the limit conditions

$$(3.21) \quad \phi \to 0(\text{mod } 2\pi), \quad \frac{d\phi}{d\xi} \to 0$$

$$\frac{du}{d\xi} \to 0, \quad \frac{dv}{d\xi} \to 0 \text{ for } |\xi| \to \infty,$$

is equivalent to the following nonlinear ordinary differential equation

$$(3.22) \quad (\hat{Q}^2 - \hat{\Omega}^2)\frac{d^2\phi}{d\xi^2} - \sin\phi + \eta(\Omega,Q)\sin 2\phi = 0,$$

where

$$(3.23) \quad \eta(\Omega,Q) = \frac{1}{2}\hat{Q}^2\left(\frac{\alpha^2}{\Omega^2-\Omega_L^2} - \frac{\beta^2}{\Omega^2-\Omega_T^2}\right)$$

and

$$\hat{\Omega} = \Omega/A \quad , \quad \hat{Q} = Q/A \quad ,$$

$$A = \left(\frac{\Omega^2 - \hat{\Omega}_L^2}{\Omega^2 - \Omega_L^2}\right)^{1/2} \quad ,$$

$$(3.24) \quad \Omega_L^2 = V_L^2 Q^2 \quad , \quad \Omega_T^2 = V_T^2 Q^2 \quad ,$$

$$\hat{\Omega}_L^2 = \hat{V}_L^2 Q^2 \quad , \hat{V}_L^2 = V_L^2(1-\epsilon_{ML}) \quad ,$$

$$\epsilon_{ML} = \alpha^2 / V_L^2 \quad ,$$

where $\hat{V}_L$ is the "magnetostrictively"reduced speed of longitudinal elastic waves [Maugin, Hakmi(1984) or Ristic (1983),p.83]. The direct integration of (3.22) under proper limit conditions yields stable solitary-wave solutions such as

$$(3.25) \quad \vartheta(\xi) = \frac{\pi}{2} + \tan^{-1}\left[\frac{\sinh\xi}{\hat{Q}\,(1-V^2)^{1/2}}\right] \quad ,$$

$V = \Omega/Q$, on the conditions that points (Ω,Q) belong to branch (a) in the disper-diagram of Figure 12. There, (Ω^+,Q^+) is the magnetoacoustic resonance in the harmonic linear case(Maugin,1980b,1981) and branches Ω_+ and Ω_- are the coupled transverse acoustic (phonon) and magnon branches. The dispersion relation of this harmonic linear case is given by

$$(3.26) \quad D_L(\Omega,Q) = (\Omega^2 - \Omega_T^2)(\Omega^2 - \Omega_S^2) - \beta^2 Q^2 = 0$$

with $\Omega_S^2 := Q^2 + 1$, while the nonlinear solitary-wave case has a pseudo dispersion relation given by

$$(3.27) \quad D_{NL}(\Omega,Q) = (\Omega^2 - \Omega_T^2)(\Omega^2 - \bar{\Omega}_S^2) + \beta^2 Q^2 = 0$$

with $\bar{\Omega}_S^2 := Q^2 - 1$. The displacement and stress fields generated by a solution (3.25) are obtained by direct integration. For instance,

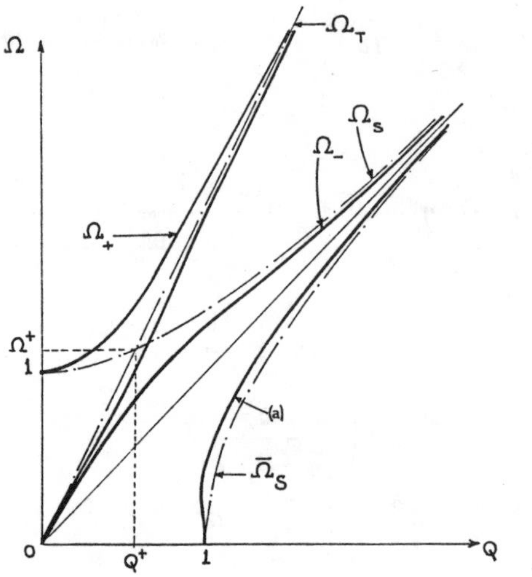

Fig. 12 Dispersion relation for solitary waves (Néel wall) in an elastic ferromagnetic thin film

$$\sigma_{xx}(\xi) = \frac{2\alpha Q(1-2\eta)}{[(1-2\eta)+\sinh^2\xi](\Omega^2-\Omega_L^2)},\tag{3.28}$$

$$\sigma_{yx}(\xi) = \mp \frac{2\beta Q(1-2\eta)^{1/2}}{(\Omega^2-\Omega_T^2)} \frac{\sinh\xi}{(1-2\eta)+\sinh^2\xi},$$

where the sign depends on whether the rotation of the magnetization in the wall is clockwise or counterclockwise.

For $\underset{\sim}{H}^{o} \neq \underset{\sim}{0}$, a perturbation scheme similar to those mentioned in Paragraph 2C has to be envisaged. This is also the case if dissipative processes such as coupled viscosity and spin-lattice relaxation (Maugin, 1975) are present.

4 MICROPOLAR ELASTIC CRYSTALS

4.1 Kinematics and field equations

In the general micromorphic theory set forth by Eringen and Suhubi in 1964, the change in the microstructure is accounted for through a general linear application. For micropolar bodies (rigid microstructure) this application reduces to an orthogonal transformation $\underset{\sim}{\chi}$, so that the generalized motion of the continuum is described by the set

$$\begin{aligned} &\underset{\sim}{x} = \underset{\sim}{\varkappa}(\underset{\sim}{X},t), \quad \underset{\sim}{\chi} = \underset{\sim}{\chi}(\underset{\sim}{X},t) \\ &\underset{\sim}{\chi}^{T} = \underset{\sim}{\chi}^{-1}, \quad \det\underset{\sim}{\chi} = +1 \end{aligned}\tag{4.1}$$

in which the first is the classical nonlinear motion of the body, $\underset{\sim}{x}$ is the Eulerian representation of a material point and $\underset{\sim}{X}$ is its Lagrangian or "material" placement. The finite deformation of a material is described by using so-called Cosserat and wryness tensors, with components (Kafadar, Eringen, 1971)

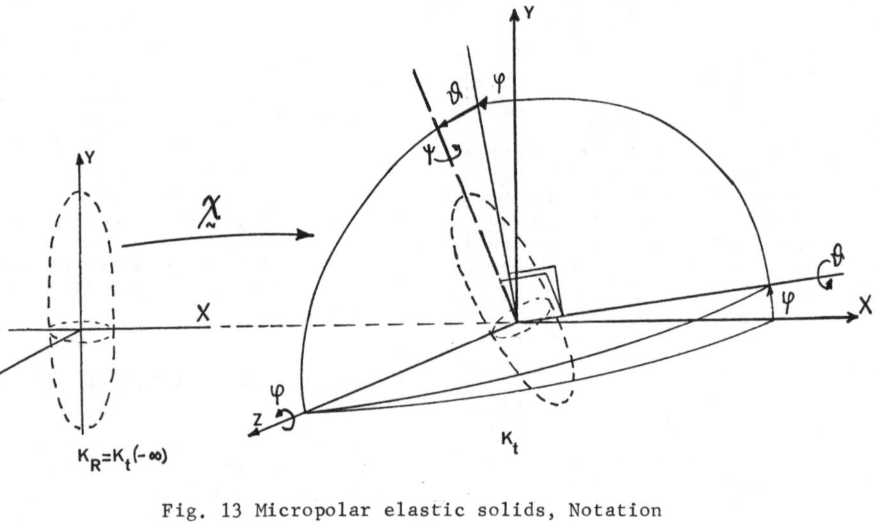

Fig. 13 Micropolar elastic solids, Notation

$$(4.2) \qquad \mathfrak{E}_{KL} := \varkappa_{k,K}\chi_{kL} , \qquad \Gamma_{KL} := \tfrac{1}{2}\varepsilon_{KMN}\chi_{kM,L}\chi_{kN} .$$

We may think of each material point as a small elongated ellipsoid which, in the reference configuration, has its longest axis along the y-direction in Figure 13 (the reference configuration may be chosen as the physical situation at $-\infty$). In this configuration of matter density ρ_o the inertia tensor of the particles is diagonal. The two basic equations of motion ,using Piola-Kirchhoff stress tensor $T_{K\ell}$ and couple-stress tensor $M_{K\ell}$, read

$$(4.3) \qquad T_{K\ell,K} + \rho_o f_\ell = \rho_o \frac{\partial^2 u_\ell}{\partial t^2} , \qquad M_{K\ell,K} + \varepsilon_{\ell mn}\varkappa_{m,K} T_{Kn} + \rho_o I_\ell = \rho_o \frac{\partial \sigma_\ell}{\partial t} ,$$

where $\underset{\sim}{f}$ and $\underset{\sim}{I}$ are external body force and couple per unit mass and σ_ℓ is the intrinsic spin vector. Let $\Sigma = \rho_o \Psi$ be the internal energy per unit volume in the reference configuration. Then we have the constitutive equations

$$(4.4) \qquad T_{K\ell} = \chi_{\ell L} \frac{\partial \Sigma}{\partial \mathfrak{E}_{KL}} , \quad M_{K\ell} = \chi_{\ell L} \frac{\partial \Sigma}{\partial \Gamma_{LK}} .$$

We select the following form of the energy:

$$(4.5) \qquad \Sigma = \tfrac{1}{2}[\lambda(\mathrm{tr}\,\underset{\sim}{\mathfrak{E}})^2 + \mu\,\mathrm{tr}\,\underset{\sim}{\mathfrak{E}}^2 + (\mu+\kappa)\,\mathrm{tr}(\underset{\sim}{\mathfrak{E}}\underset{\sim}{\mathfrak{E}}^T) + \alpha(\mathrm{tr}\,\underset{\sim}{\Gamma})^2 + \beta\,\mathrm{tr}\,\underset{\sim}{\Gamma}^2 - \gamma\,\mathrm{tr}(\underset{\sim}{\Gamma}\underset{\sim}{\Gamma}^T)]$$

with $3\lambda + 2\mu + \kappa = 0$, where $\lambda, \mu, \kappa, \alpha, \beta$ and γ are material coefficients that can be identified with those introduced in linear micropolar elasticity (Eringen,1968) by considering the case of infinitesimally small deformation and rotations of the microstructure in the following equations. The conditions relating λ, μ and κ is some kind of Stokes' hypothesis which has been found helpful in the computations (this is a constraint on Poisson's ratio).

In general $\underset{\sim}{\chi}$ is a composition of several rotations. Following Maugin and Miled(1985b), we consider only two special cases where these rotations reduce to a pure finite rotation about the x-axis (obviously, a Bloch configuration) and a pure finite rotation about the z-axis (obviously, a Néel-wall configuration - see Figure 14.

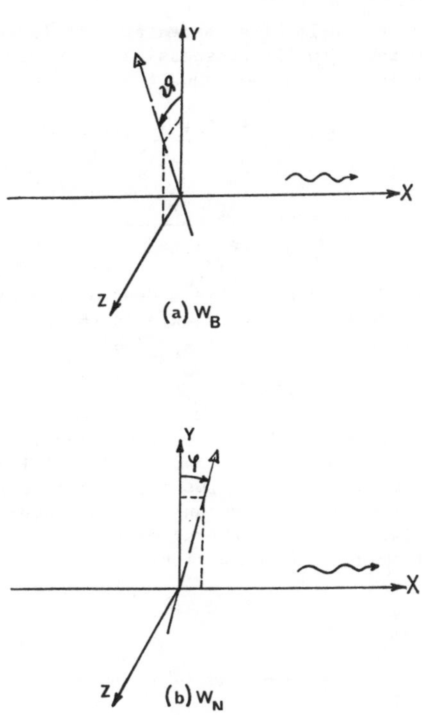

Fig.14 Micropolar elastic crystals: (a) Bloch wall;(b) Néel wall

The usual deformations are assumed to be small since the essential nonlinearity is brought in the picture through the micro-structure.

4.2 Bloch wall

In this case, in the absence of external force and torque, it is found that only the longitudinal elastic displacement and the internal rotation ϑ remain coupled. After nondimensionalization, one has the following system of partial differential equations for these two variables ($\phi = 2\vartheta$):

$$(4.6) \qquad \frac{\partial^2 u}{\partial t^2} - V_L^2 \frac{\partial^2 u}{\partial X^2} = \eta \frac{\partial}{\partial X}\left(1 + \cos\frac{\phi}{2}\right) , \qquad \frac{\partial^2 \phi}{\partial t^2} - \frac{\partial^2 \phi}{\partial X^2} - \sin\phi = e \sin\frac{\phi}{2} + \eta \frac{\partial u}{\partial X} \sin\frac{\phi}{2} ,$$

where space and time are nondimensionalized

with the help of a characteristic length δ (a kind of wall thickness) and a characteristic frequency ω_M such that

$$(4.7)\qquad \delta = \frac{1}{2}\left[\frac{\alpha+\beta+\gamma}{2(\lambda+\mu)}\right]^{1/2} , \qquad \omega_M = 2\left[\frac{2(\lambda+\mu)}{\rho_o J_1}\right]^{1/2} .$$

Here J_1 is the inertia moment of the microstructure about the x-axis in the reference configuration (say at $x = -\infty$) and

$$e = \lambda/2(\lambda+\mu) > 0$$

The system (4.6) consists in a double sine-Gordon equation for ϕ and a d'Alembert equation for u, the two being nonlinearly coupled through the coupling parameter η (related to κ). Whenever we impose the following limit conditions

$$(4.8)\qquad \phi \longrightarrow 0 \,(\text{mod } 4\pi), \quad \frac{d\phi}{d\xi} \longrightarrow 0$$

$$\frac{du}{d\xi} \longrightarrow 0 \quad \text{as } |\xi| \longrightarrow \infty ,$$

and consider propagative solutions of (4.6) that depend on a single phase variable $\xi = QX - \Omega t + \xi_o$ only, we can say that we have a representation of a 360° Bloch wall (since the real angle is $\vartheta = \phi/2$) in a micropolar elastic crystal (of which KNO_3 is an example). By eliminating u from (4.6) for such solutions satisfying (4.8), it is shown that the system (4.6) is equivalent to the following nonlinear ordinary differential equation

$$(4.9)\qquad (\hat{\Omega}^2 - \hat{Q}^2)\frac{d^2\phi}{d\xi^2} - \sin\phi - \tilde{\gamma}(\Omega, Q)\sin\frac{\phi}{2} = 0$$

with

$$\hat{\Omega} = \Omega/E , \qquad \hat{Q} = Q/E ,$$

$$(4.10)\qquad E = \left(\frac{\Omega^2 - \hat{\Omega}_L^2}{\Omega^2 - \Omega_L^2}\right)^{1/2} ,$$

$$\hat{\Omega}_L^2 = \hat{V}_L^2 Q^2 , \quad \hat{V}_L^2 = V_L^2(1-\epsilon), \quad \epsilon = \eta^2/2V_L^2 ,$$

$$\tilde{\gamma} = \frac{e}{E^2} + \frac{\Omega_L^2 - \hat{\Omega}_L^2}{\Omega^2 - \Omega_L^2} .$$

Thus the general problem (4.6)-(4.8) is cast in the framework already given in previous sections [compare eqns. (4.9) and (3.22)]. Stable solitary-wave solutions can be exhibited which satisfy a dispersion relation of the same type as the one already pictured in Figure 12 except that the coupling, if any, here is with a longitudinal elastic displacement. The thickness of this 360° Bloch wall is found to be given by (Maugin, Miled, 1985b)

$$\delta_B = 2\pi[(4-\tilde{\gamma})/(2-\tilde{\gamma})]^{1/2} .$$

4.3 Néel wall

If we consider $\vartheta = 0$ always in Figure 14, then we may say that we have a Néel wall configuration if appropriate conditions are imposed at $\pm\infty$. Indeed, for the special case where $\lambda = 0$, it can be shown (Maugin, Miled, 1985b) that both longitudinal and transverse elastic displacements (hence the elastic displacement component parallel to the (x,y)-plane) remain coupled to the internal rotation φ. Setting $\phi = 2\varphi$, we, in fact, obtain the following nondimensional system of governing equations:

$$(4.11)\qquad \frac{\partial^2 u}{\partial t^2} - V_L^2\frac{\partial^2 u}{\partial X^2} = \eta\frac{\partial}{\partial X}\left(1+\cos\frac{\phi}{2}\right) ,$$

$$\frac{\partial^2 v}{\partial t^2} - V_T^2\frac{\partial^2 v}{\partial X^2} = \eta\frac{\partial}{\partial X}\left(\sin\frac{\phi}{2}\right) ,$$

$$\frac{\partial^2 \phi}{\partial t^2} - \frac{\partial^2 \phi}{\partial X^2} - \sin\phi = \eta\left(\frac{\partial u}{\partial X}\sin\frac{\phi}{2} - \frac{\partial v}{\partial X}\cos\frac{\phi}{2}\right) ,$$

where, instead of (4.7) we have

$$(4.12)\qquad \delta = \tfrac{1}{2}(\gamma/2\mu)^{1/2} , \quad \omega_M = 2(2\mu/\rho_o J_3)^{1/2}$$

and

$$\eta = \mu(J_3/2\gamma)^{1/2} ,$$

where J_3 is the inertia moment of the microstructure about the z-axis in the reference configuration at $-\infty$. With the limit conditions

$$(4.13)\qquad \phi \longrightarrow 0 \,(\text{mod } 4\pi) \quad \frac{d\phi}{d\xi} \longrightarrow 0 ,$$

$$\frac{du}{d\xi} \longrightarrow 0, \quad \frac{dv}{d\xi} \longrightarrow 0 \text{ as } |\xi| \longrightarrow \infty ,$$

where $\xi = QX - \Omega t + \xi_o$, the system (4.11) can be shown to represents the dynamics of a 360° Néel wall in a micropolar elastic crystal . The problem encapsulated in eqns. (4.11)-(4.13) is obviously of the same type as the one stated in eqns.(3.18) - for $\underset{\sim}{H}^o = \underset{\sim}{0}$ - and (3.21), with the exception that it is $\phi/2$ that is involved in $(4.11)_{1,2}$ so that we shall not dwell on this problem any longer.

4.4 Action of an external torque

In Paragraphs 4.2 and 4.3 the microstructure (e.g., elongated ellipsoids) has no distinguishable ends, so that the configurations at minus and plus infinity are absolutely identical . As a consequence , from a distance , the wall appears as a real defect in the otherwise (i.e., outside a thickness of the order of δ) well ordered structure. But this microstructure can be endowed with a real direction if there is attached to it an electric ($\underset{\sim}{P}$) or magnetic ($\underset{\sim}{M}$) dipole (for instance, aligned with the longest principal axis of the ellipsoids). Then an external stimulus can be imposed on the microstructure in the form of a density of torque $\underset{\sim}{I}$ — see eqn.$(4.3)_2$ — if one applies an electric or magnetic field. This torque will have one of the following two forms

$$(4.14) \quad \rho_o \underset{\sim}{I} = \underset{\sim}{P} \times \underset{\sim}{E}^o \quad \text{or} \quad \rho_o \underset{\sim}{I} = \underset{\sim}{M} \times \underset{\sim}{H}^o \quad .$$

Summing over the action of such torques across the wall thickness where the dipoles are not aligned with the applied field will thus provide a motive power for the wall. If the external field is suddenly applied, then we recover the transient-motion problem of Paragraph 2C to which we refer the reader.

5 ORIENTED ELASTIC MEDIA

5.1 General comments

In the same way as liquid crystals may be phenomenologically described by means of micropolar theory à la Eringen or by means of the Leslie-Ericksen theory using one director field, certain solid elastic crystals or objects behaving elastically (e.g., long macromolecular chains) may be alternatively described by means of the micropolar theory sketched out in Section 4 or by means of a one-director theory of elastic solids as given, for instance,by Pouget and Maugin (1982) for piezoelectric powders (the latter are,per se, micromorphic media). The two descriptions should be equivalent as we envision Figure 15 where $\vec{d}$ is the director field evolving from its reference configuration $\vec{D}$ to its actual

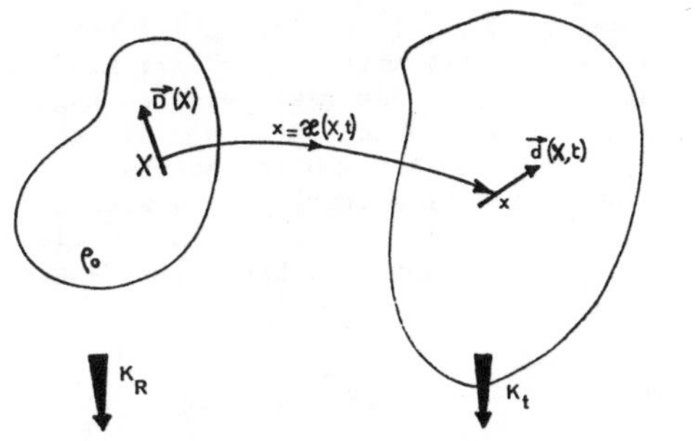

Fig.15 Kinematics of oriented elastic solids

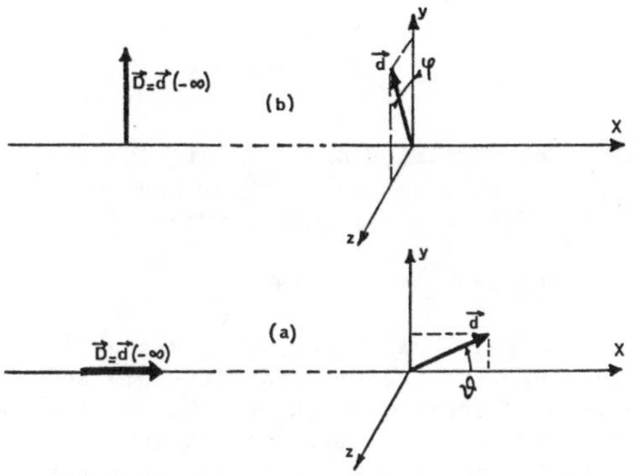

Fig.16 Walls in elastic oriented solids; (a)Néel wall;(b) Bloch wall

orientation at time t. The rotation of such a director is obviously described by an orthogonal transformation. It is nonetheless equally clear that the whole theory constructed by using a vector field $\vec{d}$, and the nonlinearities which may appear subsequently, will be quite different from those involving a tensor, let it be orthogonal like $\underset{\sim}{\chi}$. As a matter of fact, the basic laws of conservation, instead of (4.3), will read (Pouget, Maugin,1985e;also Maugin,1980a)

$$(5.1) \quad \begin{aligned} \rho \frac{d\underset{\sim}{v}}{dt} &= \text{div}\, \underset{\sim}{t} + \rho \underset{\sim}{f} \quad , \\ \rho I \frac{d^2\vec{d}}{dt^2} &= \rho(\vec{g} + \vec{G}) + \text{div}\underset{\sim}{\Pi} + (\lambda - \text{div}\, \vec{\beta})\vec{d}, \end{aligned}$$

where λ and $\vec{\beta}$ are scalar and vectorial Lagrange multipliers that account for the constraints

$$(5.2) \quad \vec{d}.\vec{d} = \text{const.} \ , \quad (\nabla\vec{d}).\vec{d} = \vec{0}$$

and I is the component of the inertia

tensor on $\vec{d}$, $\vec{g}$ is an internal force acting on $\vec{d}$ (a constitutive equation is given for this), $\vec{G}$ is an external force acting on $\vec{d}$, and the tensor $\underset{\sim}{\Pi}$ accounts for interactions within the continuum of $\vec{d}$-vectors.

By vectorial multiplication with $\vec{d}$, eqn.$(5.1)_2$ yields the equation of balance of angular momentum in the form (Maugin, 1980a)

$$(5.3)\qquad \rho \frac{d\vec{\sigma}}{dt} = \rho\vec{c} + \rho(\vec{G}\times\vec{d}) + \operatorname{div}\underset{\sim}{m},$$

where

$$m_{ij} = \varepsilon_{ipq}\Pi_{pj}d_q,\quad \vec{\sigma} = I\left(\frac{d}{dt}\vec{d}\right)\times\vec{d},$$

$$c_i = (\vec{g}\times\vec{d})_i - \rho^{-1}\varepsilon_{ipq}\Pi_{pj}d_{q,j}.$$

The thermodynamics and constitutive theory of such media for an elastic behaviour may be found in Pouget, Maugin (1985e).

5.2 Néel and Bloch walls

For a crystal of the trigonal system in class 6/m (centrosymmetric crystal) and a reference configuration selected as the limit solution at $-\infty$, one can show for infinitesimally small strains but large amplitude rotations of the microstructure that the following nondimensional systems of partial differential equations are obtained (Pouget, Maugin, 1985e):

. Néel wall (Figure 16a; $\phi = 2\vartheta$):

$$(5.4)\qquad \begin{aligned} &\frac{\partial^2 u}{\partial t^2} - \hat{V}_L^2\frac{\partial^2 u}{\partial X^2} = \alpha\frac{\partial}{\partial X}(1+\cos\phi),\\ &\frac{\partial^2 v}{\partial t^2} - \hat{V}_T^2\frac{\partial^2 v}{\partial X^2} = \beta\frac{\partial}{\partial X}(\sin\phi),\\ &\frac{\partial^2\phi}{\partial t^2} - \frac{\partial^2\phi}{\partial X^2} - \sin\phi + \alpha\frac{\partial u}{\partial X}\sin\phi\\ &\qquad + \beta\frac{\partial v}{\partial X}\cos\phi = 0; \end{aligned}$$

. Bloch wall (Figure 16b; $\phi = 2\varphi$)

$$(5.5)\qquad \begin{aligned} &\frac{\partial^2 u}{\partial t^2} - V_L^2\frac{\partial^2 u}{\partial X^2} = \gamma\frac{\partial}{\partial X}(1+\cos\phi),\\ &\frac{\partial^2 v}{\partial t^2} - V_T^2\frac{\partial^2 v}{\partial X^2} = 0,\\ &\frac{\partial^2\phi}{\partial t^2} - \frac{\partial^2\phi}{\partial X^2} - \sin\phi + \gamma\frac{\partial u}{\partial X}\sin\phi = 0, \end{aligned}$$

so that we have problems of the same type as those encountered in eqns.(3.18) - for $\underset{\sim}{H}^o = \underset{\sim}{0}$ - and (2.3), respectively. Because of these analogies we dot not pursue the matter further [we refer the reader to Pouget and Maugin (1985e)].

The present model can be used to study the dynamics of an elastic chain representing certain macromolecules (D.N.A., polyelectrolytes). An external stimulus can be applied to the microstructure since $\vec{G}$ is an external field (e.g., an electric or magnetic field) if the microstructure contains the microscopic mechanism to respond to such a stimulus (electric or magnetic dipole).

6 CONCLUSION

It is clear that a common framework emerges from the series of examples examined in Sections 2 through 5This should come as no surprise since it is the microstructure which brings both nonlinearity and dispersion (long range interactions) in the system in all caseswhile the usual strain field remains subjected to the working hypothesis of infinitesimally small strains, and the microstructure always materializes in additional"rotational" degrees of freedom although the accompanying angular momentum may have various origins (purely mechanical in many cases, of the magnetic-spin type in others) . In all, a real unifying view of moving walls generating acoustic effects in a variety of crystals and elastic bodies (ferroelectric crystals of the molecular-group type, elastic ferromagnetic crystals, crystals such as KNO_3, long elastic chains of polymer with twist) emerges. The common mathematical problem obtained consists in a sine-Gordon equation (or a double sine-Gordon equation) which is nonlinearly coupled to none, one or two wave equations. Apart from the first case, the very form of the obtained systems of dispersive nonlinear hyperbolic equations indicates that if solitary waves propagate (in which case it has been shown that the whole coupled problem is often equivalent to the one of a double sine-Gordon equation),there are no pure soliton solutions because of the accompanying radiations. An"almost" soliton-like behaviour is however exhibited. This emphasizes once more time the relevance of the sine-Gordon equation and the double sine-Gordon equations to numerous problems of mechanics and condensed-matter physics, the latter field in direct relation with phase-transition phenomena (see Paragraph 2.3, but these considerations -with appropriate alterations - could obviously be extended to the other cases examined in the present work). The solitary-wave problem, the multiple-soliton problem and the transient motion of such nonlinear waves, as well as perturbations by dissipative processes, can be studied with common mathematical tools or techniques, the inverse scattering method, Bäcklund transformations, Green functions, singular—perturbation techniques and the averaged-Lagrangian method of G.B.Whitham. These should be developed in another paper (Maugin,1985).

REFERENCES

Askar, A. 1973. A model for coupled rotation displacement modes of certain molecular crystals.Illustration for KNO_3. J.Phys.Chem.Solids 34, 1901-1907.

Auld B.A.,Fester M.M. 1981. Elastic nonlinearity and domain wall motion in ferroelastic crystals. Ferroelectrics 38, 931-934.

Bäcklund A.V. 1882. Zur Theorie der Flächentransformationen. Math.Annalen 19,387-422.

Baldokhin Yu.V., Goldanski V.I.,Makarov E.F., Mitin A.V., Povitskii V.A. 1972. The Mössbauer study of ultrasound stimulated by radiofrequency field in ferrodielectrics. J.Physique.Coll.C6,C6-145—C6-149.

Barone A., Esposito F., Magee C.J., Scott A.C. 1971. Theory and application of the sine-Gordon equation. Riv.Nuovo Cimento 1 , 227-267.

Boucher J.F., Regnault L.P.,Rossat-Mignod J., Henry J.Y., Bouillot J.,Stirling W.G., Soares E.A., Wiese J., Renard J.P. 1983. Solitons dans les chaînes magnétiques. Congress of the French Physical Society. Sept.1983, Orsay:Les Editions de Physique.

Bullough R.K., Caudrey P.J., Gibbs H.M. 1980. The double sine-Gordon equations: a physically admissible system of equations. In Solitons,Vol.17 of Topics in Current Physics. R.K.Caudrey and P.J. Bullough(eds.),p.107-141.Berlin:Springer.

Currie J.F. 1977.Aspects of Exact Dynamics for general solutions of the sine-Gordon equation with applications to domain wall. Phys.Rev. A16, 1692-1699.

Eisenhart L.P. 1960. A treatise on the differential geometry of curves and surfaces. New York:Dover(Reprint).

Enz U. 1964. Die Dynamik der Blochschen Wand. Helvetica Phys.Acta 37, 245-251.

Eringen A.C. 1964. Simple microfluids. Int.J.Engng.Sci. 2, 205-217.

Eringen A.C. 1968. Micropolar elasticity. In H.Liebowitz (ed.) Fracture-A Treatise, p.621-679. New York:Academic Press.

Feldkeller E. 1968. Magnetic domain wall dynamics. Phys.Stat.Sol. 27, 161-169.

Frenkel J.,Kontorova T. 1939. On the theory of plastic deformation and twinning . J.Phys.USSR, 1, 137-149.

Goursat E. 1925. Le problème de Bäcklund. Mém.Sci.Math.Fasc.6. Paris: Gauthier-Villars.

Kafadar C.B. 1972. On the nonlinear theory of rods. Int.J.Engng.Sci., 10, 369-391.

Kafadar C.B., Eringen A.C. 1971. Micropolar theory-I-The classical theory. Int.J.Engng. Sci. 9, 271-305.

Kittel C. 1971. Introduction to Solid State Physics. New York:J.Wiley.

Kléman M. 1983. Points, lines and walls. New York:J.Wiley.

Kunin I.A. 1982-3. Elastic media with microstructure. Two volumes. Berlin: Springer-Verlag.

Landau L.D.,Lifshitz E.M. 1935.On the theory of the dispersion of magnetic permeability in ferromagnetic bodies. Phys.Z.Sowjet, 8, 153 -reprinted In D.Ter Haar (ed.) Collected papers of L.D.Landau,p.101-116. New York:Gordon and Breach (1965).

Laval J. 1957. L'élasticité du milieu cristallin, J.Phys.Radium 18, 247-369.

Lebedev A.Yu., Ozhogin V.I., Safonov V.L., Yakubovskii A.Yu. 1983. Nonlinear Magnetoacoustics of orthoferrites near spin flip. Soviet Phys.J.E.T.P. 58(3), 616-623.

Lee J.D., Eringen A.C.1974. Relation of two continuum theories of liquid crystals. In J.F.Johnson and R.S.Porter (eds.) Liquid crystals and ordered fluids.Vol.2, p.315-330. New York: Plenum Press.

Lugovoï A.A.,Turov E.A. 1983. Magnetoelastic resonance of domain walls in ferro- and antiferromagnets. Jl Magnetism Magnetic Materials. 31-34, 693-694.

Maugin G.A. 1975. On the spin relaxation in deformable ferromagnets. Physica 81A, 454-468.

Maugin G.A. 1976. A continuum theory of deformable ferrimagnetic bodies-I-field equations. J.Math.Phys. 17, 1727-1738.

Maugin G.A. 1979. Classical magnetoelasticity in ferromagnets with defects. In Electromagnetic interactions in elastic solids, H.Parkus (ed.)p.243-324.Wien: Springer-Verlag.

Maugin G.A. 1980a. The principle of virtual power in continuum mechanics.Application to coupled fields. Acta Mechanica, 35, 1-70.

Maugin G.A. 1980b. Elastic-electromagnetic resonance couplings in electromagnetically ordered media. In F.P.J.Rimrott and B.Tabarrok (eds.) Theoretical and Applied mechanics, p.345-355. Amsterdam:North Holland.

Maugin G.A. 1981. Wave motion in magnetizable deformable solids-Recent advances. Int.J.Engng.Sci. 19, 321-388.
Maugin G.A. 1985. Solitons in microstructured elastic media-Mathematical aspects. In K.Kirchgässner and E.Kröner (eds.) Proc.Symp.Trends in the applications of pure mathematics to Mechanics (Bad Honnef, Oct.1985). Berlin:Springer-Verlag.
Maugin G.A., Drouot R. 1983. Internal variables and the thermodynamics of macromolecule solutions. Int.J.Engng.Sci. 21,705-724.
Maugin G.A.,Eringen A.C. 1972. Deformable magnetically saturated media-I,II. J.Math.Phys. 13, 143-155 , 1334-1347.
Maugin G.A., Hakmi A. 1984. Magnetoacoustic wave propagation in paramagnetic insulators exhibiting induced linear magnetoelastic couplings. J.Acoust.Soc.America 76,826-840.
Maugin G.A., Miled A. 1985a. Solitary waves in elastic ferromagnets. Phys.Rev.B (submitted for publication in).
Maugin G.A., Miled A. 1985b. Solitary waves in micropolar elastic crystals. Int.J. Engng.Sci. (submitted for publication in).
Motogi S.,Maugin G.A. 1984a. Effects of magnetostriction on vibrations of Bloch and Néel walls. Phys.Stat.Sol. 81a,519-532.
Motogi S., Maugin G.A. 1984b. Magnetoelastic oscillations of a Bloch wall in ferromagnets with dissipations. Japan.Jl.Appl. Phys. 23, 1026-1031.
Miura R.M.(Editor) 1976. Bäcklund Transformations. Vol.515 of Lecture Notes in Mathematics (eds. A.Dold and B.Eckmann). Berlin:Springer-Verlag.
Néel L. 1955. Energie des parois de Bloch dans des couches minces. C.R.Acad.Sci. Paris. 241, 533-536.
Nowacki W. 1981. Teoria niesymetrycznej Sprezystosci (Polish,2nd enlarged edition) Warsaw: P.W.N.
Pouget J. 1984. Influence de champs rémanents ou initiaux sur les propriétés dynamiques de milieux élastiques polarisables. Doctoral thesis in Mathematics. Paris:Université Pierre-et-Marie Curie.
Pouget J., Askar A.,Maugin G.A. 1985a. Lattice models for elastic ferroelectric crystals: microscopic approach (submitted for publication).
Pouget J., Askar A., Maugin G.A. 1985b. Lattice models for elastic ferroelectric crystals:continuum approximation (submitted for publication).
Pouget J., Maugin G.A. 1982. Nonlinear electroacoustic equations for piezoelectric powders. J.Acoust.Soc.America 74, 925-940.
Pouget J., Maugin G.A. 1984. Solitons and electroacoustic interactions in ferroelectric crystals-I-Single solitons and domain walls. Phys.Rev. B30, 5306-5325.
Pouget J., Maugin G.A. 1985a. Solitons and electroacoustic interactions in ferroelectric crystals-II- Interactions of solitons and radiations. Phys.Rev. B31, 4633-4651.
Pouget J., Maugin G.A. 1980. Coupled acoustic-optic modes in deformable ferroelectrics. J.Acoust.Soc.America 68, 588-601.
Pouget J., Maugin G.A. 1985b. Influence of an external electric field on the motion of a ferroelectric domain wall. Phys. Letters A (in the press).
Pouget J., Maugin G.A. 1985c. Transient Motion of a solitary wave in elastic ferroelectrics. Contr. to Intern.Conf. Nonlinear Mechanics ,Shanghai (China, Oct.28-31,1985).To appear in the proceedings.
Pouget J., Maugin G.A. 1985d. Transient motion of solitary waves in ferroelectric crystals. Phys.Rev.B (submitted for publication).
Pouget J., Maugin G.A. 1985e. Solitary waves in oriented elastic solids. Wave Motion (to appear).
Ristic V.M. 1983. Principles of acoustic devices. New York:J.Wiley-Interscience.
Seeger A. 1955. Theorie der Gitterfehlstellen.In S.Flügge (ed.) Handbuch der Physik. Bd 7, p.383-665.Berlin:Springer.
Seeger A. 1979. Solitons in crystals. In Continuum models of Discrete systems 3 (Proc.Symp.Freudenstadt,1979).Vol. 15 of Solid Mechanics Studies Series. Waterloo(Canada):Univ.of Waterloo Mechanics Division.
Seeger A.,Wesołowski Z. 1981. Standing wave solutions of the Enneper equation (sine-Gordon equation). Int.J.Engng.Sci. 19 , 1535-1549.
Soohoo R.F. 1965. Magnetic thin films. New York:Harper and Row (International Student Edition)and Tokyo:John Weatherhill.
Stokes V.K. 1984. Theories of fluids with microstructure. Berlin:Springer-Verlag.
Suzuki S., Takagi M. Topographic study of ferroelectrics $NaNO_2$ crystal I-structure of 180° domain wall. J.Phys.Soc.Japan 30, 188-195.
Takhtadzhyan L.A. 1977. Integration of the continuous Heisenberg chain through the inverse scattering method. Phys.Lett. A64, 235-237.
Tran C.D., Gerbaux X., Hadni A. 1981. Application of the pyroelectric probe technique to the study of domain wall motion in Ferroelectric $NaNO_2$ and TGS. Ferroelectrics 33, 31-35.
Turov E.A. 1984. Symmetry breaking and magnetoacoustic effects in ferro- and antiferromagnets. In G.A.Maugin(ed.) The Mechanical behavior of Electromagnetic

of solid continua ,p.255-267.Amsterdam: North-Holland.
Turov E.A., Tchavrov V.G. 1983. Symmetry breaking and magnetoacoustic effects in ferro- and antiferromagnets. Progress in Physical Sciences (in Russian:Uspeki Fiz.Nauk), 140 , 429-462.
Wadas R.S. 1974. Magnetism in spinels, garnets and perovskites. Warsaw :P.W.N.
Wesołowski Z. 1983. Dynamics of a bar of asymmetric cross section.Jl of Engineering mathematics 17, 315-321.
Whitham G.B. 1974. Linear and Nonlinear waves. New York:J.Wiley-Interscience.
Winters J. 1961. Bloch wall excitation: Applications to nuclear resonance in a Bloch wall. Phys.Rev. 124, 452-459.

5th International Symposium on Continuum Models of Discrete Systems / Nottingham / 14-20 July 1985

A continuum model for polyelectrolytes in solution

A.Morro
Biophysical & Electronic Engineering Department, University of Genova, Italy
R.Drouot & G.A.Maugin
Mécanique Théorique, Université Pierre-et-Marie Curie, Paris, France

ABSTRACT: The polyelectrolyte solution is modelled as a mixture; the solvent is a fluid, the solute merely consists of polyelectrolyte macroions along with some types of counterions. The motion of counterions is described through electric current densities which need not be conserved also because of possible ionization and recombination reactions. Both the solvent and the solute are supposed to undergo isochoric motions. First the balance equations are considered. Then, starting from the Clausius-Duhem inequality (in a general form) as the expression for the second law, noticeable thermodynamic restrictions on the constitutive relations are derived. Finally, it is shown how a free energy function consisting of an elastic term and an electrostatic one gives reason for the mechanochemical effect.

1 INTRODUCTION

As is well-known, polyelectrolytes consist of macromolecules whose monomers contain an ionic group. We are now accustomed to a large variety of macromolecules such as synthetic and natural polimers, proteins, nucleic acids. This large variety results in a wide phenomenology, the most interesting phenomena being related to polyelectrolytes in solution.

When dissolved in fluids, the macromolecules are dissociated into polyvalent macroions and a large number of small ions of opposite charge (counterions). These macroions have a structure which depends heavily on the interaction of the ionized groups. It is then easy to realize that the conformation and the volume of the macromolecule may be strongly influenced by the distribution of counterions and the action of an external electric field.

Consider the mechanochemical effect which may be exemplified by looking at the molecule of polyacrilic acid (constituted by CH_2-CHCOOH monomers). On the addition of alkali the carboxyl groups are dissociated and the molecule gains a number of negative charges meanwhile producing positive counterions. Of course the number of dissociated groups depends on the amount of the added alkali and then it increases with the pH of the solution. With increasing charge the molecule changes its shape from the spherical one to a fully extended one. Thus the macromolecule elongates with increasing pH (and viceversa).

As to macroscopic phenomena like the mechanochemical effect, a handy description of polyelectrolyte behaviour needs a continuum model which then should be based on the general principles of continuum thermodynamics. With the purpose of elaborating a model allowing also the description of phenomena due to the interaction with an external electric field, in this note we develop a thermodynamic theory of polyelectrolyte solutions acted upon by an electric field. As an application we show how the theory accounts for the mechanochemical effect.

2 A MODEL FOR POLYELECTROLYTE SOLUTIONS

The phenomena concerning polyelectrolytes

in solution are intimately connected with the structures these macromolecules take on in the dissolved state. Long molecular chains are randomly coiled in the solution. Molecules with hydrogen-bonded helical structures maintain such a structure in solvents that have little tendency to form hydrogen bonds between solvent and solute molecules. Globular proteins are supposed to have a compact and symmetrical structure at least in water solution.

In dealing with theoretical considerations of macromolecular solutions it is advantageous to model the macromolecules as material particles undergoing deformations. Two possible representations are a sphere and a long thin rod; the former might be a good representation of globular proteins, the latter of helical rod-shaped molecules. Due to mathematical convenience and to generality of the model, dissolved molecules are often represented as ellipsoids of revolution.

With this in mind, we describe the deformable configuration of macromolecules by a symmetric second-order tensor $\mathbf{C}$. In the case of ellipsoidal models it is natural to define $\mathbf{C}$ as

$$\mathbf{C} = a\, \mathbf{e}_1 \otimes \mathbf{e}_1 + b(\mathbf{e}_2 \otimes \mathbf{e}_2 + \mathbf{e}_3 \otimes \mathbf{e}_3) \ ,$$

$\mathbf{e}_1$ being the unit vector along the axis of revolution. As to randomly coiled molecules we may still describe the configuration by a symmetric tensor $\mathbf{C}$ which then would be defined via the average over all possible entanglements of the polymer chain. In any case we may set

$$\mathbf{C} = \tfrac{1}{3}\kappa \mathbf{1} + \mathbf{K}$$

with $\operatorname{tr}\mathbf{K} = 0$ and hence $\kappa = \operatorname{tr}\mathbf{C} > 0$. So κ describes the change in volume while $\mathbf{K}$ describes the deviation from the spherical shape. Accordingly we regard $\mathbf{K}$ as an internal variable whose evolution equation describes the conformational relaxation toward equilibrium (Maugin & Drouot, 1983; Bampi & Morro, 1984).

The polyelectrolyte solution is modelled as a mixture, the solvent being a fluid (carrier fluid). The solute merely consists of the polyelectrolyte macromolecules, or rather macroions. Moreover, different types of ions are dissolved into the fluid. For the sake of simplicity the solvent and the solute are supposed to undergo isochoric motions. The motion of the different types ions give rise to electric current densities which need not be conserved also because of possible ionization and recombination reactions; the volume charge density as a whole is assumed to vanish. So the polyelectrolyte solution is modelled as a reacting mixture of several fluid components. In order to avoid the cumbersome relations which would result from the direct application of the standard theory of mixtures (Müller, 1975; Bowen, 1976), as a consequence of the complexity of the polyelectrolyte solution, we introduce some approximations which, while appearing physically sound, reduce the number of effective unknown fields. Let ρ_c, ρ_m be the mass densities of the carrier fluid and the macroion fluid. Owing to the assumption of isochoric motions we have $\rho_c, \rho_m =$ = constant and hence the total mass density $\rho = \rho_c + \rho_m$ is constant. The corresponding velocity fields $\mathbf{v}$, $\mathbf{v}_m$ satisfy $\nabla\cdot\mathbf{v} = \nabla\cdot\mathbf{v}_m = 0$, ∇ denoting the spatial gradient operator. We ascribe material properties to the carrier fluid and the macroion fluid only; so ρ is the mass density of the mixture. The local electric properties are described through the electric polarization per unit volume, $\mathbf{P}$, or unit mass, $\boldsymbol{\Pi} = \mathbf{P}/\rho$. Maxwell's equations are considered within the framework of quasi-electrostatics. The carrier fluid in itself is regarded as neutral.

3 BALANCE EQUATIONS

The different types of ions existing in the solution are described through a set of charge densities q_α, q_m being the charge density of macroions. The balance of charge for the ions in the solution may be written as

$$\partial q_\alpha/\partial t + \nabla\cdot(q_\alpha \mathbf{v}_\alpha) = \gamma_\alpha \qquad (3.1)$$

γ_α being the charge supply and $\mathbf{v}_\alpha$ the velocity of the α-th type of ions; of course the conservation of charge implies that

$$\Sigma_\alpha\, \gamma_\alpha = 0 \ .$$

Letting $\mathbf{j}_\alpha = q_\alpha(\mathbf{v}_\alpha - \mathbf{v})$, in view of the condition $\nabla\cdot\mathbf{v} = 0$ we have

$$\nabla\cdot(q_\alpha \mathbf{v}_\alpha) = \nabla\cdot\mathbf{j}_\alpha + \mathbf{v}\cdot\nabla q_\alpha \ .$$

Then the equation of balance of charge may be given the form

$$\dot{q}_\alpha + \nabla\cdot\mathbf{j}_\alpha = \gamma_\alpha, \qquad (3.2)$$

the superposed dot denoting the material time derivative (relative to the fluid). It is worth emphasizing the particular behaviour of the macroion charge, namely

$$\mathbf{j}_m = \mathbf{0}, \qquad \dot{q}_m = \gamma_m .$$

The equation of balance of linear momentum for the whole solution is considered in the form

$$\rho\dot{\mathbf{v}} = \nabla\cdot\mathbf{t} + \mathbf{f} + \mathbf{f}^{em} \tag{3.3}$$

the divergence of the stress tensor $\mathbf{t}$ being taken on the second index and $\mathbf{f}$ denoting the mechanical body force (per unit volume). In view of the quasi-electrostatics approximation the volume electromagnetic force $\mathbf{f}^{em}$ is given by (Maugin & Eringen, 1975)

$$\mathbf{f}^{em} = (\mathbf{P}\cdot\nabla)\mathbf{E}$$

while $\mathbf{t}$ is written as (Maugin, 1980)

$$\mathbf{t} = \boldsymbol{\sigma} + \text{skw}(\mathbf{P}\otimes\mathbf{E})$$

$\boldsymbol{\sigma}$ being the symmetric mechanical intrinsic stress tensor and skw denoting the skew-symmetric part. Let $\mathbf{J} = \Sigma_\alpha\, \mathbf{j}_\alpha$ be the total current density in the solution. Moreover, let e be the internal energy density, $\mathbf{h}$ the heat flux, and r the heat supply (per unit mass). On account of the balance equation (3.3) and objectivity arguments the energy balance equation is taken in the form

$$\rho\dot{e} = \text{tr}(\boldsymbol{\sigma}\,\mathbf{D}) + \mathbf{J}\cdot\mathbf{E} + \rho\mathbf{E}\cdot D_J\boldsymbol{\Pi} - \nabla\cdot\mathbf{h} + \rho r \tag{3.4}$$

where $\mathbf{D}$ is the rate-of-strain tensor and D_J denotes the objective Jaumann derivative such that (for a vector field $\boldsymbol{\Pi}$)

$$D_J\boldsymbol{\Pi} = \dot{\boldsymbol{\Pi}} - \boldsymbol{\Omega}\times\boldsymbol{\Pi} , \quad \boldsymbol{\Omega} = \tfrac{1}{2}\nabla\times\mathbf{v} .$$

As to the entropy inequality, we follow Müller's viewpoint that the entropy flux need not be the heat flux times the inverse of the absolute temperature θ. Then, letting η be the entropy density, we adopt the entropy inequality as

$$\rho\dot{\eta} \geq \theta^{-1}\rho r - \nabla\cdot(\theta^{-1}\mathbf{h} + \mathbf{k})$$

$\mathbf{k}$ being the entropy extra-flux. Hence, in view of the energy balance equation (3.4), the entropy inequality becomes

$$-\rho(\dot{\psi} + \eta\dot{\theta}) + \text{tr}(\boldsymbol{\sigma}\,\mathbf{D}) + \mathbf{J}\cdot\mathbf{E} - \rho\mathbf{E}\cdot D_J\boldsymbol{\Pi} + \theta\nabla\cdot\mathbf{k} - \theta^{-1}\mathbf{h}\cdot\mathbf{g} \geq 0$$

ψ being the free energy density and $\mathbf{g}$ the temperature gradient. For later developments it is convenient to introduce the quantity $\phi = \psi - \mathbf{E}\cdot\boldsymbol{\Pi}$; then we have

$$-\rho(\dot{\phi} + \eta\dot{\theta}) + \text{tr}(\boldsymbol{\sigma}\,\mathbf{D}) + \mathbf{J}\cdot\mathbf{E} - \rho\boldsymbol{\Pi}\cdot D_J\mathbf{E} + \theta\nabla\cdot\mathbf{k} - \theta^{-1}\mathbf{h}\cdot\mathbf{g} \geq 0 . \tag{3.5}$$

We end this section with some comments about the above formulation of the balance equations. The main approximations, with respect to a genuine application of the theory of mixtures, are summarized as follows. First, except for the macroions, the ions have been modelled as electric currents but not as components of the mixture. Second, because of the condition $\mathbf{v}_m = \mathbf{v}$, we have considered only one balance equation for the linear momentum as well as for the energy. Third, we have neglected the exchanges of linear momentum and energy between the ions and the fluid (carrier fluid and macroions) but the contribution $\mathbf{J}\cdot\mathbf{E}$ (Joule effect). Finally, it is worth remarking that our model resembles, in a sense, the one elaborated by Tiersten (1984) in connection with deformable semiconductors. In this respect it turns out that we have disregarded the pressures of the different ions on the fluid.

4 CONSTITUTIVE EQUATIONS AND THERMODYNAMIC RESTRICTIONS

Most of the effects revealed by polymer solutions are accounted for through the dependence on the temperature θ, the temperature gradient $\mathbf{g}$, the electric field $\mathbf{E}$, the rate-of-strain tensor $\mathbf{D}$, the charge densities q_α, and the conformation tensor $\mathbf{C}$. Since we are interested here in volume-preserving motions of macroions, we let κ be constant and then we confine the attention to the traceless part $\mathbf{K}$ only. Because $\mathbf{K}$ is regarded as an internal variable, we write the whole set of constitutive equations as - see the general methodology in (Maugin & Drouot, 1983; Bampi & Morro, 1984) -

$$\phi = \phi(\theta, \mathbf{E}, q_\alpha, \mathbf{K}, \mathbf{g}, \mathbf{D}) \tag{4.1}$$

along with analogous equations for η, $\boldsymbol{\sigma}$, the $\mathbf{j}_\beta$'s, $\boldsymbol{\Pi}$, $\mathbf{h}$, $\mathbf{k}$ while

$$D_J\mathbf{K} = \hat{\mathbf{K}}(\theta, \mathbf{E}, q_\alpha, \mathbf{K}, \mathbf{g}, \mathbf{D}); \tag{4.2}$$

the constitutive function ϕ is assumed continuously differentiable with respect to its argument while η, $\boldsymbol{\sigma}$, $\mathbf{j}_\beta$, $\boldsymbol{\Pi}$, $\mathbf{h}$, $\mathbf{k}$, and $\hat{\mathbf{K}}$ are continuous. It is worth remarking that a possible dependence on the mass density ρ is disregarded in view of the incompressibility of the fluid. Meanwhile the incompressibility allows $\boldsymbol{\sigma}$ to be identified with the deviatoric part of the mechanical stress thus leaving the pressure p thermodynamically undetermined.

Substitution of (4.1), (4.2) into (3.5)

and account of (3.2) give

$$-\rho(\eta+\partial\phi/\partial\theta)\dot{\theta} - \rho(\Pi+\partial\phi/\partial\mathbf{E})\cdot D_J\mathbf{E}$$
$$-\rho(\partial\phi/\partial\mathbf{g})\cdot D_J\mathbf{g} - \mathrm{tr}[(\partial\phi/\partial\mathbf{D})D_J\mathbf{D}]$$
$$- \mathrm{tr}[(\partial\phi/\partial\mathbf{K})\hat{\mathbf{K}}] + \mathrm{tr}(\boldsymbol{\sigma}\mathbf{D}) \quad (4.3)$$
$$+ \Sigma_\alpha\, \mathbf{j}_\alpha\cdot\mathbf{E} + \rho\Sigma_\alpha(\partial\phi/\partial q_\alpha)\nabla\cdot\mathbf{j}_\alpha$$
$$-\rho\Sigma_\alpha(\partial\phi/\partial q_\alpha)\gamma_\alpha + \theta\nabla\cdot\mathbf{k} - \theta^{-1}\mathbf{h}\cdot\mathbf{g} \geqslant 0.$$

The inequality (4.3) holds identically provided that

$$\partial\phi/\partial\mathbf{g} = \mathbf{0}\,, \qquad \partial\phi/\partial\mathbf{D} = \mathbf{0}\,,$$

and

$$\eta = -\partial\phi/\partial\theta\,, \qquad \Pi = -\partial\phi/\partial\mathbf{E}\,. \quad (4.4)$$

Hence, because of the identity

$$(\partial\phi/\partial q_\alpha)\nabla\cdot\mathbf{j}_\alpha = \nabla\cdot[(\partial\phi/\partial q_\alpha)\mathbf{j}_\alpha] - \mathbf{j}_\alpha\cdot\nabla(\partial\phi/\partial q_\alpha),$$

the entropy inequality (4.3) reduces to

$$\mathrm{tr}(\mathbf{F}\hat{\mathbf{K}} + \boldsymbol{\sigma}\mathbf{D}) + \Sigma_\alpha\, \mathbf{j}_\alpha\cdot[\mathbf{E} - \rho\nabla(\partial\phi/\partial q_\alpha)]$$
$$- \Sigma_\alpha(\partial\phi/\partial q_\alpha)\gamma_\alpha + \rho\nabla\cdot[\Sigma_\alpha(\partial\phi/\partial q_\alpha)\mathbf{j}_\alpha] \quad (4.5)$$
$$+ \theta\nabla\cdot\mathbf{k} - \theta^{-1}\mathbf{h}\cdot\mathbf{g} \geqslant 0$$

where we have set

$$\mathbf{F} = -\rho\,\partial\phi/\partial\mathbf{K}\,.$$

As with usual chemical reactions, the charge supplies γ_α are not independent of one another. Specifically, if n distinct reactions between atoms and ions are possible we may write

$$\gamma_\alpha = \Sigma_r\, \nu_{\alpha r}\, \Lambda_r\,,$$

r running from 1 to n; the $\nu_{\alpha r}$'s play the role of stoichiometric coefficients and the Λ_r's the role of reaction rates. Then, on applying the standard reasoning we conclude that (4.5) holds only if

$$\Sigma_\alpha\, \nu_{\alpha r}\, \partial\phi/\partial q_\alpha = 0,\ r = 1,..,n. \quad (4.6)$$

The relation (4.6) governs the equilibrium between the various types of ions in the solution. If, as it seems natural, we regard ϕ as the sum of two parts, one concerning the fluid and one the ions, then (4.6) may be viewed as a relation between the reduced chemical potentials (relative to the potential of the fluid).

So far we have derived necessary conditions for the validity of the entropy inequality. To exhaust the set of necessary conditions is a formidable task which, at least, would lead to very cumbersome relations. That is why here we content ourselves with sufficient conditions which enable us to describe the phenomena we are interested in.

First we observe that the identity

$$\theta\nabla\cdot\mathbf{k} = \nabla\cdot(\theta\mathbf{k}) - \mathbf{k}\cdot\mathbf{g}$$

and the result (4.6) allow (4.5) to be written as

$$\mathrm{tr}(\mathbf{F}\hat{\mathbf{K}} + \boldsymbol{\sigma}\mathbf{D}) + \Sigma_\alpha\mathbf{j}_\alpha\cdot[\mathbf{E} - \rho\nabla(\partial\phi/\partial q_\alpha)]$$
$$+ \nabla\cdot[\rho\Sigma_\alpha(\partial\phi/\partial q_\alpha)\mathbf{j}_\alpha + \theta\mathbf{k}] - (\mathbf{k} + \theta^{-1}\mathbf{h})\cdot\mathbf{g} \geqslant 0\,.$$

This expression for the entropy inequality strongly suggests that we take

$$\mathbf{k} = -\rho\theta^{-1}\Sigma_\alpha(\partial\phi/\partial q_\alpha)\mathbf{j}_\alpha$$

so that we are left with the reduced entropy inequality

$$\mathrm{tr}(\mathbf{F}\hat{\mathbf{K}} + \boldsymbol{\sigma}\mathbf{D}) \quad (4.7)$$
$$+ \Sigma_\alpha\, \mathbf{j}_\alpha\cdot[(\mathbf{E} - \rho\nabla(\partial\phi/\partial q_\alpha)] - \mathbf{S}\cdot\mathbf{g} \geqslant 0$$

where

$$\mathbf{S} = \theta^{-1}[\mathbf{h} - \rho\Sigma_\alpha(\partial\phi/\partial q_\alpha)\mathbf{j}_\alpha]$$

is the entropy flux.

According to the purpose of this note and for the sake of definiteness, we consider now constitutive equations involving linear dependences on the "forces" **F**, **D**, **g**, and **E** while θ, **K** are regarded as parameters. Specifically, as to $\hat{\mathbf{K}}$ and $\boldsymbol{\sigma}$ we set - cf. (Maugin & Drouot, 1983) -

$$\hat{\mathbf{K}} = \xi\mathbf{F} - \lambda\mathbf{D} + \beta(\mathbf{K}\mathbf{D} + \mathbf{D}\mathbf{K})\,,$$
$$\boldsymbol{\sigma} = \lambda\mathbf{F} + \mu\mathbf{D} - \beta(\mathbf{K}\mathbf{F} + \mathbf{F}\mathbf{K})\,. \quad (4.8)$$

As to the constitutive relations for the current densities $\mathbf{j}_\alpha$ and the entropy flux **S** it is convenient to write them in the form

$$\mathbf{E} - \rho\nabla(\partial\phi/\partial q_\alpha) = \mathbf{R}_\alpha\mathbf{j}_\alpha + A_\alpha\mathbf{g}\,,$$
$$\mathbf{S} = -\Sigma_\alpha A_\alpha\mathbf{j}_\alpha - B\mathbf{g}\,. \quad (4.9)$$

The scalar coefficients ξ, λ, μ, β, A_α, B are viewed as functions of $\mathrm{tr}\,\mathbf{K}^2$, θ, and the q_α's. The second-order symmetric tensors $\mathbf{R}_\alpha$ account for Ohm's law and then they are taken as non-singular.

As an aside, a simple way of describing the possible anisotropy effects on the electric conduction is obtained by setting

$$\mathbf{R}_\alpha = r'_\alpha\,\mathbf{1} + r''_\alpha\,\mathbf{K}\,,$$

r'_α, r''_α being functions of $\mathrm{tr}\,\mathbf{K}^2$, θ, and the q_β,s; the term $r''_\alpha\mathbf{K}$ represents the resistance due to the conformation of the macromolecules (elasto-resistance).

Upon substitution of (4.8) and (4.9) into the reduced dissipation inequality (4.7) we see at once that compatibility with thermodynamics implies

$$\xi \geqslant 0,\quad \mu \geqslant 0,\quad B \geqslant 0\,,$$
$$\mathbf{R}_\alpha \text{ positive definite},$$
$$A_\alpha^2 \leqslant B\,\bar{R}_\alpha\,,$$

$\bar{R}_\alpha(>0)$ being the smallest eigenvalue of $\mathbf{R}_\alpha$, and the λ and β terms contributing nothing to the dissipation in reason of the gyroscopic behaviour (Maugin & Drouot, 1983).

5 CONFORMATIONAL RELAXATION AND MECHANOCHEMICAL EFFECT

Confine our attention to the equilibrium conformation of macroions and to their time evolution toward the equilibrium conformation. Accordingly let the carrier fluid be at rest (and hence $\mathbf{D} = \mathbf{0}$) and let the external electric field vanish. Then the evolution function $(4.8)_1$ leads to the evolution equation

$$\dot{\mathbf{K}} = \xi \mathbf{F} \quad , \tag{5.1}$$

which means that the evolution of $\mathbf{K}$ is determined as soon as we specify $\mathbf{F}$ in terms of $\mathbf{K}$ and, possibly, θ and the q_α's.

The equilibrium conformation of macroions is influenced by the temperature and the charge densities q_α; obviously the dependence on the charge density of macroions q_m ($q_m < 0$) is likely to play a dominant role. Then, letting $\ell = (\mathrm{tr}\,\mathbf{K}^2)^{1/2}$ be a scalar measure for the deviation of the conformation from the spherical form, we denote by $\ell_e(\theta\,,\,q_\alpha)$ its equilibrium value. At the simplest level of approximation the free energy of macroions may be viewed as resulting from the deformation energy and the interaction between deformation and electric charge. Since $\mathrm{tr}\,\mathbf{K} = 0$, this suggests that we write the free energy ϕ as

$$\phi = \tfrac{1}{2}\chi\ell^2 + \nu\ell^{-1} q_m \tag{5.2}$$

the coefficients χ, ν being functions of and the charge densities but q_m. It seems natural, if not imperative, to assume that χ is positive so that ϕ is minimum in the undeformed spherical conformation; note that the macromolecule is neutral in the spherical conformation ($q_m = 0$).

Owing to (5.2) we have

$$\mathbf{F} = -\rho(\chi - \nu\ell^{-3}q_m)\,\mathbf{K} \tag{5.3}$$

and hence $\mathbf{F} = \mathbf{0}$ if either

$$\mathbf{K} = \mathbf{0} \quad , \tag{5.4}$$

whence $\ell = 0$, or

$$\ell = (\nu\chi^{-1} q_m)^{1/3} . \tag{5.5}$$

On account of (5.1) it follows that (5.4) and (5.5) single out equilibrium conformations. Indeed, (5.5) provides just the function $\ell_e(\theta\,,\,q_\alpha)$. Moreover, because ℓ is non-negative by definition, (5.5) implies that $\nu < 0$. As we will see shortly, the negativeness of ν provides an immediate explanation for the mechanochemical effect.

Substitution of (5.3) into (5.1), contraction with $\mathbf{K}$ and some rearrangement yield

$$\tau\,\dot{\ell} = -(\ell - \ell_e) \tag{5.6}$$

where

$$\tau = \ell^2\,[\rho\xi\chi(\ell^2 + \ell_e\ell + \ell_e^2)]^{-1} .$$

Since $\tau > 0$, eq. (5.6) tells us that, when θ and the q_α's are constant, ℓ tends eventually to the equilibrium value ℓ_e. So eq. (5.6) is a relaxation-type equation for the deviation $\ell - \ell_e$ and τ plays the role of relaxation time.

Experimentally, it is possible (Ouibrahim, 1981) that τ depends on ℓ with τ decreasing as the shear of macromolecules increases. Theoretically this may be represented by a (probably weakly) ℓ-dependent ξ or, equivalently, by assuming a weakly nonlinear elastic shear behaviour of macromolecules. In this case eq. (5.5) may be replaced by an expression of the form

$$\nu\ell^3\,[1 + f(\ell)] - \nu q_m = 0$$

where $f(\ell)$ accounts for the nonlinearity in the elastic behaviour. A perturbation method then yields an equilibrium value which differs only slightly from the result (5.5).

The number of dissociated groups in the macroions increases (and hence q_m decreases) with increasing the pH value p of the solution, namely

$$(dq_m/\partial p) < 0 .$$

On the other hand, in view of (5.5) we have

$$(\partial\ell_e/\partial q_m) = (\nu/3\chi\ell^2) < 0 .$$

This implies that

$$(\partial\ell_e/\partial p) > 0 ,$$

which is the content of the mechanochemical effect (Volkenshtein, 1983).

6 REMARKS

A very simple model allows also an account of the conformational phase transitions, namely sudden volume changes induced, for example, by an external electric field. This phenomenon however is not investigated here; the interested reader is referred to our paper (Morro, Drouot & Maugin, 1985).

In ending this note we observe that our model provides detailed results inasmuch as the free energy ϕ is known in detail. This suggests that we turn our attention to the determination of ϕ in terms of suitable microscopic parameters connected with the main properties of the macromolecule. This study is under way.

REFERENCES

Maugin, G.A., Drouot, R. 1983. Internal variables and the thermodynamics of macromolecule solutions. Int. J. Engng Sci. 21, 705-724.

Bampi, F., Morro, A. 1984. Nonequilibrium thermodynamics: a hidden variable approach. In J. Casas-Vazquez, D. Jou, G. Lebon (eds.), Recent Developments in Nonequilibrium Thermodynamics, Lecture Notes in Physics, 199, Springer, Berlin.

Müller, I. 1975. Thermodynamics of mixtures of fluids. J. Mécan. 14, 267-303.

Bowen, R.M. 1976. Theory of mixtures. In A.C. Eringen (ed.), Continuum Physics III, Academic Press, New York.

Maugin, G.A., Eringen, A.C. 1975. On the equations of the electrodynamics of deformable bodies of finite extent. J. Mécan. 16, 101-147.

Maugin, G.A. 1980. The principle of virtual power in continuum mechanics: Application to coupled fields. Acta Mech. 35, 1-70.

Tiersten, H.F. 1984. Electric fields, deformable semiconductors and piezoelectric devices. In G.A. Maugin (ed.), The Mechanical Behavior of Electromagnetic Solid Continua, North Holland, Amsterdam.

Ouibrahim, A. 1981. Thesis of Doct. ès Sciences Physiques, Univ. Paris VI, mimeographed.

Volkenshtein, M.V. 1983. Biophysics, Mir, Moscow.

Morro, A., Drouot, R., Maugin, G.A. 1985. Thermodynamics of polyelectrolyte solutions in an electric field. J. NonEquilib. Thermodyn. 10.

5th International Symposium on Continuum Models of Discrete Systems / Nottingham / 14-20 July 1985

Nonlinear diffusion and Schrödinger equation

A.Trzesowski & R.Kotowski
Institute of Fundamental Technological Research, Polish Academy of Sciences, Warsaw

INTRODUCTION

The problem of explanation the quantum phenomena with the help of statistical physics has attained the interest almost from the beginning of the quantum theory [1]. Recently a great number of papers has appeared in the connection with the theory of E. Nelson [2], [3] of the so-called 'stochastic quantization method'. The approach of Nelson to quantum mechanics was criticised by Mielnik and Tengstrand [4]. In our previous paper [5] we tried to incorporate certain Nelson's ideas into the scheme of (not quantum) description of motion of point defects in crystal lattices.

There exist two main continual methods of description of crystal lattice with defects. The first one, which we call 'physical', is based on the use of Euclidean coordinate systems and material coefficients ('constants') depending on the position in the space. The second, 'geometrical' approach, makes use of the differential geometry language and treats the crystal with defects as an ideal one but immersed in the non-Euclidean space and to describe the curvature and torsion a certain affine connection is used [6], [7], [8].

In this paper we wish to generalise our previous results on a Riemannian manifold, i.e. we want to put the 'geometrical' description to profit. We recall briefly at the beginning the case when the concentration of defects does not influence the value of the diffusion coefficient. More details one can find in [5].

1 THE CASE OF CONSTANT DIFFUSION COEFFICIENT

Let us consider a diffusion motion of point crystal lattice defects with mass which we treat as identical but distinguishable particles with negligible interactions. We restrict our discussion to the infinite body which can be identified with the configuration space of a single diffusing particle and to which it is possible, according to the above assumptions, to reduce considerations concerning the configuration space of all diffusing particles. In a macroscale such a diffusion process is characterised by volume concentration of matter taking part in the diffusion process

$$n = n(\underline{x},t), \qquad [n] = \mathrm{cm}^{-3},$$

$$\underline{x} \in R^3, \qquad t \geqslant 0$$

and subjected to the condition that

$$\forall t \geqslant 0, \qquad N(t) = \int_{R^3} n(\underline{x},t)\, dV(\underline{x}) \tag{1}$$

$$dV(\underline{x}) = dx^1\, dx^2\, dx^3.$$

(N.B. In paper [5] $\dot{N}(t) = 0$ is not assumed.)

If $dp_t(\underline{x})$ is the probability of observing at instant t a diffusing particle in an infinitesimal neighbourhood dV of the point $\underline{x}$, then it follows from our assumptions that [9]

$$dp_t(x) = p(\underline{x},t)\,dV(\underline{x})$$
$$= \lim_{N(t)\to\infty} n(\underline{x},t)\,dV(\underline{x})/N(t) \tag{2}$$

so that the approximation

$$p(\underline{x},t) = n(\underline{x},t)/N(t) \tag{3}$$

is the better, the greater N(t) is.

Diffusion can be regarded in many cases as a random motion of a particle. Such an approach is justified for example for crystal solid bodies if one considers a chemical diffusion with small concentrations.

A mathematical model which relatively well describes a random walk of a diffusing particle in the presence of an external force is the so-called diffusion Markov process. We constrain ourselves considering the diffusion Markov process defined by stochastic Ito equation of the form

$$d\underline{x}(t) = \underline{b}(\underline{x},t)\,dt + d\underline{W}(t)\ , \quad \underline{x} \in R^3,\ t \geqslant 0,$$
$$\underline{W}(0) = 0, \qquad E\{\underline{W}(t)\} = 0,$$
$$E\{\underline{W}(t) \times \underline{W}(t)\} = \sigma^2 tI, \qquad \sigma^2 > 0, \tag{4}$$
$$I = (\delta_{ij};\ i,j = 1,2,3),$$

where $\underline{W}$ is the Wiener process in the configuration space R^3, E is the expectation value operator, I is the unit matrix, and σ^2 is the variance parameter. In this case the probabilistic density p defined by equation (3) fulfils the Fokker-Planck equation

$$\partial_t p = \tfrac{1}{2}\sigma^2 \Delta p - \operatorname{div}(p\underline{b}). \tag{5}$$

When the proper assumptions about the function $\underline{b}$ are made, the solution of the Ito equation with the following initial condition

$$\underline{x}(0) = \underline{x}_0 \in R^3$$

exists, it is continuous and unique defined with probability one. This solution is in general not differentiable and we need a substitute for the derivative operator along the trajectory of the Ito equation. One can generalise the notion of the one-side derivative in the following way [2], [3]:

$$D_+ f(\underline{x}(t),t) = \lim_{h \to 0_+} E_{\underline{x}(t)} \{(f(\underline{x}(t+h),t+h) - f(\underline{x}(t),t))/h\}$$
$$D_- f(\underline{x}(t),t) = \lim_{h \to 0_-} E_{\underline{x}(t)} \{(f(\underline{x}(t),t) - f(\underline{x}(t-h),t-h))/h\}, \tag{6}$$

where $E_{\underline{x}(t)}$ is a conditional expectation value operator and f is a smooth vector or scalar function. Particularly, at every instant of time there exist two mean velocities of the wandering particle:

velocity $\underline{b}_*$ with which the particle arrives at the point $\underline{x}$

$$\underline{b}_*(\underline{x}(t),t) = D_- \underline{x}(t); \tag{7}$$

velocity $\underline{b}$ with which the particle starts from the point $\underline{x}$

$$\underline{b}(\underline{x}(t),t) = D_+ \underline{x}(t). \tag{8}$$

The quantity $\underline{b}$ is strictly this quantity which occurs in Ito and Fokker-Planck equations.

If we introduce two new velocities $\underline{u}$ and $\underline{v}$

$$\underline{v} = \tfrac{1}{2}(\underline{b}_* + \underline{b}), \qquad \underline{u} = \tfrac{1}{2}(\underline{b}_* - \underline{b}), \tag{9}$$

then we obtain

$$\underline{u} = \underline{v} - \underline{b} = -\frac{\sigma^2}{2} \frac{\nabla p}{p}, \qquad (p = \frac{n}{N}), \tag{10}$$

where σ^2 is the variance parameter and ∇ is the gradient operator.

In the literature concerning the diffusion theory, the velocity $\underline{b}$ is called the mean velocity while the velocity $\underline{b}_*$ is not considered at all. In the mixture theory velocities analogous to $\underline{v}$ and $\underline{u}$ are discussed. They are called peculiar and diffusion velocities respectively.

One can also relate to the Ito equation an operator of the mean second derivatives [2], [3]

$$A x(t) = \tfrac{1}{2}(D_+ D_- + D_- D_+) x(t). \tag{11}$$

This operator is invariant with respect to the symmetry operation $t \to -t$

$$A\underline{x}(-t) = A\underline{x}(t) \tag{12}$$

and has the form

$$A\underline{x}(t) = a(\underline{x}(t),t) \tag{13}$$

where

$$\underline{a} = \partial_t \underline{v} + (\underline{v}.\nabla)\underline{v} - (\underline{u}.\nabla)\underline{u} + \tfrac{1}{2}\sigma^2 \Delta \underline{u}. \tag{14}$$

We call the field $\underline{a}$ the peculiar acceleration field of the diffusion process.

The influence of an external field on a point defect is described in the classical formulation of the diffusion theory by the constitutive relation in the form of the so-called Stokes relation

$$\underline{F} = \zeta \underline{b} \tag{15}$$

which connects the force $\underline{F}$ acting on a point defect with its mean velocity $\underline{b}$. In our paper we do not assume the Stokes relation; it means that the force $\underline{K}$ of the form

$$\underline{K}(\underline{x},t) = -\zeta\underline{b}(\underline{x},t) + \underline{F}(\underline{x},t) \qquad (16)$$

does not vanish identically.

We assume, following the Nelson procedure of the stochastic quantization method, the constitutive equation in the form of the following Langevin-like equation (m - mass of point defect)

$$\underline{K}(\underline{x},t) = m\underline{a}(\underline{x},t) . \qquad (17)$$

We call this equation the Nelson relation, although in Nelson's papers occurs the force $\underline{F}$ only.

Now, if we assume the following relation between the variance parameter σ^2 and the diffusion coefficient D

$$\tfrac{1}{2}\sigma^2 = D \qquad (18)$$

then our diffusion process is described by the three equations

$$\partial_t \rho = D\Delta\rho - \operatorname{div}(\rho(\underline{v}-\underline{u})) + \beta(t) , \qquad (19)$$

$$\partial_t \underline{v} = -\frac{1}{\tau}(\underline{v}-\underline{u}) + \nabla\{D \operatorname{div}\underline{u} + \tfrac{1}{2}(\underline{v}^2-\underline{u}^2)\} + \frac{1}{m}\underline{F} , \qquad (20)$$

$$\underline{u} = -D\frac{\nabla\rho}{\rho} , \qquad (21)$$

where

$$\beta(t) = \dot{N}(t)/N(t) = \dot{M}(t)/M(t) ,$$

$$M(t) = N(t)m , \qquad (22)$$

$$\rho = \rho(\underline{x},t) = n(\underline{x},t)m/M(t) ,$$

and τ is a constant of a dimension of time and by definition it has the form

$$\tau = m/\zeta . \qquad (23)$$

Eqn. (19) is the diffusion equation; it is equivalent to the Fokker-Planck equation. Eqn. (20) is equivalent to the Nelson relation. In statistical physics the constant τ is called the kinetic relaxation time.

Let us observe that if τ_* is an interval of time which defines the scale of time observations of the diffusion process, then for

$$\tau \ll \tau_* , \qquad (\tau \to 0,\ \zeta = \text{constant})$$

the Nelson relation reduces to the Stokes relation and our system of equations reduces to the classical linear diffusion equation. If

$$\tau \geqslant \tau_*$$

then according to statistical physics, the use of the Stokes relation is not justified. In this case we propose to use the Nelson relation in place of the Stokes relation.

We see that Eqns. (19) - (21) describe nonlinear diffusion process with finite relaxation time.

In the particular case when the peculiar velocity $\underline{v}$ and the external force $\underline{F}$ are potential functions

$$\underline{v}(\underline{x},t) = 2D\nabla_{\underline{x}} S(\underline{x},t)$$
$$\underline{F}(\underline{x},t) = -\nabla_{\underline{x}} U(\underline{x},t) \qquad (24)$$

our system of equations can be written down in the following form

$$i\hbar\partial_t\psi = -\frac{\hbar^2}{2m}\Delta\psi + V\psi \qquad (25)$$

$$\psi = e^{R+iS} ,$$

$$R = \tfrac{1}{2}\ln\frac{\rho}{\rho_0} , \qquad (u = -2D\nabla R)$$

$$V = U + \frac{\hbar}{2}(\ln|\psi|^2 + i\ln\frac{\psi^*}{\psi}) , \qquad \hbar = 2mD ,$$

$$[\hbar] = [\hbar] = gcm^2s^{-1} , \qquad [D] = cm^2s^{-1} .$$

It is the nonlinear Schrödinger-like equation which fuses the collective description of atoms behaviour, manifesting itself in the existence of the diffusion flux, with Newtonian description of the dynamics of a single particle. It is not the Schrödinger equation because, generally speaking, the parameter h is not equal to the reduced Planck constant, that is

$$\hbar \neq \hbar = h/2\pi . \qquad (26)$$

2 INFLUENCE OF THE POINT DEFECTS DISTRIBUTION ON THE VALUE OF THE DIFFUSION COEFFICIENT

It is known that diffusion of point defects is the "structural sensitive" process. This means that disturbances in the geometrically perfect crystal lattice influence the value of the diffusion coefficient [10]. We assume that these disturbances originate from the existence in a body a certain amount of extra-matter only. By extra-matter we understand these atoms that do not exist in the perfect crystal lattice, i.e. they are both intrinsic and extrinsic mass point defects

[11]. We assume, moreover, that but a part of the extra-matter with a prescribed mass density $\rho = \rho(x,t) = mn(x,t)$ diffuses. Here m denotes the mass of a point defect. If the concentration of defects is small, then it can be accepted that the diffusion coefficient does not change in time t and depends on the stationary distribution of the not diffusing extra-matter only. The assumption of the stationary distribution of the not diffusing extra-matter in the unbounded space and not existance of the chemical reactions means that the total mass of diffusing matter should be constant, i.e.,

$$M(t) = \int_{R^3} \rho(\underline{x},t)\, dV(\underline{x}) = M_0 , \tag{27}$$

where $dV(\underline{x})$ is the volume element of the configuration space of the diffusing particle.

It follows from the fundamental considerations that existence of point defects in a body produces the appearance of the curvature of the material space [8]. The geometrical model for such a body with defects is a Riemannian manifold $M = (R^3, g_{ij})$. We constrain ourselves to an unbounded body. The metric tensor of the mentioned manifold has a form [12]

$$g_{ij} = B^k_i B^\ell_j a_{k\ell} , \tag{28}$$

where $a_{k\ell}$ it is the so called elastic metric tensor which in the Cartesian coordinate system x^k on R^3 has a form

$$a_{k\ell} = \chi^p_k \chi^q_\ell \delta_{pq} , \tag{29}$$

$$\chi^p_k = \frac{\partial \chi^p}{\partial x^k} ,$$

where $\chi = (\chi^p) : M \to R^3$ is a diffeomorphism on R^3.

The non-orthogonal matrices B^k_ℓ can be considered, for example, in the form

$$B^k_\ell = \lambda_\ell \delta^k_\ell , \qquad \lambda_\ell > 0 . \tag{30}$$

The case $0 < \lambda_\ell < 1$ is interpreted as describing the influence of the existence of vacancies on the deformation of the crystal lattice [13].

The mathematical model of the considered process in the presented approach is the Markov process with values in the manifold M described in a certain coordinate system (x^i) by the stochastic Ito equation

$$dx^i(t) = b^i(\underline{x}(t),t)dt + dW^i(t) \tag{4$_i$}$$

but now [14]

$$\begin{aligned} E_{\underline{x}(t)}\{dW^i(t)\} &= E\{dW^i(t)\,|\,\underline{x}(t) = \underline{x}\} \\ &= \alpha^i(x)dt \\ E_{\underline{x}(t)}\{dW^i(t)\ dW^j(t)\} &= 2Dg^{ij}(\underline{x})dt , \end{aligned} \tag{31}$$

where

$$\begin{aligned} \alpha^i(\underline{x}) &= -D\Gamma^i_{jk}(\underline{x})g^{jk}(\underline{x}) \\ \Gamma^i_{jk}(\underline{x}) &= g^{i\ell}\Gamma_{\ell jk} \\ \Gamma_{ijk}(\underline{x}) &= \tfrac{1}{2}(\partial_j g_{ki} + \partial_k g_{ij} - \partial_i g_{jk}) \\ \partial_i &= \frac{\partial}{\partial \underline{x}^i} \end{aligned} \tag{32}$$

and

$$D = \frac{\hbar}{2m} \tag{33}$$

is the diffusion coefficient of the crystal in a flat space. Matrix

$$D^{ij}(\underline{x}) = Dg^{ij}(\underline{x}) \tag{34}$$

is interpreted as a representation of the matrix of diffusion coefficients in the coordinate system (x^i) [15].

The Fokker-Planck equation has its obvious form

$$\partial_t p = D\Delta p - \mathrm{div}(p\underline{b}) \tag{35}$$

but Δ is now the Laplace-Beltrami operator on the manifold M [16]

$$\Delta = \frac{1}{\sqrt{g}}\, \partial_i (\sqrt{g}\ g^{ij}\ \partial_j) . \tag{36}$$

Here $g = \det(g_{ij})$ and (cf. Eqns (3), (26) and observe that $M = mN$)

$$p(\underline{x},t) = \rho(\underline{x},t)/M_0 \tag{37}$$

is the probability density of finding at the instant t the diffusing particle in the infinitesimal neighbourhood $dV(\underline{x})$ of the point $\underline{x}$

$$P(\underline{x}(t) \in \mathcal{P}) = \int_{\mathcal{P}} p(\underline{x},t)\, dV(\underline{x})$$

$$dV(\underline{x}) = \sqrt{g(\underline{x})}\, dx^1 dx^2 dx^3 \qquad (38)$$

$$\mathcal{P} \in \text{top } M .$$

It follows from Eqn. (37) that Eqn. (35) is equivalent to the diffusion equation of the form

$$\partial_t \rho = D\Delta\rho - \mathrm{div}(\rho \underline{b}) . \qquad (39)$$

If we denote

$$u^i = 2D\nabla^i R(\underline{x},t) \qquad (40)$$

where

$$R(\underline{x},t) = \ell n(\rho(\underline{x},t)/2\rho_0) \qquad (41)$$

$$\nabla^i = g^{ij}\nabla_j$$

and ρ_0 is an arbitrary constant with the dimension of ρ, and because

$$\underline{v} = \underline{u} + \underline{b} , \qquad (42)$$

then Eqn. (39) can be transformed into the continuity equation

$$\partial_t \rho + \nabla_i(\rho v^i) = 0 \qquad (43)$$

and $\underline{J} = \rho\underline{v}$ is interpreted as the flux of matter diffusing in the manifold M with the diffusion velocity $\underline{v}$.

Generalization of the Nelson procedure on the Riemannian space needs new definitions of the mean left and right derivatives D_+ and D_- along the trajectories of the Ito equation, because we request Eqns (7) - (12) to obey. In this case Equations (13), (16) and (17) can be accepted too.

To define the derivatives along the trajectories of the Ito equation we need to construct in a proper manner the notion of transport of vectors along these trajectories. The most natural definition of transport, the Levi-Civita parallel transport, does not give satisfactory results [16]. In the paper [14] the new definition, which takes into account the so-called geodesic correlations to parallel transport, was proposed. Consequently, (for details, see [14]) we obtain the following formula for the peculiar acceleration (cf. Eqn. (14))

$$a^i = \partial_t v^i + (\underline{v}.\nabla)v^i - (\underline{u}.\nabla)u^i + D\Delta u^i + DR^i_j u^j , \qquad (44)$$

where R^i_j is the Ricci curvature tensor for the Levi-Civita connection Γ^i_{jk} on the Riemannian manifold M. Thus the discussed process of diffusion with the finite relaxation time τ is described by the non-linear system of equations (39) and (42) and (cf. Eqns (16) and (17))

$$\partial_t v^i = -(\underline{v}.\nabla)v^i + (\underline{u}.\nabla)u^i - D\Delta u^i - DR^i_j u^j - \frac{1}{\tau}(v^i - u^i) + \frac{1}{m}F^i , \qquad (45)$$

where τ is given by Eqn. (23). If we now assume additionally that formulae (24) are valid (but where ∇ is the covariant derivative on the Riemannian manifold M) then this system of equations can be transformed into the form (25) but with Δ given by (36) (cf. [14]). Let us note that the Schrödinger-like equation (25) (with Δ defined by Eqn. (36)) does not contain, in contrast to Eqn. (45), the term depending on the curvature tensor. This fact that the governing equation preserves its formal shape in both Euclidean and non-Euclidean spaces reminds us the situation appearing in the Noll's method of description the bodies with defects [17]. This circumstance causes that also analysis of stationary solutions

$$\psi(\underline{x},t) = \psi_0(\underline{x})\exp(-iFt/\hbar) \qquad (46)$$

carried out in our previous paper [5] can be repeated once more without changes. Also without any changes we can repeat the discussion of the dependence of the diffusion coefficient D on the relaxation time

$$D = D(\Lambda_\tau) = k\Lambda_\tau/\zeta$$
$$\Lambda_\tau = \theta/2k\alpha(\tau) , \qquad (47)$$

where $\theta > 0$ is the characteristic energy of a stationary state of a diffusing matter, e.g. $\theta = kT$ for a thermodynamically equilibrium state (k - Boltzmann constant, T - absolute temperature) and $\alpha(t) > 0$ is such a function that

$$\lim_{\substack{\tau\to 0 \\ \zeta = \text{const.}}} \alpha(\tau) = \alpha(0) = \tfrac{1}{2} . \qquad (48)$$

Thus we can repeat also our argumentation

that the Schrödinger-like equation is connected with the existence of stationary states of diffusing matter far from the thermodynamically equilibrium states. So, it suggests that the mentioned equation can be connected with the theory of dissipative structures. It needs however the further generalization of our considerations on the case of the Riemannian manifold M with a boundary (cf. discussion of that problem in [5]).

REFERENCES

1. A Szczepanski, A possible physical model for quantum objects, Bull. Acad. Polon. Sci., Sér. Sci. Techn., 24, 71-73, 1976.
2. E. Nelson, Derivation of the Schrödinger equation from Newtonian mechanics, Phys. Rev. 150, 1079-1085, 1966
3. E. Nelson, Dynamical theories of Brownian motion, Mathematical Notes, Princeton University Press, Princeton, New Jersey, 1967.
4. B. Mielnik, G. Tengstrand, Nelson-Brown motion: some question marks, Int. J. Theor. Phys. 19, 239-250, 1980
5. A. Trzesowski, R. Kotowski, Nonlinear diffusion and Nelson-Brown movement, Int. J. Theor. Phys. 24, 533-556, 1985
6. E. Kröner, Continuum theory of defects, in Physics of Defects, Les Houches Session XXXV, 1980, R. Balian et al. Eds., North-Holland, 1981
7. E. Kröner, Differential geometry of defects in condensed system of particles with only translational mobility, Int. J. Engng. Sci. 19, 1507-1515, 1981
8. E. Kröner, Basic results of the field theory of defects, in Gauge Field Theories of Defects in Solids, E. Kröner, Ed., Conf. Proc., Max-Planck-Institut für Festkörperforschung, Stuttgart, 1982
9. J. L. Klimontovitch, Statistical Physics, Nauka, Moscow, 1982 (in Russian)
10. J. W. Christian, The Theory of Transformations in Metals and Alloys, Pergamon Press, Oxford, 1975. Part I, p.75
11. A. Seeger, Point defects in metals, in Theory of Crystal Defects, F. Kroupa, Ed., Academic, Prague, 1966
12. H. Günther, M. Zórawski, On geometry of point defects and dislocations, Ann. d. Physik 42, 41-46, 1985
13. M. Zórawski, Certain problems in imperfected crystal. Applications of Weyl and Minkowskian geometry, Bull. Acad. Polon. Sci., Sér. Sci. Techn. 13, 313-320, 1965
14. D. Dohrn, F. Guerra, Nelson's stochastic mechanics on Riemannian manifolds, Lett. Nuovo Cimento 22, 121-127, 1978
15. S. A. Moltschanov, Diffusion processes and Riemannian geometry, Usp. Fiz. Nauk. 30, 3-59, 1975
16. T. G. Dankel, Jr., Mechanics on manifolds and the incorporation of spin into Nelson's stochastic mechanics, ARMA 37, 192-221, 1971
17. C. C. Wang, Mathematical Principles of Mechanics and Electromagnetism, Part A: Analytical and Continuum Mechanics, Plenum Press, New York, 1979

5th International Symposium on Continuum Models of Discrete Systems / Nottingham / 14-20 July 1985

Jump conditions in magnetodiffusion in solids

J.Stefaniak
Technical University of Poznań, Poland

ABSTRACT: In this report a simple case of diffusion of uncharged components in a solid consisting of two parts of different mechanical, thermal and electromagnetic properties is considered. The solid is assumed to be a magnetic and a conductor of electricity. On the basis of laws of balance of mass, balance of energy, Maxwell equations, second law of thermodynamics and phenomenological equations the equations of motion, heat conduction equation and equations of diffusion, as well as jump conditions in the considered body are derived. Finally, the equations and conditions in linear approximation and a particular case of a solid surrounded by a one-component gas are presented.

1 INTRODUCTION

The equations, which describe the interaction of electromagnetic, thermal and mechanical fields are presented in works of many authors. Also the problems of diffusion in solids have been investigating for a long time. In last years have appeared some papers devoted to diffusion in solids with respect of action of the mentioned fields. In several papers are also presented, in general form, the jump conditions concerning these problems. In the present report we consider a particular case of diffusion in solid. We assume

1. The physical properties of the body vary jump-like at the surface S, which devides the body in two parts, V_1 and V_2. ∂V denotes the boundary of the body.
2. The body is a magnetic and a conductor of electricity.
3. In the body the diffusion of electrically insensitive particles, which are also chemically neutral /electrically and chemically neutral gases/ is possible.

The start point of our considerations are balance equations, Maxwell equations, second law of thermodynamics, invariance under superposed rigid-body motion. Particulary, S may be identical with the boundary of the body ∂V. Then, some jump conditions pass into the boundary conditions.

2 BALANCE OF MASS

We take under consideration the difussion of n gas-components in the solid. Therefore we can assume, that

$$\rho^k \ll \rho^o \quad , \quad k = 1,2,\ldots,n,$$

where: ρ^k - density of component k,

ρ^o - density of the body, i.e. $\rho^o = \rho_o^1$ in V_2 and $\rho^o = \rho_o^2$ in V_1. For both indices the inequality is valid. As consequence we can assume

$$\rho = \sum_{k=0}^{n} \rho^k \approx \rho^o \ .$$

The law of conservation of mass for each of components takes the form /de Groot and Mazur 1962/

$$\int_V \frac{\partial \rho^k}{\partial t} dV' + \int_{\partial V} \rho^k v_i n_i d\,\partial V' = 0, \qquad /1/$$
$$k=0,1,\ldots,n$$

where V' denotes an arbitrary material volume, $V' \subset V$. The global equations of conservation of mass /1/ are equivalent to the local equations

$$\frac{\partial \rho^k}{\partial t} + (\rho^k v_i^{\,k})_{,i} = 0, \quad k=o,1,2,\ldots,n \ /2/$$

If the concentrations /mass fractions/ $c^k = \rho^k/\rho$ are employed, equations /2/ take the forms

$$\rho^k \frac{dc^k}{dt} = -\eta^k_{i,i} , \qquad /3/$$

where $\eta^k_i = \rho^k (v^k_i - v^s_i)$ denotes diffusion flow of component k, $v^s_i = \sum_{k=o}^{n} \rho^k v^k_i / \rho$ is the barycentric velocity. We can postulate $v^s_i \approx v^o_i = v_i$, where v_i are components of the velocity of solid. Then, for k=o /solid/ we have

$$\eta^o_i \approx \rho(v^o_i - v_i) \equiv 0 .$$

From relation /1/ there follow also the jump conditions on the surface S:

$$[\![\eta^k_i n_i]\!] = 0 \qquad /4/$$

3 MAXWELL EQUATIONS

The electromagnetic phenomena are described by the following equatiors of balance in the global form /Van de Ven 1975/

$$\frac{d}{dt} \int_S B_i n_i dS = - \oint_C (E_i + e_{ijk} v_j B_k) dl_i ,$$

$$\oint_S B_i n_i dS = 0 ,$$

$$\frac{d}{dt} \int_S D_i n_i dS = \oint_C (H_i - e_{ijk} v_j D_k) dl_i +$$

$$- \int_S (j_i - \rho_e v_i) n_i dS,$$

$$\oint_S D_i n_i dS - \int_V \rho_e dV = 0,$$

$$\frac{d}{dt} \int_V \rho_e dV = - \oint_S (j_i - \rho_e v_i) n_i dS,$$

where: E_i - electric field density, D_i - electric displacement, j_i - electric current density, H_i - magnetic field density, B_i - magnetic induction, ρ_e - free charge.

From these equations we obtain
- the system of local balance equations /Maxwell equations/

$$\frac{\partial B_i}{\partial t} = - e_{ijk} E_{k,j}, \quad B_{i,i} = 0$$

$$\frac{\partial D_i}{\partial t} + j_i = e_{ijk} H_{k,j}, \quad D_{i,i} = 0 \qquad /5/$$

$$\frac{\partial \rho_e}{\partial t} + j_{i,i} = 0 ,$$

- jump conditions on the surface S

$$[\![e_{ijk} E_j n_k + B_i v_k n_k]\!] = 0,$$

$$[\![e_{ijk} H_j n_k - D_i v_k n_k]\!] = 0, \qquad /6/$$

$$[\![D_i n_i]\!] = 0, \quad [\![B_i n_i]\!] = 0,$$

$$[\![j_i n_i - \rho_e v_i n_i]\!] = 0.$$

According to the assumption of conductivity of the body we have

$$D_i = \varepsilon_o E_i , \quad B_i = \mu_o (H_i + M_i), \qquad /7/$$

where: μ_o - permeability of free space, ε_o - permittivity of free space, M_i - - magnetization per unit of mass.

4 BALANCE OF ENERGY

The global equations of balance of energy may be postulated in the following form /Parkus 1972, Stefaniak 1982/:

$$\frac{d}{dt} \int_{V'} [\rho (U + \frac{v^2}{2}) + U_e] dV' = \int_{V'} X_i v_i dV' +$$

$$+ \int_{\partial V'} (\sigma_{ij} v_j - e_{ijk} E_j H_k - q_i + \qquad /8/$$

$$- \sum_{k=1}^{n} \mu^k \eta^k_i + U_e v_i) n_i d\partial V',$$

where: U specific internal energy /internal energy per unit of mass/, U_e - electromagnetic energy density, X_i - body force, σ_{ij} - stress tensor, q - heat flux, μ^k - chemical potential of component k.

Applying the Gauss theorem we can write equation /8/ in the form

$$\int_{V'} (\rho v_i \dot{v}_i + \rho \frac{dU}{dt} + \frac{\partial U_e}{\partial t}) dV' = \int_{V'} [X_i v_i +$$

$$+ (\sigma_{ij} v_j)_{,i} - e_{ijk} (E_j H_k)_{,i} - q_{i,i} +$$

$$- \sum_{k=1}^{n} \mu^k_{,i} \eta^k_i - \sum_{k=1}^{n} \mu^k \eta^k_{i,i} \; dV'$$

or in local form

$$\rho \frac{dU}{dt} + \frac{\partial U_e}{\partial t} + (\rho \dot{v}_i - X_i) v_i +$$

$$-(\sigma_{ij} v_j)_{,i} + e_{ijk} (E_j H_k)_{,i} + q_{i,i} + \qquad /9/$$

$$+ \sum_{k=1}^{n} \mu^k_{,i} \eta_i + \sum_{k=1}^{n} \mu^k \eta^k_{i,i} = 0.$$

Let us postulate, that the specific free energy F = U - TS depends on

- gradient of deformation $x_{i,A} = \frac{\partial x_i}{\partial X_A}$, where x_i and X_A are spatial and material coordinates, respectivelly
- temperature T
- concentrations /mass fractions/ c^k, $k=1,2,\ldots,n$
- specific magnetization m_i :

$F = F(x_{i,A}, T, c^k, m_i)$,

S denotes the specific entropy.

Then on the one hand we can write

$$\rho\frac{dF}{dt} = \rho\frac{\partial F}{\partial x_{i,A}}\dot{x}_{i,A} + \rho\frac{\partial F}{\partial T}\dot{T} + \rho\sum_{k=1}^{n}\frac{\partial F}{\partial c^k}\dot{c}^k + \rho\frac{\partial F}{\partial m_i}\dot{m}_i, \qquad /10/$$

and on the other hand

$$\frac{dF}{dt} = \frac{dU}{dt} - T\dot{S} - S\dot{T} \qquad /11/$$

Let us notice, that

$$\rho\frac{\partial F}{\partial x_{i,A}}\dot{x}_{i,A} = \rho\frac{\partial F}{\partial x_{i,A}}\dot{x}_{i,j}x_{j,A} = \left(\rho\frac{\partial F}{\partial x_{i,A}}x_{j,A}v_i\right)_{,j} - \left(\rho\frac{\partial F}{\partial x_{i,A}}x_{j,A}\right)_{,j}v_i \qquad /12/$$

Combining /9/ - /12/ we obtain

$$\rho T\dot{S} + \rho S\dot{T} + \rho\frac{\partial F}{\partial T}\dot{T} + \left(\rho\frac{\partial F}{\partial x_{i,A}}x_{j,A}\right)_{,j}v_i + (\rho\dot{v}_i - X_i)v_i + \left(\rho\frac{\partial F}{\partial x_{j,A}}x_{j,i} + {}\right.$$
$$- (\sigma_{ij}v_j)_{,i} + e_{ijk}(E_jH_k)_{,i} + q_{i,i} + \frac{\partial U_e}{\partial t} + \rho\frac{\partial F}{\partial m_i}\dot{m}_i + \sum_{k=1}^{n}\mu^k_{,i}\eta^k_i + \sum_{k=1}^{n}\mu^k\eta^k_{i,i} + \rho\sum_{k=1}^{n}\frac{\partial F}{\partial c^k}\dot{c}^k = 0. \qquad /13/$$

Now let us consider the term $\frac{\partial U_e}{\partial t}$. We have

$U_e = 1/2\ (\varepsilon_o E^2 + \mu_o H^2)$.

Applying the Maxwell equations /5/ and relations /7/ we find

$$\frac{\partial U_e}{\partial t} = \varepsilon_o E_i\frac{\partial E_i}{\partial t} + \mu_o H_i\frac{\partial H_i}{\partial t} = - e_{ijk}(E_iH_k)_{,i} - E_ij_i - \mu_o H_i\frac{\partial M_i}{\partial t}. \qquad /14/$$

Taking into account the relation between specific magnetization and magnetization per unit of mass: $\rho m_i = M_i$, mass continuity equation

$$\frac{\partial\rho}{\partial t} = -(\rho v_i)_{,i}$$

and the relation

$$\frac{dm_i}{dt} = \frac{\partial m_i}{\partial t} + m_{i,j}v_j$$

we obtain

$$H_i\frac{\partial M_i}{\partial t} = \rho H_i\frac{dm_i}{dt} - (H_im_i\rho v_k)_{,k} + \rho m_iH_{i,k}v_k.$$

Then, the equation /14/ may be written in the form

$$\frac{\partial U_e}{\partial t} = - e_{ijk}(E_jH_k)_{,i} - \mu_o\rho H_i\frac{dm_i}{dt} + \mu_o(H_jm_j\rho v_i)_{,i} - \mu_o m_jH_{j,i}v_i - E_ij_i. \qquad /15/$$

Using relations /3/ and /15/ we can finally write the equation /13/ in the form

$$\rho\left(S + \frac{\partial F}{\partial T}\right)\dot{T} + \Big[\rho\dot{v}_i - X_i - \left(\rho\frac{\partial F}{\partial x_{i,A}}x_{j,A}\right)_{,j} - \mu_o\rho m_jH_{j,i} + \left(\rho\frac{\partial F}{\partial x_{i,A}}x_{j,A}\right)_{,j} - \sigma_{ji,j} + \mu_o(\rho H_km_k\delta_{ij})_{,j}\Big]v_i + \left(\rho\frac{\partial F}{\partial x_{i,A}}x_{j,A} + {}\right.$$
$$\left. - \sigma_{ji} + \mu_o\rho H_km_k\delta_{ij}\right)v_{i,j} + \rho\left(\frac{\partial F}{\partial m_i} - \mu_o H_j\right)\dot{m}_i + \rho\sum_{k=1}^{n}\left(\frac{\partial F}{\partial c^k} - \mu^k\right)\dot{c}^k + \rho T\dot{S} + q_{i,i} - E_ij_i + \sum_{k=1}^{n}\mu^k_{,i}\eta^k_i = 0. \qquad /16/$$

From the global form of balance of energy /8/ we also obtain, in the usual way /Truesdell and Toupin 1960/ the jump conditions on the surface S

$$\Big[\sigma_{ij}v_jn_i - e_{ijk}E_jH_kn_i - q_in_i - \sum_{k=1}^{n}\mu^k\eta^k_in_i\Big] = 0. \qquad /17/$$

5 EQUATIONS OF MOTION AND HEAT CONDUCTION EQUATIONS

Assuming the invariance under superposed rigid-body motion we obtain from equation /16/
- equations of motion

$$\left(\rho\frac{\partial F}{\partial x_{i,A}}x_{j,A}\right)_{,j}+\mu_0 M_j H_{j,i}+X_i=\rho\dot{v}_i \qquad /18/$$

- constitutive equations

$$\sigma_{ji}=\rho\frac{\partial F}{\partial x_{i,A}}x_{j,A}+\mu_0 M_k H_k\delta_{ji} \qquad /19/$$

and from jump condition /17/

$$[\![\sigma_{ij}n_i]\!]=0,$$

$$[\![e_{ijk}E_j H_k n_i+q_i n_i-\sum_{k=1}^{n}\mu^k\eta_i^k n_i]\!]=0. \qquad /20/$$

The functions T, c^k, m_i are independent variables, since the coefficients at $\dot{T}$, $\dot{c}_k$, $\dot{m}_i$, as well as the free term, in equation /16/ must vanish. This yields to
- next constitutive equations

$$H_i=\frac{1}{\mu_0}\frac{\partial F}{\partial m_i},\quad \mu^k=\frac{\partial F}{\partial c^k},\quad S=-\frac{\partial F}{\partial T} \qquad /21/$$

- heat conduction equation

$$\rho T\dot{S}=-q_{i,i}-\sum_{k=1}^{n}\mu^k_{,i}\eta_i^k+E_i j_i. \qquad /22/$$

$\rho\frac{\partial F}{\partial x_{i,A}}x_{j,A}$ is the Cauchy's stress tensor.

6 PHENOMENOLOGICAL EQUATIONS

The heat conduction equation /22/ one can write in the form

$$\rho\dot{S}=-(q_i/T)_{,i}+\Omega,$$

where

$$\Omega=-\frac{q_i}{T^2}+\frac{j_i E_i}{T}-\sum_{k=1}^{n}\frac{\eta_i}{T}\mu^k_{,i} \qquad /23/$$

is the entropy produation. This corresponds to the general expression for entropy production /de Groot and Mazur/

$$\Omega=\frac{1}{T}\sum_{l=1}^{m}\underset{\sim}{J}_l\underset{\sim}{X}_l,$$

where $\underset{\sim}{J}_l$ are independent fluxes and $\underset{\sim}{X}_i$ are independent thermodynamic forces. In general, the fluxes are complex fuctions of forces

$$\underset{\sim}{J}_l=\underset{\sim}{J}_l(\underset{\sim}{X}_1,\dots,\underset{\sim}{X}_m),$$

but we postulate them in the linear form

$$\underset{\sim}{J}_l=\sum_{j=1}^{m}L_{lj}\underset{\sim}{X}_j. \qquad /24/$$

In our case we have

$$\underset{\sim}{J}_1=\underset{\sim}{q},\quad \underset{\sim}{J}_2=\underset{\sim}{j},\quad \underset{\sim}{J}_3=\eta^1,\dots,\underset{\sim}{J}_{n+2}=\eta^n,$$

$$\underset{\sim}{X}_1=-\frac{1}{T}\nabla T,\quad \underset{\sim}{X}_2=\underset{\sim}{E},\quad \underset{\sim}{X}_3=-\nabla\mu^1,$$

$$\dots\ \underset{\sim}{X}_{n+2}=-\nabla\mu^n$$

Then, the relations /24/ take the form

$$q_i=-\frac{L_{11}}{T}T_{,i}+L_{12}E_i-\sum_{k=1}^{n}L_{1\,2+k}\mu^k_{,i}$$

$$j_i=-\frac{L_{21}}{T}T_{,i}+L_{22}E_i-\sum_{k=1}^{n}L_{2\,2+k}\mu^k_{,i},$$

$$\eta^1_i=-\frac{L_{31}}{T}T_{,i}+L_{32}E_i-\sum_{k=1}^{n}L_{3\,2+k}\mu^k_{,i},$$

$$\dots \qquad /25/$$

$$\eta^n_i=-\frac{L_{n+2\,1}}{T}T_{,i}+L_{n+2\,2}E_i+$$

$$-\sum_{k=1}^{n}L_{n+2\,2+k}\mu^k_{,i}.$$

These relations state the generalization of Fourier's, Fick's and Ohm's laws. It is remarkable, that the Ohm's law was not assumed before, but is a result of thermodynamic considerations. Now, with help of equations /25/, the equation /23/ may be written in the form

$$T\Omega=\frac{T_{,i}}{T}\Big(L_{11}\frac{T_{,i}}{T}-L_{12}E_i+$$

$$+\sum_{k=1}^{n}L_{1\,2+k}\mu^k_{,i}+E_i\Big(-L_{21}\frac{T_{,i}}{T}+$$

$$+L_{22}E_i-\sum_{k=1}^{n}L_{2\,2+k}\mu^k_{,i}+$$

$$+\mu^1_{,i}\Big(L_{31}\frac{T_{,i}}{T}-L_{32}E_i+ \qquad /26/$$

$$+\sum_{k=1}^{n}L_{3\,2+k}\mu^k_{,i}+\dots+$$

$$+\mu_{,i}^{n}\left(L_{n+2\ 1}\frac{T_{,i}}{T}-L_{n+2\ 2}E_i+\right.$$ /26/ cont

$$+\sum_{k=1}^{n} L_{n+2\ 2+k}\overset{k}{\mu}_{,i}\ .$$

According to the second law of thermodynamic $\Omega \geqslant 0$. Ω is zero for reversible processes and positive for irreversible processes, so the quadratic form on the right-hanside of equation /26/ is non-negative defined. Therefore for irreversible processes the coefficients satisfy the conditions

$$\begin{vmatrix} L_{pp} & \frac{1}{2}\left(L_{pq}+L_{qp}\right) \\ \frac{1}{2}\left(L_{qp}+L_{pq}\right) & L_{qq} \end{vmatrix} > 0,$$

$L_{pp} > 0$, for $p,q = 1,2,\dots,n+2$.

7 LINEAR APPROXIMATION

We have assumed the free energy F as an arbitrary function of $x_{j,A}$, T, m_i, c^k. Let us expand it in the power series at the neighbourhood of equilibrium state, described by $T = T_o$, $x_{i,A} = 0$, $m_i = 0$, $c^k = 0$. We restrict the terms to the second powers. The free energy at the equilibrium state reachs minimum, thus, the coefficients at linear terms vanish. Furthermore we assume, that the strain is sufficiently small. Then we can write

$$\frac{\partial F}{\partial x_{i,A}}x_{j,A} \approx \frac{\partial F}{\partial u_{i,j}}, \quad \text{and}$$

$$E_{ij} \approx e_{ij} \approx 1/2\left(u_{i,j}+u_{j,i}\right), \qquad /27/$$

where: E_{ij} - Green's strain tensor, e_{ij} - Almansi's strain tensor.

Then, the free energy depends on $u_{i,j}$, T, m_i, c^k. Regarding the symmetry of strain tensor $e_{ij} = e_{ji}$ we can introduce as independent variables e_{ij} instead of $u_{i,j}$. So, for an isotropic body, but with properties varying jump-like at the surface S, we can write

$$\overset{\alpha}{\varrho}\overset{\alpha}{F} = \overset{\alpha}{\varrho}\overset{\alpha}{F}_o + \overset{\alpha}{\mu}\overset{\alpha}{e}_{ij}\overset{\alpha}{e}_{ij} + \overset{\alpha}{\lambda}/2\,\overset{\alpha}{e}_{pp}\overset{\alpha}{e}_{qq} +$$

$$-\overset{\alpha}{\gamma}_o\overset{\alpha}{e}_{pp}\overset{\alpha}{\theta} - \overset{\alpha}{e}_{pp}\sum_{k=1}^{n}\overset{\alpha}{\gamma}{}_c^k\,\overset{\alpha}{c}{}^k +$$

$$+\overset{\alpha}{\varrho}/2\sum_{k=1}^{n}\sum_{l=1}^{n}\overset{\alpha}{a}{}^{kl}\overset{\alpha}{c}{}^k\overset{\alpha}{c}{}^l - \overset{\alpha}{n}/2\,\overset{\alpha}{\varrho}\overset{\alpha}{\theta}{}^2 + \qquad /28/$$

$$-\overset{\alpha}{\varrho}\theta\sum_{k=1}^{n}\overset{\alpha}{d}{}^k\overset{\alpha}{c}{}^k - \frac{\overset{\alpha}{\alpha}}{2}m_pm_p\ ,$$

where $\theta = T - T_o$, $\alpha = 1,2$.

The index α denotes, that all quantities /coefficients and functions/ may be different in each part of the body. Substituting /28/ into /18/, /19/ and /21/ we find

$$\sigma^*_{ij,j} + \mu_oM_jH_{j,i} + X_i = \varrho\dot{v}_i\ , \qquad /29/$$

$$\sigma_{ij} = \sigma^*_{ij} + \mu_o M_kH_k\delta_{ij} = 2\mu e_{ij} +$$

$$+\lambda e_{pp} - \gamma_\theta\,\theta - \sum_{k=1}^{n}\gamma_c^k\,c^k + \mu_o M_pH_p\delta_{ij},$$

$$\mu^k = -\frac{\gamma_c^k}{\varrho}e_{pp} - d^k\theta + \sum_{k=1}^{n}a^{kl}c^l, \qquad /30/$$

$$S = \frac{\gamma_\theta}{\varrho}e_{pp} + n\theta + \sum_{k=1}^{n}d^k c^k\ ,$$

$$H_i = \frac{\alpha}{\mu_o}M_i\ ,$$

$$\text{where}\quad \sigma^*_{ij} = \varrho\frac{\partial F}{\partial e_{ij}}\ .$$

In these relations the index α is omited, but we have to remember, that for each part of the body /on the one handside of the surface S and on the other handside of it/ the coefficients and the functions can be different. Now let us combine /22/, /25/, $/30/_2$ and $/30/_3$. We find

$$\gamma_\theta\,\dot{e}_{pp} + \varrho n\dot{\theta} + \varrho\sum_{k=1}^{n}d^k\dot{c}^k =$$

$$= \frac{L_{11}}{T}\left(\frac{\theta_{,i}}{T}\right)_{,i} - \frac{L_{12}}{T}E_{i,i} + \qquad /31/$$

$$-\sum_{k=1}^{n}\frac{L_{1\ 2+k}}{T}\left(\frac{\gamma_c^k}{\varrho}\nabla^2e_{pp} + d^k\nabla^2\theta +\right.$$

$$\left.-\sum_{l=1}^{n}a^{lk}\nabla^2c^l - \dots\ ,\right.$$

Considering the equations of conservation of mass /3/, with help of $/25/_3$

and /30/$_2$ we obtain

$$\rho\dot{c}^k = L_{2+k\;1}\left(\frac{\theta_{,i}}{T}\right)_{,i} - L_{2+k\;2}\,E_{i,i} +$$
$$-\sum_{l=1}^{n} L_{2+k\;2+l}\,\frac{\gamma_c}{\rho}\,\nabla^2 e_{pp} + d^l\,\nabla^2\,\theta +$$
$$-\sum_{m=1}^{n} a^{ml}\,\nabla^2\,c^m\Big),\qquad k=1,2,\dots,n. \qquad /32/$$

The free energy F we have expanded in power series at the neighbourhood of the temperature T_o. Assuming, that $\theta/T_o \ll 1$ we replace in equations /31/ and /32/ T by T_o. Furthermore, omiting the nonlinear terms in equation /31/ we obtain

$$\gamma_0\,\dot{e}_{pp} + \rho n\dot{\theta} + \rho\sum_{k=1}^{n} d^k\dot{c}^k = \frac{L_{11}}{T_o^2}\,\nabla^2\theta +$$
$$-\frac{L_{12}}{T_o}\,E_{i,i} - \sum_{k=1}^{n}\frac{L_{1\;2+k}}{T_o}\Big(\frac{\gamma_c^k}{\rho}\,\nabla^2 e_{pp} +$$
$$+\, d^2\,\nabla^2\,\theta - \sum_{l=1}^{n} a^{lk}\,\nabla^2\,c^l\Big). \qquad /33/$$

Using relations /30/$_3$, /30/$_4$ and /25/ we can eliminate the functions S, M_i, μ_k, q_i, j_i, η_i^k. Then, the functions to be find are E_i, H_i, B_i, D_i, ρ_e, θ, c^k, σ_{ij}, e_{ij}, u_i. To find them, we have the following sets of equations and jump conditions

1. Maxwell equations /5/ and relations /7/
2. Equations of motion /29/, relations /7/
3. Heat conduction equation /33/
4. Set of equations of diffusion /32/
5. Jump conditions /4/, /6/ and /20/.

If we consider a body surrounded by one - component gas /n=1/ of a constant concentration /$\overset{2}{c}$=1 and constant temperature /$\overset{2}{\theta}$=const/ and assume the surroundings as a free space from the electromagnetic point of view, our equations reduce to

$$\frac{\partial B_i}{\partial t} = -\,e_{ijk}\,E_{k,j}\,,\qquad B_{i,i} = 0,$$

$$\frac{\partial D_i}{\partial t} = e_{ijk}\,H_{k,j} - j_i\,,\qquad D_{i,i} = {}_e\,,$$

$$\frac{\partial \rho_e}{\partial t} + j_{i,i} = 0,$$

$$\frac{\partial \overset{2}{B}_i}{\partial t} = -\,e_{ijk}\,\overset{2}{E}_{k,j}\,,\qquad \overset{2}{B}_{i,i} = 0,$$

$$\frac{\partial \overset{2}{D}_i}{\partial t} = e_{ijk}\,\overset{2}{H}_{k,j}\,,\qquad \overset{2}{D}_{i,i} = 0\,.$$

$$\left(\sigma^*_{ji}\right)_{,j} + \mu_o\,M_j\,H_{j,i} + X_i = \rho\dot{v}_i\,,$$

$$\sigma_{ij} = 2\mu e_{ij} + \lambda e_{pp} - \gamma_0\,\theta - \gamma_c\,c +$$
$$+\,\mu_o\,M_l\,H_l\,\delta_{ij}\,,$$

$$e_{ij} = \frac{1}{2}\left(u_{i,j} + u_{j,i}\right),$$

$$H_i = \frac{\alpha}{\mu_o}\,M_i\,,\qquad \sigma^*_{ij} = \rho\,\frac{\partial F}{\partial e_{ij}}\,,$$

$$\rho\dot{c} = \frac{L_{31}}{T_o}\,\nabla^2\,\theta - L_{32}\,E_{i,i} +$$
$$-\,L_{33}\left(\frac{\gamma_c}{\rho}\,\nabla^2\,e_{pp} + d\,\nabla^2\,\theta - a\,\nabla^2\,c\right),$$

$$\gamma_0 e_{pp} + \rho n\,\theta + \rho dc = \frac{L_{11}}{T_o^2}\,\nabla^2\,\theta +$$
$$-\frac{L_{12}}{T_o}\,E_{i,i} - \frac{L_{13}}{T_o}\left(\frac{\gamma_c}{\rho}\,\nabla^2 e_{pp} + d\,\nabla^2\theta - a\nabla^2 c\right),$$

$$[\![e_{ijk}\,E_j\,n_k + B_i\,v_k\,n_k]\!] = 0\,,$$

$$[\![e_{ijk}\,H_j\,n_k - D_i\,v_k\,n_k]\!] = 0\,,$$

$$[\![D_i\,n_i]\!] = 0\,,\qquad [\![B_i\,n_i]\!] = 0\,,$$

$$[\![j_i\,n_i + \rho_e\,v_i\,n_i]\!] = 0\,,\qquad \sigma_{ij}\,n_j = 0,$$

$$[\![e_{ijk}\,E_j\,H_k\,n_i]\!] + [\![q_i\,n_i]\!] - [\![\mu^1 \eta_i]\!] n_i = 0,$$

$$[\![e_{ijk}E_jH_kn_i]\!] + \Big[\overset{2}{L}_{12}\overset{2}{E}_i + \frac{L_{11}}{T_o}\,\theta_{,i} - L_{12}E_i +$$
$$+\,L_{13}\Big(\frac{\gamma_c}{\rho}\,e_{pp} - d\theta + ac\Big)_{,i}\Big]\,n_i - \Big(-\overset{2}{d}\,\overset{2}{\theta} +$$
$$+\,\overset{2}{a}\overset{2}{c} + \frac{\gamma_c}{\rho}\,e_{pp} + d\theta - ac\Big)\Big[-\frac{L_{31}}{T_o}\,\theta_{,i} + L_{32}E_i +$$
$$-\,L_{33}\Big(-\frac{\gamma_c}{\rho}\,e_{pp} - d\theta + ac\Big)\Big]\,n_i = 0$$

In these relations, the functions and

coefficients without top indices refer to the body and with indices 2 to the surroundings.

The conditions, which refer to electromegnetic field are jump conditions, but refering to stress, temperature of the body and concentration inside of the body are, under assumed conditions /$\overset{2}{c}$ = const, $\overset{2}{\theta}$ = const/, boundary conditions. One can notice, that not only the equations which describe the considered phenomena, but also the jump and boundary conditions are coupled.

REFERENCES

de Groot S.R. and P.Mazur 1962. Non--equilibrium thermodynamics. Amsterdam: North-Holland Publishing Company.

Parkus H. 1972. Magneto-thermoelasticity. Wien-New York: Springer Verlag.

Stefaniak J. 1982. The influence of electromagnetic field on thermodiffusion in an izotropic solid /in polish/. Warszawa-Poznań: PWN.

Truesdell C. and R.A. Toupin 1960. Handbuch der Physic B.III/1, 226-790. Berlin-Göttingen-Heidelberg: Springer -Verlag.

Ven A.A.F van de 1975. Interaction of electromagnetic and elastic fields in solids. University of Technology Eindhoven.

5th International Symposium on Continuum Models of Discrete Systems / Nottingham / 14-20 July 1985

Latent microstructure

G.Capriz
University of Pisa, Italy

ABSTRACT: An appropriate concept of latent microstructure may help in explaining deviations from standard thermomechanical behaviour of continua.

1. INTRODUCTION

Proposals for changes in (or generalizations of) the standard form of the balance equations of continuum thermodynamics are numerous.

It has been suggested, for instance, that:

(i) The Cauchy stress tensor need not be symmetric;

(ii) the entropy flux need not be equal to the heat flux divided by the absolute temperature or, rather, the energy flux, stress power apart, need not be equal to the heat flux;

(iii) certain requirements of objectivity may have to be relaxed.

In a recent paper (Capriz 1985) I have shown that a concept of 'latent' microstructure may provide a background within which a certain incompatibility between constitutive prescriptions involving displacement gradients of higher order and the Clausius-Duhem inequality can be circumvented.

I show here that, within the same background, also assertions (i), (ii), (iii) may apply naturally. In the quoted paper the concept is related to a general definition of microstructure introduced earlier. Here I would like to avoid back references; I base my considerations on the special and well accepted case of the one-director theory (which is traditionally the starting point for a continuum description of the behaviour of liquid crystals, for instance); on the other hand, I exploit certain ideas even more radically.

2. BALANCE EQUATIONS

I begin by writing the balance equations for mass, momentum, moment of momentum, energy and entropy in an ordinary continuum, so that the notation adopted is clear and the contrast with the equations which are quoted later and are valid in a more general context is immediate:

$$\dot{\rho} + \rho \operatorname{div} v = 0 , \qquad (2.1)$$

$$\rho \dot{v} = \rho b + \operatorname{div} T , \qquad (2.2)$$

$$\operatorname{skw} T = 0 , \qquad (2.3)$$

$$\rho \dot{\varepsilon} = T \cdot \operatorname{grad} v + \operatorname{div} q + \rho \chi , \qquad (2.4)$$

$$\rho \dot{\eta} = \operatorname{div} \left(\frac{q}{\theta}\right) + \rho \frac{\chi}{\theta} ; \qquad (2.5)$$

in (2.3) 'skw' stands for 'antisymmetric part of'.

The alternative case I consider here is when the microstructure can be described by a vector field, the director d. Such is the case in certain crystalline bodies when the director is used to specify polarization and also in liquid crystals (where, however, d is usually required to be of unit length). Then the equations (2.1), (2.2) and (2.5) do not change, whereas the microstructure influences the balances of moment of momentum and of energy; besides, an equation of balance for micromomentum is required. The latter equation mirrors the Cauchy equation (2.2) with a body force c, a microstress V and an inertia per unit mass $\gamma\ddot{d}$ (where γ is an appropriate constant):

$$\rho\gamma\ddot{d} = \rho c + \operatorname{div} V - z \,. \qquad (2.6)$$

In (2.6), also a resultant internal force -z per unit volume appears, due to internal actions because such actions on the microstructure need not be separately balanced. The power density per unit volume of microstresses turns out to be equal to $V \cdot \operatorname{grad}\dot{d} + z \cdot \dot{d}$, and the equation of balance of energy must be altered through the addition of such a term; thus eqn (2.4) is substituted by

$$\rho\dot{\varepsilon} = T \cdot \operatorname{grad} v + V \cdot \operatorname{grad}\dot{d} + z \cdot \dot{d} + \operatorname{div} q + \rho\chi \,. \qquad (2.4)'$$

In classical continua eqn (2.3) is equivalent to the condition that the internal power density vanish for all rigid body velocity distributions. As in our context that density equals the sum of the first three terms in the right-hand side of (2.4'), and, in a rigid body motion with angular velocity ω, one has $\operatorname{grad} v = e\,\omega$ (e, Ricci's third order permutation tensor) and $\dot{d} = (e\,d)\omega$, eqn (2.3) must be substituted by

$$\operatorname{skw} T = \operatorname{skw}\,(d \otimes z + (\operatorname{grad} d)V^T). \qquad (2.3)'$$

3. INTERNAL CONSTRAINT

The idea in (Capriz 1985) is to think of situations where the microstructure (represented here by the vector d) becomes, in some sense, latent. A purely formal step in this direction is to cross out from the list the extra equation (2.6) by using it to eliminate z, so that (2.3)', (2.4)' change into

$$\operatorname{skw} T = \operatorname{skw}\,(\rho d \otimes (c - \gamma\ddot{d}) + \operatorname{div}(d \otimes V)), \qquad (2.3)''$$

$$\rho\dot{\tilde{\varepsilon}} = T \cdot \operatorname{grad} v + \operatorname{div}\tilde{q} + \rho\tilde{\chi} \,, \qquad (2.4)''$$

where

$$\tilde{\varepsilon} = \varepsilon + \tfrac{1}{2}\gamma\,\dot{d}^2 \,,$$

$$\tilde{q} = q + V^T\dot{d} \,, \qquad (3.1)$$

$$\tilde{\chi} = \chi + c \cdot \dot{d} \,.$$

Now, from a purely formal point of view, the set of equations (2.1), (2.2), (2.3)", (2.4)", (2.5) appears as a set of balance equations not dissimilar from (2.1)÷(2.5) but where the possibilities (i), (ii), (iii) quoted in the introduction may be realized.

The second, substantial, step is to imagine the microstructure constrained to follow the macromotion; this constraint can be rendered explicit in many different ways. Here I explore the consequences of a very special choice; alternative choices could be easily dealt with, at the cost of similar manipulations. The choice is

$$d = F\,d^* \,, \qquad (3.2)$$

where F is the position gradient in the macromotion and d* is a vector field defined on the reference placement.

The consequences are

$$T^c + z^c \otimes d + V^c(\operatorname{grad} d)^T = 0,$$

$$d_i V^c_{jk} + d_k V^c_{ji} = 0 \,, \qquad (3.3)$$

where latin indices indicate Cartesian components.

It follows further that

$$\operatorname{skw} T^c = \operatorname{skw}\,(d \otimes z^c + (\operatorname{grad} d)(V^c)^T),$$

$$\operatorname{div} T^c = \operatorname{div}\,((-z^c + \operatorname{div} V^c) \otimes d),$$

with the conclusion that the balance equations for moment of momentum (2.3)' and energy (2.4)' apply with the substitution of T^a, V^a, z^a for T, V, z respectively.

The equations of balance of momentum and micromomentum together give also rise to a single 'pure' equation (i.e., an equation where only active stresses appear):

$$\rho\dot{v} = \rho b + \operatorname{div}(T^a + (z^a - \operatorname{div} V^a + \rho(\gamma\ddot{d} - c)) \boxtimes d). \tag{3.4}$$

If we define now a global Cauchy stress $\tilde{T}$ as

$$\tilde{T} = T^a + (z^a - \operatorname{div} V^a + \rho(\gamma\ddot{d} - c)) \boxtimes d),$$

then (3.4) takes up again the standard form

$$\rho\dot{v} = \rho b + \operatorname{div}\tilde{T}\ ; \tag{2.2)'}$$

the same occurs for the energy equation

$$\rho\dot{\tilde{\varepsilon}} = \tilde{T}\cdot \operatorname{grad} v + \operatorname{div}\tilde{q} + \rho\tilde{\chi}\ , \tag{2.4)'''}$$

where $\tilde{\varepsilon}$, $\tilde{q}$, $\tilde{\chi}$ are defined as in (3.1), or more precisely as

$$\tilde{\varepsilon} = \varepsilon + \tfrac{1}{2}\gamma((\operatorname{grad} v)d)^2\ ,$$

$$\tilde{q} = q + V^{aT}((\operatorname{grad} v)d),$$

$$\tilde{\chi} = \chi + (\operatorname{grad} v)\cdot(c \boxtimes d)\ .$$

Finally the equation of balance of moment of momentum becomes

$$\operatorname{skw}\tilde{T} = \operatorname{skw}\ (\rho d \boxtimes (c - r(\operatorname{grad}\dot{v})d) + \operatorname{div}(d \boxtimes V^a)). \tag{2.3)'''}$$

Thus, the 'pure' equations of balance in our continuum with latent microstructure are of the type predicated in the introduction.

4. FINAL REMARKS AND ACKNOWLEDGEMENT

Ideas akin to those presented here occur in the literature repeatedly, even if in general only implicitly. I hope readers find appealing the simplicity and directness of the example provided above; of course, the process of reduction could be applied in much more comprehensive circumstances.

The paper reflects work done while the author was visiting the Institute for Mathematics and its Applications at the University of Minnesota. The support of that Institute and of the Italian C.N.R. is acknowledged.

REFERENCE

Capriz, G. 1985. Continua with latent microstructure. Arch. Rat. Mech. Analysis (to appear).

5th International Symposium on Continuum Models of Discrete Systems / Nottingham / 14-20 July 1985

Graph-theoretical approach to the mechanics of granular materials

M.Satake
Tohoku University, Sendai, Japan

ABSTRACT: This paper summarizes the recent studies (Satake 1978a,1978b,1983) of the graph-theoretical approach to the mechanics of granular materials. Firstly the replacement of an assembly of grains to graphs is explained and the relationship between redundancy number, which describes the looseness of packing, and the coodination number is derived. The mechanical quantities and their fundamental equations are given by the aid of fundamental matrices in the graph theory. Lastly, the fabric tensor and branch and contact tensors, which describe the anisotropy of an assembly of grains, are introduced. The definition of stress and strain for granular materials is proposed and the necessity of two stress tensors in granular materials is explained. The so-called induced-anisotropy is analyzed by virtue of the fabric tensor and some results obtained by photoelastic experiments are shown.

1 INTRODUCTION

As a granular material is considered to consist of discrete particles, a discrete treatment for force and deformation may be fundamentally necessary in the study of the mechanics of granular materials.

In this paper, the replacement of an assembly of grains to graphs is proposed and the graph theory is applied to the expressions of mechanical quantities and their fundamental equations. It is of interest to note that the expressions of the fundamental equations by use of the graph theory are very analogous to the corresponding expressions in the mechanics of generalized continua.

For the practical use, however, such discrete expressions of mechanical quantities in granular materials are to be modified into global and continuous ones. For this purpose, some tensors which describe the anisotropy of fabric in an assembly of grains are introduced by virtue of the graph representation and applied to the definition of stress and strain in granular materials. It is observed that two kinds of stress tensor related each other through the contact tensor become necessary in granular materials. The induced-anisotropy, which is a special feature of granular materials, can be expressed by using the fabric tensor and some considerations on the induced-anisotropy are given through the performance of photoelastic experiments.

For simplicity, only two dimensional analysis is given in this paper, and each particle is assumed as circular and rigid. The symbolic notation is used for vector and tensor throughout this paper.

2 REPLACEMENT INTO GRAPHS

We begin with the replacement of an assumbly of grains to an oriented graph. The proposed replacement is briefly explained in Fig.1. The obtained graph is called the replaced graph of the assembly of grains and the correspondence between the replaced graph and the original assembly is shown in Table 1.

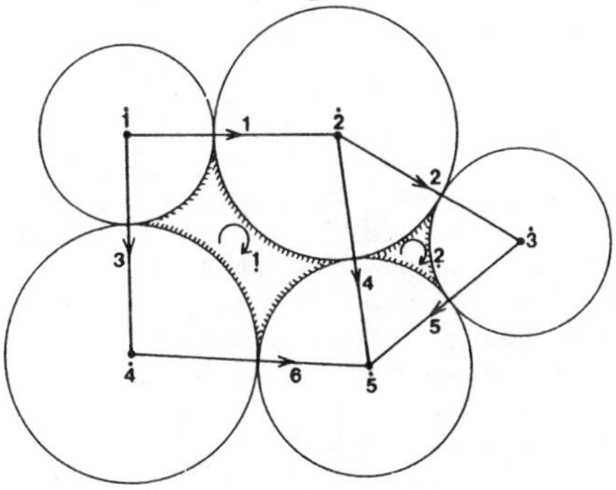

Fig. 1 The replaced graph

Table 1. Correspondence between assembly of grains and replaced graph

assembly of grains	replaced graph	index	number
grain	point	$\dot{i}$	$\dot{n}$
contact point	branch	j	n
void (in 2-dimensions)	loop	k	$\underset{\cdot}{n}$

Next we define the modified replaced graph (or briefly modified graph) by introducing virtual branches. A virtual branch corresponds with an imaginary contact point, which is not a real contact point at present but is considered to have a high possibility for changing to a real contact point by a small deformation. To introduce the virtual branches and to obtain the modified graph, the following procedure is applied:

(1) Find a polygon having more than three sides in the given replaced graph (or in a tentatively-modified graph), and choose a diagonal with the shortest length in the polygon as a new virtual branch. Adding the virtual branches thus introduced, we obtain a succesive tentatively-modified graph.

(2) Repeat this procedure succesively till one can find no other polygon than triangles in the graph.

Examples of virtual branches are shown by dotted lines in Fig.2(b). Then as is shown in Fig.2(c), the dual graph of the modified graph having the following properties can easily be defined:

(1) The numbers of points and loops in the dual graph are equal to those of loops and points respectively in the corresponding modified graph.

(2) The number of branches is equal in both graphs, and corresponding branches intersect perpendicularly each other.

(3) The dual graph is an oriented graph as well as the modified graph, and the orientation of branch $\bar{j}$ in the dual graph is determined as

$$n_{\bar{j}} = \hat{n}_{j'} \qquad (2.1)$$

where $n_{j'}$ and $n_{\bar{j}}$ are the branch directions, which are unit vectors denoting the direction of branches j' and $\bar{j}$ in the modified and the dual graphs respectively.

Indices with prime and bar, as j' and $\bar{j}$, denote elements in the modified and the dual graphs respectively and $\wedge$ (and $\wedge\wedge$) indicate the dual vector (and the dual tensor) explained in Appendix.

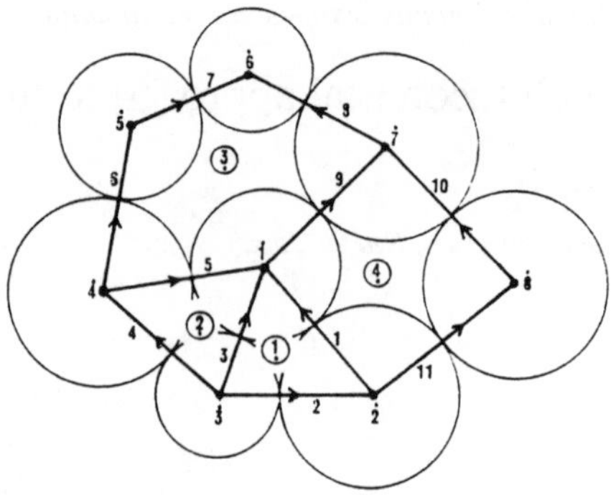

(a) Replaced graph ($\dot{n}=8$, $n=11$, $\underset{\cdot}{n}=4$)

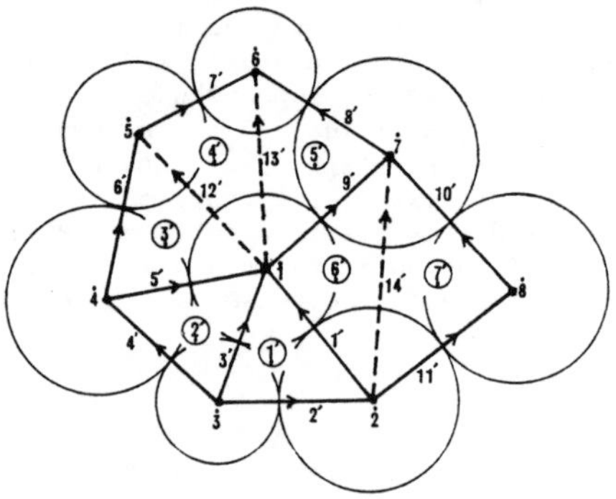

(b) Modified graph ($\dot{n}=8$, $n'=14$, $\underset{\cdot}{n}'=7$)

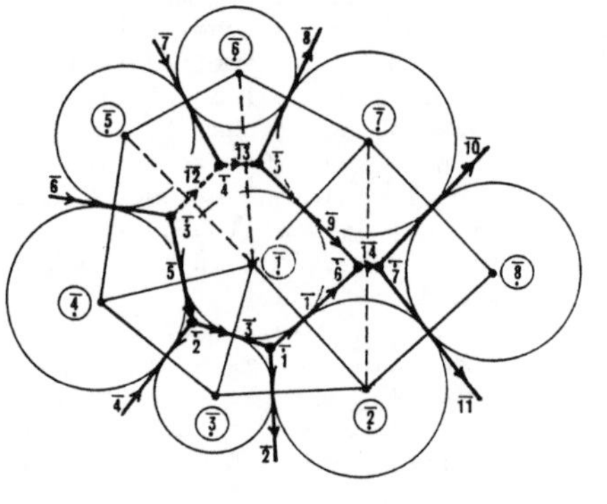

(c) Dual graph ($\dot{n}=7$, $\bar{n}=14$, $\bar{\underset{\cdot}{n}}=8$)

Fig. 2 Graph representations of a granular assembly

3 RELATIONSHIP BETWEEN REDUNDANCY NUMBER AND COORDINATION NUMBER

A void k in the assembly of grains, which contains (r_k+ 1) points of the dual graph,

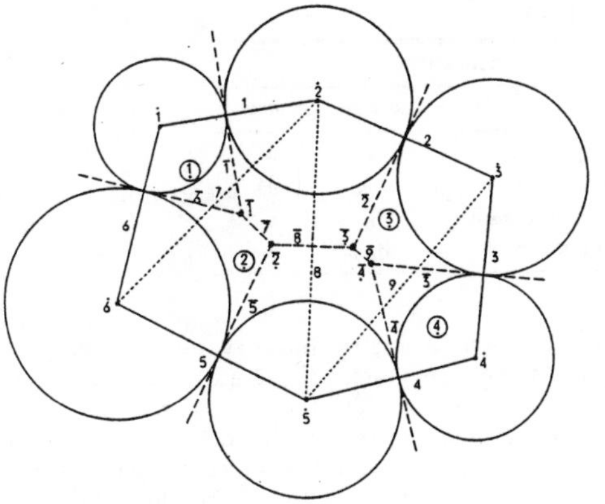

Fig. 3 A void with redundancy number $r_k = 3$

includes r_k virtual branches in it (Fig. 3). From this reason, r_k is called the redundancy number of void k.

Putting

$$r = \sum r_k \qquad (3.1)$$

we have

$$\left.\begin{aligned} \bar{\dot{n}} &= \dot{\eta}' = \dot{\eta} + r \\ \bar{n} &= n' = n + r \\ \bar{\dot{\eta}} &= \dot{n}' = \dot{n} \end{aligned}\right\} \qquad (3.2)$$

where $\dot{n}'$, n' and $\dot{\eta}'$ are numbers of points, branches and loops in the modified graph respectively and $\bar{\dot{n}}$, $\bar{n}$ and $\bar{\dot{\eta}}$ are similar numbers in the dual graph. Euler equations with respect to the replaced, modified and dual graphs are expressed respectively as

$$\left.\begin{aligned} \dot{n} - n + \dot{\eta} &= 1 \\ \dot{n}' - n' + \dot{\eta}' &= 1 \\ \bar{\dot{n}} - \bar{n} + \bar{\dot{\eta}} &= 1 \end{aligned}\right\} \qquad (3.3)$$

The mean value of r_k given by

$$R = \frac{r}{\dot{\eta}} = \frac{1}{\dot{\eta}} \sum r_k \qquad (3.4)$$

is called the (mean) redundancy number of the assembly of grains and is considered to represent the degree of looseness of packing. R is also considered to correspond with the redundant void ratio defined by $(e - e_{min})$, which is an important measure in soil mechanics, where e denotes the void ratio of soil.

On the other hand, the coordination number n_i with respect to a grain i is given by the number of branches connected with point i.

The mean value of n_i given by

$$N = \frac{2n}{\dot{n}} = \frac{\sum n_i}{\dot{n}} \qquad (3.5)$$

is called the (mean) coordination number of the assembly of grains. Quite similarly the dual coordination number, which denotes the mean number of branches composing loop in the replaced graph is given by

$$\bar{N} = \frac{2n}{\dot{\eta}} \qquad (3.6)$$

By virtue of Euler equation (3.3), it follows approximately that

$$N = \frac{2\bar{N}}{\bar{N} - 2} \qquad (3.7)$$

As the modified graph is composed of only triangles, the dual coordination number with respect to the modified graph $\bar{N}'$ is written as

$$\bar{N}' = \frac{2n'}{\dot{\eta}'} = 3 \qquad (3.8)$$

As the coordination number with respect to the modified graph N' has a similar relation to Eq.(3.7), we have

$$N' = \frac{2\bar{N}'}{\bar{N}' - 2} = 6 \qquad (3.9)$$

From Eqs.(3.2),(3.4),(3.6)and(3.8), we have

$$\bar{N} = 3 + R \qquad (3.10)$$

and the substitution of Eq.(3.10) into Eq. (3.7) leads

$$N = \frac{2(3+R)}{1+R}, \; or \; R = \frac{6-N}{N-2} \qquad (3.11)$$

Eq.(3.11) gives a relationship between redundancy number R and coordination number N, which is illustrated in Fig.4.

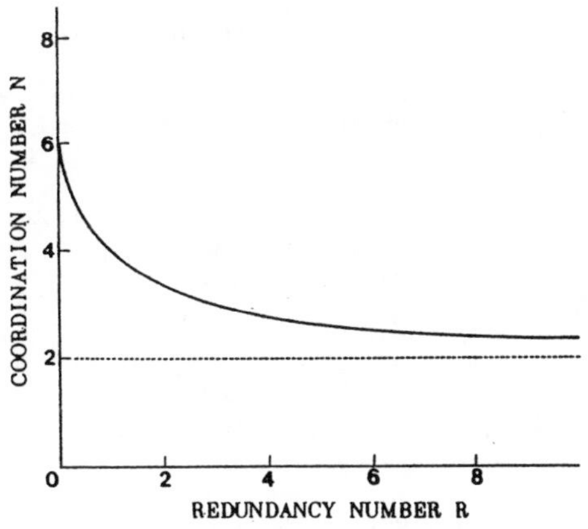

Fig. 4 Relationship between coordination number and redundancy number

Table 2. Basic mechanical quantities

force			deformation		
	notation	number		notation	number
body force	$\boldsymbol{F}_i$	$\dot{n}$	grain displacement	$\boldsymbol{u}_i$	$\dot{n}$
body couple	$\boldsymbol{N}_i$	$\dot{n}$	grain rotation	$\boldsymbol{w}_i$	$\dot{n}$
contact force	$\boldsymbol{S}_j$	n	relative displacement	$\Delta\boldsymbol{u}_j$	n
contact couple	$\boldsymbol{M}_j$	n	relative rotation	$\Delta\boldsymbol{w}_j$	n

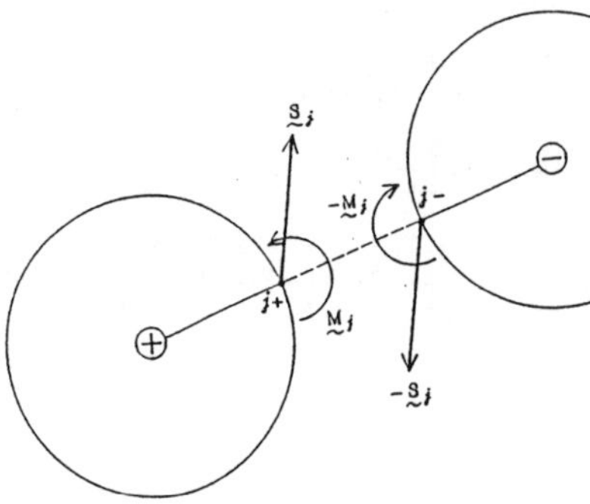

Fig.5 Definitions of contact force and couple

4 FUNDAMENTAL MATRICES IN GRAPHS

Fundamental matrices in the replaced graph are defined as follows:

(1) Incidence Matrix D_{ij}

$$D_{ij} = \begin{cases} 1, & \text{if branch } j \text{ is incident at point } i \text{ and is oriented away from point } i \\ -1, & \text{if branch } j \text{ is incident at point } i \text{ and is oriented toward point } i \\ 0, & \text{otherwise} \end{cases} \quad (4.1)$$

(2) Loop Matrix L_{kj}

$$L_{kj} = \begin{cases} 1, & \text{if branch } j \text{ is in loop } k \text{ and the orientation of the branch and that of the loop coincide} \\ -1, & \text{if branch } j \text{ is in loop } k \text{ and the orientation of the branch and that of the loop do not coincide} \\ 0, & \text{otherwise} \end{cases} \quad (4.2)$$

(3) Simplified Incidence Matrix D'_{ij}

$$D'_{ij} = \begin{cases} 1, & \text{if branch } j \text{ is incident at point } i \\ 0, & \text{otherwise} \end{cases} \quad (4.3)$$

(4) Simplified Loop Matrix L'_{kj}

$$L'_{kj} = \begin{cases} 1, & \text{if branch } j \text{ is in loop } k \\ 0, & \text{otherwise} \end{cases} \quad (4.4)$$

We can easily derive the identities

$$D_{ij}L_{jk} = 0 \quad (4.5)$$

$$L_{kj}D_{ji} = 0 \quad (4.6)$$

and we have similar identities to Eqs.(4.5) and (4.6) for fundamental matrices in the modified and the dual graphs, where we write, for example, the incidence matrices in the modified and the dual graphs as $D_{ij'}$ and $D_{\bar{k}\bar{j}}$ respectively. From the duality of the modified and the dual graphs, we have

$$D_{\bar{j}\bar{k}} = -L_{j'k'} \quad (4.7)$$

$$L_{\bar{i}\bar{j}} = D_{ij'} \quad (4.8)$$

where the orientation of loops is always taken as clockwise.

5 MECHANICAL QUANTITIES AND FUNDAMENTAL EQUATIONS

5.1 Mechanical quantities

The mechanical quantities quoted in this paper are listed in Table 2. For the most generality, the body couple and the contact couple are taken into consideration. It is noted that the contact force and couple are to be defined for grains in positive contact, as is shown in Fig.5, and that the relative displacement and rotation are also defined with respect to grains in positive contact, where positive or negative contact means that the corresponding component of the incidence matrix D_{ij} is 1 or -1 respectively.

5.2 Modified relative displacement

The modified relative displacement $\mathring{\Delta}\boldsymbol{u}_j$ denotes the relative displacement at contact j, as is explained in Fig.6, and is expressed as

$$\mathring{\Delta}\boldsymbol{u}_j = \Delta\boldsymbol{u}_j + \boldsymbol{n}_j \times (r_1\boldsymbol{w}_1 + r_2\boldsymbol{w}_2) = \Delta\boldsymbol{u}_j + l_j\boldsymbol{n}_j \times \boldsymbol{w}_j \quad (5.1)$$

where r_i is radius of grain i,

$$l_j = r_1 + r_2 = D'_{ji}r_i \quad (5.2)$$

denotes length of branch j and

$$\boldsymbol{w}_j = \frac{1}{l_j} D'_{ji}r_i\boldsymbol{w}_i \quad (5.3)$$

is called the branch rotation. It is noted that, as index i in Eqs.(5.2) and (5.3),

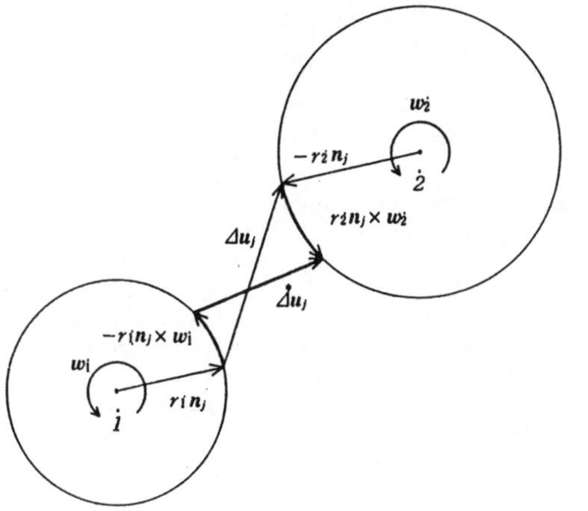

Fig. 6 Illustration of $\mathring{\Delta} u_j$

for twice repeated indices after a fundamental matrix in the same term, summation is to be applied.

5.3 Fundamental equations

The equilibrium equations for force and couple with respect to grain i are expressed, by virtue of the incidence and simplified incidence matrices, as

$$D_{ij} S_j + \Delta A_i F_i = 0 \qquad (5.4)$$

$$D_{ij} M_j + r_i D'_{ij} (n_j \times S_j) + \Delta A_i N_i = 0 \qquad (5.5)$$

where ΔA_i is area of grain i.

On the other hand, the relative displacement and rotation are written as

$$\Delta u_j = -D_{ji} u_i \qquad (5.6)$$

$$\Delta w_j = -D_{ji} w_i \qquad (5.7)$$

From the identity (4.6), it follows that

$$L_{kj} \Delta u_j = 0 \qquad (5.8)$$

$$L_{kj} \Delta w_j = 0 \qquad (5.9)$$

As the modified relative displacement is expressed as

$$\mathring{\Delta} u_j = -D_{ji} u_i + l_j n_j \times w_j \qquad (5.10)$$

we have

$$L_{kj} \mathring{\Delta} u_j = L_{kj} (l_j n_j \times w_j) \qquad (5.11)$$

Eqs.(5.8),(5.9) and (5.11) are considered as the compatibility equations for Δu, Δw and $\mathring{\Delta} u$ respectively.

5.4 Void force

Though the equilibrium equations (5.4) and (5.5) are given for the replaced graph, it is noticed that these equations are also applicable to the modified graph, because no contact force and couple exists along the virtual branches. Thus we can write as

$$D_{ij'} S_{j'} + \Delta A_i F_i = 0 \qquad (5.12)$$

$$D_{ij'} M_{j'} + r_i D'_{ij'} (n_{j'} \times S_{j'}) + \Delta A_i N_i = 0 \qquad (5.13)$$

where index j' indicates branches in the modified graph and $S_{j'}$ and $M_{j'}$ vanish, if j' corresponds with a virtual branch.

In the case where body force F_i is absent, Eq.(5.12) reduces to

$$D_{ij'} S_{j'} = 0 \qquad (5.14)$$

From Eq.(4.8),Eq.(5.14) is rewritten as

$$L_{\bar{i}\bar{j}} S_{\bar{j}} = 0 \qquad (5.15)$$

where contact force $S_{\bar{j}}$ is considered to be distributed along branch $\bar{j}$ in the dual graph. Eq.(5.15) is automatically satisfied, if we assume that

$$S_{\bar{j}} = D_{\bar{j}\bar{k}} X_{\bar{k}} \qquad (5.16)$$

where $X_{\bar{k}}$ denotes an imaginary force applied at point $\bar{k}$ in the dual graph. As $X_{\bar{k}}$ with respect to points $\bar{k}$ included in a common void k is expected to reduce to the same force X_k defined with respect to the void k, $X_k (= X_{\bar{k}})$ is called the void force. X_k is considered to be a quantity having a similar property to the stress function in continuum mechanics.

5.5 Analogy between granular materials and generalized continua

In the mechanics of generalized continua (Satake1970), the equilibrium equations are written as

$$\nabla \cdot \sigma + F = 0 \qquad (5.17)$$

$$\nabla \cdot \mu + I \cdot \times \sigma + N = 0 \qquad (5.18)$$

where σ and μ are stress and couple stress, and F and N are body force and body couple, respectively, and ∇ denotes the vectorial differential operator and I is the unit tensor.

On the other hand, the equations which define gradients are written as

$$\gamma = \nabla u + I \times w \qquad (5.19)$$

$$\alpha = \nabla w \qquad (5.20)$$

where γ and α are gradient of polar displacement and gradient of rotation, and u and w are displacement and rotation, respectively.

Further, introducing Schaefer's differential operators (Schaefer 1967) defined by

$$\text{Grad} = \begin{pmatrix} \nabla & 0 \\ I\times & \nabla \end{pmatrix} \tag{5.21}$$

$$\text{Rot} = \begin{pmatrix} \nabla\times & 0 \\ I\times\times & \nabla\times \end{pmatrix} \tag{5.22}$$

$$\text{Div} = \begin{pmatrix} \nabla\cdot & 0 \\ I\cdot\times & \nabla\cdot \end{pmatrix} \tag{5.23}$$

we have following simplified expressions for Eqs.(5.17) and (5.18):

$$\text{Div}\begin{pmatrix} \sigma \\ \mu \end{pmatrix} + \begin{pmatrix} F \\ N \end{pmatrix} = 0 \tag{5.24}$$

and for Eqs.(5.19) and (5.20):

$$\begin{pmatrix} \alpha \\ \gamma \end{pmatrix} = \text{Grad}\begin{pmatrix} w \\ u \end{pmatrix} \tag{5.25}$$

As we have identities

$$\text{Rot Grad} = 0 \tag{5.26}$$

$$\text{Div Rot} = 0 \tag{5.27}$$

we obtain from Eq.(5.24), for the case where $\boldsymbol{F}$ and $\boldsymbol{N}$ vanish,

$$\begin{pmatrix} \sigma \\ \mu \end{pmatrix} = -\text{Rot}\begin{pmatrix} \varphi \\ \chi \end{pmatrix} \tag{5.28}$$

and from Eq.(5.25)

$$\text{Rot}\begin{pmatrix} \alpha \\ \gamma \end{pmatrix} = 0 \tag{5.29}$$

where φ and χ are force and couple considered to be distributed along the boundary of the surface to which σ and μ are applied.

It is easy to see that Eqs.(5.4) and (5.5) are analogous to Eqs.(5.17) and (5.18), and that Eqs.(5.1) and (5.7) are also analogous to Eqs.(5.19) and (5.20). Further, if we introduce operators expressed as

$$\widetilde{D}_{ij} = \begin{pmatrix} D_{ij} & 0 \\ r_i D'_{ij} n_j \times & D_{ij} \end{pmatrix} \tag{5.30}$$

$$\widetilde{D}_{ji} = \begin{pmatrix} D_{ji} & -n_j \times D'_{ji} r_i \\ 0 & D_{ji} \end{pmatrix} \tag{5.31}$$

Eqs.(5.4) and (5.5) and Eqs.(5.7) and (5.10) are rewritten as

$$\widetilde{D}_{ij}\begin{pmatrix} S_j \\ M_j \end{pmatrix} + \Delta A_i \begin{pmatrix} F_i \\ N_i \end{pmatrix} = 0 \tag{5.32}$$

$$\begin{pmatrix} \mathring{\Delta} u \\ \Delta w_j \end{pmatrix} = -\widetilde{D}_{ji}\begin{pmatrix} u_i \\ w_i \end{pmatrix} \tag{5.33}$$

and Eqs.(5.32) and (5.33) are quite similar to Eqs.(5.24) and (5.25). It is noted, however, that equations for granular assemblies corresponding with Eqs.(5.28) and (5.29) are not generally obtained. Table 3 shows the above-mentioned analogy between granular materials and generalized continua.

The supplementary correspondence between $n_j n_j$ (or fabric tensor ϕ) and unit tensor I will be explained in 6 and 7.

Table 3. Correspondence between granular materials and generalized continua

granular materials	generalized continua
S_j	σ
M_j	μ
Δw_j	α
$\mathring{\Delta} u_j$	γ
D_{ij} $(\widetilde{D}_{ij})$	$\nabla\cdot$ (Div)
$-D_{ji}$ $(-\widetilde{D}_{ji})$	∇ (Grad^T)
L_{jk} , L_{kj}	$\nabla\times$
$n_j n_j$ or $\phi = 2\int_0^\pi f(\theta) n_j n_j d\theta$	I

6 ANISOTROPY TENSORS AND STATISTIC EXPRESSIONS OF MECHANICAL QUANTITIES

The anisotropy of fabric is a special feature of granular materials, and has a remarkable affection to the mechanical characteristics of the materials. To express such anisotropy of fabric, first we introduce the fabric tensor defined by

$$\phi = \frac{1}{n}\sum_R n_j n_j \tag{6.1}$$

where n_j is branch direction, which is the same to the contact normal of grains in positive contact and n is the number of branches in R. R is a mesodomain, i.e. a circular region which consists of a suf-

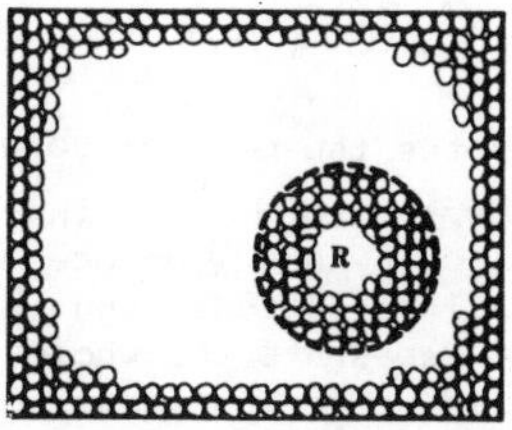

Fig. 7 Mesodomain in a granular material

ficient number of grains to apply a statistical analysis and is small enough to calculate an average value for a local measure, such as stress and strain.

As we assume that R contains a relatively large number of grains, we can also write as

$$\boldsymbol{\phi} = 2\int_0^{\pi} f(\theta)\, \boldsymbol{n}_j \boldsymbol{n}_j \, d\theta \tag{6.2}$$

where $f(\theta)$ is the probability density function of contact normals or briefly contact normal density and θ denotes inclination angle of $\boldsymbol{n}_j$, where $0 \leq \theta \leq \pi$ is assumed. From the prescribed condition

$$2\int_0^{\pi} f(\theta)\, d\theta = 1 \tag{6.3}$$

we have

$$\mathrm{tr}\, \boldsymbol{\phi} = 1 \tag{6.4}$$

Similarly, the modified fabric tensor and the dual fabric tensor are defined respectively as

$$\boldsymbol{\phi}' = 2\int_0^{\pi} f'(\theta)\, \boldsymbol{n}_{j'} \boldsymbol{n}_{j'} \, d\theta \tag{6.5}$$

$$\bar{\boldsymbol{\phi}} = 2\int_0^{\pi} f'(\theta)\, \boldsymbol{n}_{\bar{j}} \boldsymbol{n}_{\bar{j}} \, d\theta \tag{6.6}$$

where $f'(\theta)$ is the modified contact normal density defined for the modified graph in which imaginary contacts corresponding to virtual branches are taken into consideration. From Eq.(2.1), it follows that

$$\bar{\boldsymbol{\phi}} = \hat{\hat{\boldsymbol{\phi}'}} \tag{6.7}$$

Thus $\bar{\boldsymbol{\phi}}$ becomes the dual tensor(see Appendix) of $\boldsymbol{\phi}'$.

Next we introduce two tensors which also describe the anisotropy of the assembly of grains, expressed as

$$\boldsymbol{B} = \frac{\sum_R l_{j'} \boldsymbol{n}_{j'} \boldsymbol{n}_{j'}}{\sum_R l_{j'}} \tag{6.8}$$

$$\boldsymbol{C} = \frac{\sum_R s_{\bar{j}} \boldsymbol{n}_{j'} \boldsymbol{n}_{j'}}{\sum_R s_{\bar{j}}} \tag{6.9}$$

where $l_{j'}$ and $s_{\bar{j}}$ denote length of branches j' and $\bar{j}$ respectively.

Using modified contact normal density $f'(\theta)$ we can write as

$$\boldsymbol{B} = \frac{2}{\bar{l}} \int_0^{\pi} f'(\theta)\, \bar{l}(\theta)\, \boldsymbol{n}_{j'} \boldsymbol{n}_{j'} \, d\theta \tag{6.10}$$

$$\boldsymbol{C} = \frac{2}{\bar{s}} \int_0^{\pi} f'(\theta)\, \bar{s}(\theta)\, \boldsymbol{n}_{j'} \boldsymbol{n}_{j'} \, d\theta \tag{6.11}$$

where $\bar{l}(\theta)$ and $\bar{s}(\theta)$ are mean values of $l_{j'}$ and $s_{\bar{j}}$ with inclination angle θ respectively, and

$$\bar{l} = \frac{1}{n'} \sum_R l_{j'} = 2\int_0^{\pi} f'(\theta)\, \bar{l}(\theta)\, d\theta \tag{6.12}$$

$$\bar{s} = \frac{1}{n'} \sum_R s_{\bar{j}} = 2\int_0^{\pi} f'(\theta)\, \bar{s}(\theta)\, d\theta \tag{6.13}$$

are mean values of $l_{j'}$ and $s_{\bar{j}}$ with respect to all direcitons respctively and n' is the number of branches of the modified graph in $\boldsymbol{R}$. $\boldsymbol{B}$ and $\boldsymbol{C}$ are called the branch tensor and contact tensor respectively, and in the special case where $\bar{l}(\theta) = \bar{l}$ and $\bar{s}(\theta) = \bar{s}$, both $\boldsymbol{B}$ and $\boldsymbol{C}$ reduce to $\boldsymbol{\phi}'$.

The definition of stress for an assembly of grains is given in the form

$$\boldsymbol{\sigma} = \frac{2\sum_R \boldsymbol{n}_{j'} \boldsymbol{S}_{j'}}{\sum_R s_{\bar{j}}} = \frac{2}{n'\bar{s}} \sum_R \boldsymbol{n}_{j'} \boldsymbol{S}_{j'} \tag{6.14}$$

Eq.(6.14) is quite analogous to the definition of stress in a continuum given by

$$\boldsymbol{\sigma} = \lim_{a \to 0} \frac{2\oint \boldsymbol{n}\boldsymbol{S}\, ds}{\oint ds} \tag{6.15}$$

where $\boldsymbol{S}$ denotes the traction vector and the integral is applied along a circumference with radius a. The contact force $\boldsymbol{S}_{j'}$ is conversely expressed as

$$\boldsymbol{S}_{j'} = \bar{s}(\theta)\, \boldsymbol{n}_{j'} \cdot \overset{*}{\boldsymbol{\sigma}} + \delta \boldsymbol{S}_{j'} \tag{6.16}$$

where $\overset{*}{\boldsymbol{\sigma}}$ is the co-stress given by

$$\overset{*}{\boldsymbol{\sigma}} = \frac{1}{2} \boldsymbol{C}^{-1} \cdot \boldsymbol{\sigma} \tag{6.17}$$

and $\delta \boldsymbol{S}_{j'}$ denotes the residual contact force with the property

$$\sum_R \boldsymbol{n}_{j'} \delta \boldsymbol{S}_{j'} = 0 \tag{6.18}$$

In the case where contact couple and body couple are absent, we have from Eq.(5.13)

$$\sum_R \boldsymbol{n}_{j'} \times \boldsymbol{S}_{j'} = 0 \tag{6.19}$$

and consequently $\boldsymbol{\sigma}$ becomes symmetric. It is noted however that $\overset{*}{\boldsymbol{\sigma}}$ is not generally symmetric and becomes symmetric only when both principal directions of $\boldsymbol{\sigma}$ and $\boldsymbol{C}$ coincide.

In a quite similar manner as above, we define the displacement gradient in an assembly of grains by the form

$$\gamma=\frac{2\sum_R n_{j'}\overset{*}{\Delta} u_{j'}}{\sum_R l_{j'}}=\frac{2}{n'\bar{l}}\sum_R n_{j'}\overset{*}{\Delta} u_{j'} \tag{6.20}$$

Eq.(6.19) is also quite analogous to the definition of the gradient of polar dis-placement in the generalized continua (Satake 1970), given by

$$\gamma=\lim_{a\to 0}\frac{2\oint n(du)^*}{\oint dl} \tag{6.21}$$

The modified relative displacement is conversely expressed as

$$\overset{*}{\Delta} u_{j'}=\bar{l}(\theta)n_{j'}\cdot\overset{*}{\gamma}+\delta\overset{*}{\Delta} u_{j'} \tag{6.22}$$

where $\overset{*}{\gamma}$ is the co-displacement gradient given by

$$\overset{*}{\gamma}=\frac{1}{2}B^{-1}\cdot\gamma \tag{6.23}$$

and $\delta\overset{*}{\Delta} u_{j'}$ denotes the residual modified relative displacement with the property

$$\sum_R n_{j'}\delta\overset{*}{\Delta} u_{j'}=0 \tag{6.24}$$

It is easy to show that the incremental work DW in R due to an incremental grain displacement Du_i and an incremental grain rotation Dw_i is expressed as

$$DW=-\sum_R(S_{j'}\cdot\overset{*}{\Delta} Du_{j'}+M_{j'}\cdot\Delta Dw_{j'}) \tag{6.25}$$

For the case where contact couple is negligible, the substitution of Eqs.(6.16) and (6.22) leads

$$\begin{aligned}DW&=-\sum_R S_{j'}\cdot\overset{*}{\Delta} Du_{j'}\\&=-\sum_R \bar{s}(\theta)\bar{l}(\theta)n_{j'}n_{j'}\cdot\cdot(\overset{*}{\sigma}\cdot\overset{*}{\gamma}^T)+\delta DW\\&=-2AJ\cdot\cdot(\overset{*}{\sigma}\cdot\overset{*}{\gamma}^T)+\delta DW\end{aligned} \tag{6.26}$$

where

$$A=\frac{1}{2}n'\bar{s}\bar{l} \tag{6.27}$$

$$J=\frac{2}{\bar{s}\bar{l}}\int_0^\pi f'(\theta)\bar{s}(\theta)\bar{l}(\theta)n_{j'}n_{j'}d\theta \tag{6.28}$$

$$\delta DW=-\sum_R \delta S_{j'}\cdot\delta\overset{*}{\Delta} Du_{j'} \tag{6.29}$$

J is called the anisotropy tensor of the modified graph and δDW is called the re-sidual incremental work. In the case where $\bar{s}(\theta)$ and $\bar{l}(\theta)$ are regarded to have no cor-relation, we can write as

$$J=\frac{2}{\bar{A}}\int_0^\pi f'(\theta)\bar{A}(\theta)n_{j'}n_{j'}d\theta \tag{6.30}$$

where $\bar{A}(\theta)$ denotes the mean value of $A_{j'}=\frac{1}{2}s_j l_{j'}$ with respect to θ, the inclination angle of branch j', and $\bar{A}$ denotes that with respect to all directions, and in this case $A=n'\bar{A}$ repesents the whole area of R. In the case where we can assume that $\bar{s}(\theta)=\bar{s}$ and $\bar{l}(\theta)=\bar{l}$, J reduces to ϕ' as well as B and C. In such a case, we can write DW, using Eq.(6.17), as

$$\begin{aligned}DW&=-2A(C\cdot\overset{*}{\sigma})\cdot\cdot\overset{*}{\gamma}+\delta DW\\&=-A\sigma\cdot\cdot\varepsilon+\delta DW\end{aligned} \tag{6.31}$$

where ε denotes the symmetric part of $\overset{*}{\gamma}$ and is called the strain in the assembly of grains. Further, in the case where $\delta S_{j'}$ and $\delta\overset{*}{\Delta} Du_{j'}$ are regarded to have no corre-lation, the term of residual incremental work δDW can be omitted.

7 CONSIDERATION ON INDUCED-ANISOTROPY

It has been observed and noticed that the fabric in granular materials shows the so-called induced-anisotropy under appli-cation of stresses and that such anisotropy proportionally increases as stress increases. Considering the above-mentioned fact, we assume that

$$\phi=\frac{\sigma}{tr\,\sigma} \tag{7.1}$$

and this assumption means that the number of contact points increases in the direction of the major principal stress and decreases in that of the minor one. Assuming $\phi=\phi'=C$, the substitution of Eq.(7.1) into Eq.(6.17)leads

$$\overset{*}{\sigma}=pI \tag{7.2}$$

where p denotes the mean stress and I is the unit tensor. From Eq.(7.2), it may be observed that the co-stress $\overset{*}{\sigma}$ tends to a hydrostatic state due to the induced-aniso-tropy. In practice, however, such a change of fabric may be obstructed by the constraint of surrounding particles and the real induced-anisotropy seems to be smaller than that given by Eq.(7.1).

Thus, we introduce the more general expres-sion for the induced-anisotropy written as

$$\phi=\frac{\sigma^\alpha}{tr\,\sigma^\alpha} \tag{7.3}$$

where σ^α is a tensor which has the same principal directions to σ and principal values of $(\sigma_i)^\alpha$, and $0<\alpha<1$ is assumed. α is called the inductivity degree of anisotropy and is measured by use of the relation

$$\alpha \doteq \frac{\log \phi_1 - \log \phi_2}{\log \sigma_1 - \log \sigma_2} \qquad (7.4)$$

where σ_1 and σ_2 are the major and the minor principal stresses respectively.

To investigate the actual induced-anisotropy, we performed photoelastic experiments by bicompressive and simple shear apparatus using two dimensional random packing of epoxy resin disks as shown in Figs.8 and 9. For the calculation of principal values of fabric tensor ϕ as well as stress tensor σ, we use Eqs.(6.1) and (6.14) with respect to a mesodomain which includes about 200 particles.

Fig.10 shows an experimental result for the relationship between stress ratio $\frac{\sigma_1}{\sigma_2}$ and major principal strain ε_1 in bicompressive test and Fig.11 shows the relationship between $\log\sqrt{\frac{\sigma_1}{\sigma_2}}$ and $\log\lambda$ (where $\lambda = \sqrt{\frac{\varphi_1}{\varphi_2}}$). Fig. 12 also shows experimental results for the relationship between applied shear force F and shear strain γ in simple shear test and Fig.13 shows the similar relationship as Fig.11 in the case of simple shear. From Figs.11 and 13, it is observed that the value of inductivity degree of anisotropy α is about 0.4-0.6. It is of interest to note that the modified stress proposed by Nakai, which is very usefully applied to the elastoplastic models of soils, shows an inductivity degree of anisotropy with value 0.5(Nakai & Mihara 1984).

Fig. 8 Apparatus of photoelastic experiment for simple shear test

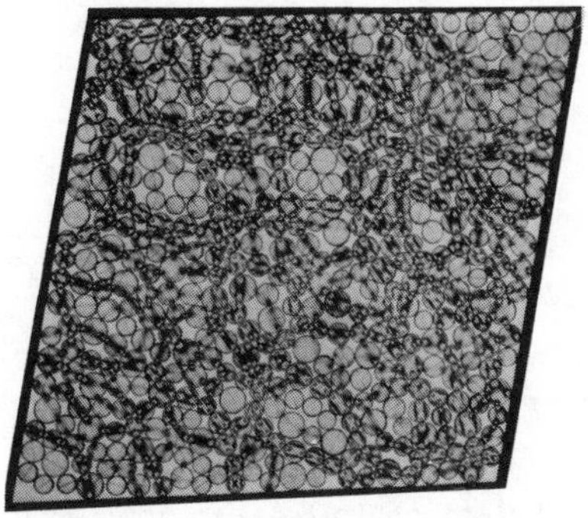

Fig. 9 Photoelastic force lines in simple shear test

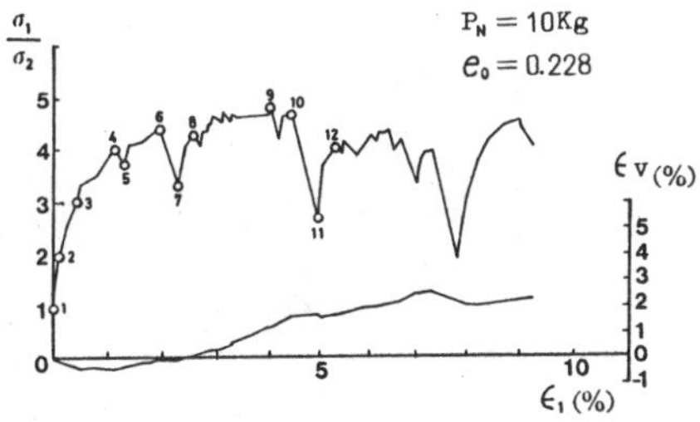

Fig. 10 Relationship between stress ratio $\frac{\sigma_1}{\sigma_2}$ and major principal strain ε_1 in bicompressive test

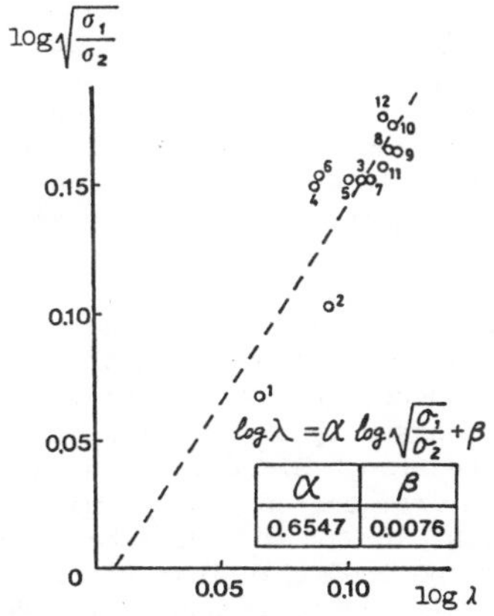

Fig. 11 Relationship between $\log\sqrt{\frac{\sigma_1}{\sigma_2}}$ and $\log\lambda$ in bicompressive test

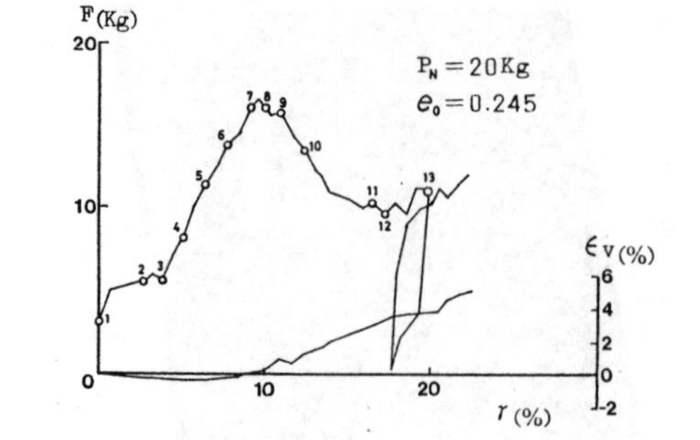

Fig. 12 Relationship between applied shear force F and shear strain γ in simple shear test

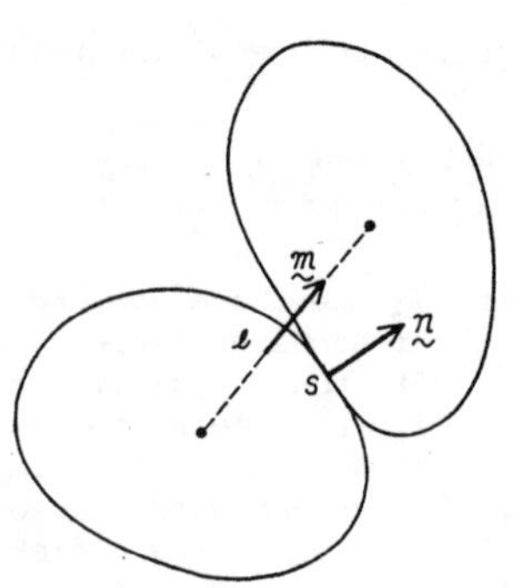

Fig. 14 Branch direction and contact normal

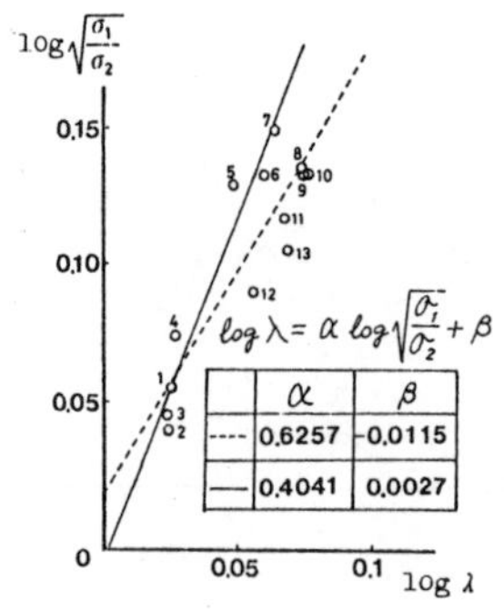

Fig. 13 Relationship between $\log\sqrt{\frac{\sigma_1}{\sigma_2}}$ and $\log\lambda$ in simple shear test

8 CONCLUDING REMARKS

In this paper the graph-theoretical approach to the mechanics of granular materials is briefly summarized. It may be seen that the fundamental equations in the microscopic and discrete viewpoint can be given in a compact form by use of the graph theory. However, to extend the graph-theoretical analysis to more general cases a lot of modifications may become necessary. For instance, in the case of an assembly of grains with arbitrary shapes, the branch direction should be distinguished from the contact normal, as shown in Fig.14(Nemat-Nasser & Mehrabadi 1983). It is also noticed that in the compatibility equations of a granular assembly a supplementary equation describing that none of two particles in the assembly can overlap each other during deformation is to be added. For these purposes, it seems necessary that the graph theory with a consideration on more quantitative properties is to be developed.

REFERENCES

Schaefer, H. 1967. Analysis der Motrofelder im Cosserat-Kontinuum. ZAMM 47: 319-328

Satake, M. 1970. On mechanical quantities in generalized continua. Technology Reports, Tohoku Univ. 35: 15-37.

Satake, M. 1978a. Constitution of mechanics of granular materials through graph representation. Theoretical and Applied Mechanics. 26: 257-266.

Satake, M. 1978b. Constitution of mechanics of granular materials through the graph theory. In S.C.Cowin & M.Satake (eds.), Continuum-mechanical and statistical approaches in the mechanics of granular materials, p.47-62. Tokyo: Gakujutsu Bunken Fukyukai.

Satake, M. 1983. Fundamental quantities in the graph approach to granular materials. In J.T.Jenkins and M. Satake (eds.), Mechanics of granular materials: New models and constitutive relations, p.9-19. Amsterdam: Elsevier.

Nemat-Nasser, S & M. Mehrabadi 1983. Stress and fabric in granular masses. In J.T.Jenkins & M. Satake (eds.), Mechanics of granular materials: New models and constitutive relations, p.1-8. Amsterdam: Elsevier.

Nakai, T & Y. Mihara. 1984 A new mechanical quantity for soils and its application to elastoplastic constitutive models. Soils and Foundations. 24-2: 82-93.

APPENDIX

For a vector $\boldsymbol{v} = (v_1 v_2)$, we define the dual vector

$$\hat{\boldsymbol{v}}=(v_2, -v_1) \tag{A.1}$$

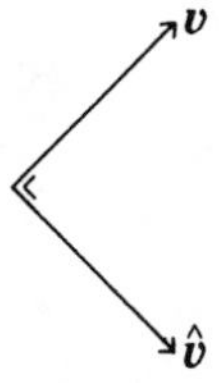

Fig. 15 Dual vector

$\hat{\boldsymbol{v}}$ is expressed symbolically as

$$\hat{\boldsymbol{v}}=\boldsymbol{I}\times\boldsymbol{v} \tag{A.2}$$

and has a property

$$\boldsymbol{v}\cdot\hat{\boldsymbol{v}}=0 \tag{A.3}$$

For a tensor $\boldsymbol{\tau}=\begin{pmatrix} t_{11} & t_{12} \\ t_{21} & t_{22} \end{pmatrix}$, we define the dual tensor

$$\hat{\hat{\boldsymbol{\tau}}}=\begin{pmatrix} t_{22} & -t_{21} \\ -t_{12} & t_{11} \end{pmatrix} \tag{A.4}$$

$\hat{\hat{\boldsymbol{\tau}}}$ is expressed symbolically as

$$\hat{\hat{\boldsymbol{\tau}}}=-(\boldsymbol{I}\times\boldsymbol{\tau})\times\boldsymbol{I} \tag{A.5}$$

and has a property

$$\boldsymbol{\tau}\cdot\cdot\hat{\hat{\boldsymbol{\tau}}}=2|\boldsymbol{\tau}| \tag{A.6}$$

5th International Symposium on Continuum Models of Discrete Systems / Nottingham / 14-20 July 1985

T matrices for elastodynamic scattering by circular discs

M.M.Carroll
University of California, Berkeley, USA
M.F.McCarthy
University College, National University of Ireland, Galway

ABSTRACT: The diffraction of time-harmonic elastic waves by a rigid circular disc in an isotropic elastic matrix is examined. The transition matrix is obtained through the use of a boundary integral technique. It is shown how symmetry considerations lead to a natural separation of the fields into even and odd components which results in a simplification of the form of the transition matrix.

1 INTRODUCTION

While the problem of scattering of an elastic wave by a rigid circular disc has been treated by many authors, the transition matrix, which describes the response of the disc to any arbitrary incident wave, has not been calculated. The corresponding problem for scalar and electromagnetic waves has been treated by Kristensson and Waterman (1982), who treat the disc as the limit of an oblate spheroid. A somewhat different approach to the scalar problem has been applied by Van den Berg (1980), who employed a boundary integral technique of the type he also used to calculate the transition matrix for two dimensional elastodynamic scattering by strips and cracks (Van den Berg, 1982).

Here we follow in the spirit of Van den Berg by taking the boundary integral (Pao and Varatharajulu, 1976) as a point of departure. The symmetry associated with the circular disc is exploited in order to effect a separation of the even and odd parts of the elastodynamic field. A detailed calculation is presented when the elastodynamic field is odd. By expanding the unknown quantities at the disc in a complete sequence of associated Legendre polynomials, the elastodynamic conditions at the edge of the disc are satisfied and the T matrix is calculated.

2 BASIC EQUATIONS

We consider an infinite, homogeneous, isotropic, linear elastic material of uniform density ρ and Lamé constants λ,μ, in which is embedded a smoth rigid inclusion which is bounded by a smooth surface S.

A monochromatic wave of frequency ω is incident on S. The displacement vector associated with the incident wave is $\underset{\sim}{u}^i$ and when this is scattered by the inclusion it gives rise to a scattered wave $\underset{\sim}{u}^s$. All field quantities will have the common time factor $\exp(-i\omega t)$ which will be suppressed. The equation of motion for $\underset{\sim}{u} = \underset{\sim}{u}^i + \underset{\sim}{u}^s$ is

$$(1/k_p^2)\nabla(\nabla\cdot\underset{\sim}{u}) - (1/k_s^2)\nabla\times\nabla\times\underset{\sim}{u} + \underset{\sim}{u} = \underset{\sim}{0}\ , \tag{2.1}$$

with $k_p^2 = \omega^2/c_p^2 = \rho\omega^2/(\lambda+2\mu)$ and $k_s^2 = \omega^2/c_s^2 = \rho\omega^2/\mu$, respectively. Corresponding to (2.1), the Green's dyadic $\underset{\sim}{G} = \underset{\sim}{G}(\underset{\sim}{x}|\underset{\sim}{x}')$ satisfies the equation

$$(1/k_p^2)\nabla(\nabla\cdot\underset{\sim}{G}) - (1/k_s^2)\nabla\times\nabla\times\underset{\sim}{G} + \underset{\sim}{G} = -\frac{\kappa}{\rho\omega^2}\delta(\underset{\sim}{x}-\underset{\sim}{x}')\underset{\sim}{I} \tag{2.2}$$

and the radiation condition at infinity. In (2.2) $\kappa = 1/\rho\omega^2$.

We are interested in determining the displacement vector $\underset{\sim}{u}$ at points outside S. If the inclusion is rigid and fixed then (Pao, Varatharajulu, 1976)

$$\underset{\sim}{u}^s(\underset{\sim}{x}) = -\int_S \underset{\sim}{t}(\underset{\sim}{x}')\cdot\underset{\sim}{G}(\underset{\sim}{x}|\underset{\sim}{x}')dS \tag{2.3}$$

where $\underset{\sim}{t}(\underset{\sim}{x}') = \underset{\sim}{n}\cdot\underset{\sim}{\tau}(\underset{\sim}{x}')$ is the surface traction at $\underset{\sim}{x}'$, $\underset{\sim}{n}$ being the unit outward normal to S and $\underset{\sim}{\tau}(\underset{\sim}{x}')$ the stress tensor. In an infinite region the Green's dyadic is given by

$$\underset{\sim}{G}(\underset{\sim}{x}|\underset{\sim}{x}') = \frac{\kappa}{4\pi}\{k_s^2 \underset{\sim}{I} g_s(\underset{\sim}{x}|\underset{\sim}{x}') + \nabla[g_p(\underset{\sim}{x}|\underset{\sim}{x}') - g_s(\underset{\sim}{x}|\underset{\sim}{x}')]\nabla'\} \quad , \qquad (2.4)$$

where

$$g_p(\underset{\sim}{x}|\underset{\sim}{x}') = \exp(ik_p r)/r \quad ,$$

$$g_s(\underset{\sim}{x}|\underset{\sim}{x}') = \exp(ik_s r)/r \quad , \qquad (2.5)$$

$$r = |\underset{\sim}{x} - \underset{\sim}{x}'| \quad .$$

In what follows we need expressions for g_p and g_s in cylindrical polar coordinates (ρ,θ,z) and then we have (Morse, Feshbach, 1953)

$$g(\underset{\sim}{x}|\underset{\sim}{x}') = \sum_{m=0}^{\infty} \varepsilon_m \cos m(\phi-\phi')\int_0^{\infty} J_m(\lambda\rho) J_m(\lambda\rho')\gamma(k)^{-1}\lambda \exp[i\gamma(k)|z-z'|]d\lambda \qquad (2.6)$$

where $k = k_p$ or k_s, $\varepsilon_m = 2 - \delta_{om}$ and $\gamma(k) = \sqrt{k^2-\lambda^2}$, with $\mathrm{Re}[\gamma(k)] \geqslant 0$, $\mathrm{Im}[\gamma(k)] \geqslant 0$.

The scatterer is assumed to be finite so that it may be circumscribed by a sphere of finite radius outside of which the scattered wave will have the form of an outgoing spherical wave. Thus, it seems natural to employ spherical polar coordinates (r,θ,ϕ), with origin inside S, and to introduce the basis functions

$$\underset{\sim}{\psi}_{1\sigma mn} = k_p \Lambda^{1/2}\xi_{mn}\underset{\sim}{L}^{\sigma}_{mn} \quad ,$$

$$\underset{\sim}{\psi}_{2\sigma mn} = k_s \eta_{mn}\underset{\sim}{M}^{\sigma}_{mn} \quad , \qquad (2.7)$$

$$\underset{\sim}{\psi}_{3\sigma mn} = k_s \eta_{mn} N^{\sigma}_{mn} \quad ,$$

where

$$\underset{\sim}{L}^{\sigma}_{mn} = h_n'(k_p r)\underset{\sim}{A}^{\sigma}_{mn} + (n^2+n)^{1/2}[h_n/k_p r]\underset{\sim}{B}^{\sigma}_{mn} \quad ,$$

$$\underset{\sim}{M}^{\sigma}_{mn} = (n^2+n)^{1/2} h_n(k_s r)\underset{\sim}{C}^{\sigma}_{mn} \quad , \qquad (2.8)$$

$$\underset{\sim}{N}^{\sigma}_{mn} = (n^2+n)[h_n/k_s r]\underset{\sim}{A}^{\sigma}_{mn} + [(n^2+n)^{1/2}/k_s r][k_s r h_n]'\underset{\sim}{B}^{\sigma}_{mn}$$

with

$$\underset{\sim}{A}^{\sigma}_{mn} = Y^{\sigma}_{mn}\hat{r} \quad , \quad \underset{\sim}{B}^{\sigma}_{mn} = (n^2+n)^{-1/2} r\nabla Y^{\sigma}_{mn} \quad ,$$

$$C^{\sigma}_{mn} = -\hat{r}\times \underset{\sim}{B}^{\sigma}_{mn} \quad ,$$

$$Y^{\sigma}_{mn} = P^m_n(\cos\theta)\,{\cos \atop \sin} m\phi \{{\sigma=e \atop \sigma=0}\} \quad , \qquad (2.9)$$

$$\xi_{mn} = [\varepsilon_n \frac{(2n+1)(n-m)!}{4\pi(n+m)!}]^{1/2} \quad ,$$

$$\eta_{mn} = [n(n+1)]^{-1/2}\xi_{mn} \quad ,$$

and $\Lambda = k_p/k_s$. $h_n(\cdot)$ is the spherical Hankel function of the first kind of order n and a prime denotes differentiation with respect to its argument. In (2.7), the first index is the mode index, $\sigma = e,0$ specifies the azimuthal parity, $m = 0,\ldots n$ specifies the rank and $n = 0,1\ldots$, specifies the order of the spherical harmonics. We use the abbreviation $\underset{\sim}{\psi}_{j\sigma mn} = \underset{\sim}{\psi}$ as long as no confusion arises.

Corresponding to the basis vectors $\{\underset{\sim}{\psi}_n\}$, we also need the regular set of basis vectors $\{\hat{\underset{\sim}{\psi}}_n\}$ which is obtained by replacing the spherical Hankel functions in ψ_n by the spherical Bessel functions j_n. The Green's dyadic may be expanded in terms of the basis vectors as follows (Varatharajulu, Pao, 1976):

$$\underset{\sim}{G}(\underset{\sim}{x}|\underset{\sim}{x}') = i\kappa \sum_n k_s \underset{\sim}{\psi}_n(\underset{\sim}{r}_>)\hat{\underset{\sim}{\psi}}_n(\underset{\sim}{r}_<) \quad , \qquad (2.10)$$

where $\underset{\sim}{r}_>$ and $\underset{\sim}{r}_<$ refer to the greater and lesser of $\underset{\sim}{x}$ and $\underset{\sim}{x}'$, respectively.

We are concerned with the scattering of elastic waves by a rigid circular disc which is located at $0 \leqslant r < a$, $0 \leqslant \phi < 2\pi$, $\theta = \pi/2$. The boundary condition which must be satisfied at the disc is

$$\lim_{\theta\to\pi/2} \underset{\sim}{u}(r,\theta,\phi) = \underset{\sim}{0} \quad , \quad 0 \leqslant r < a \ . \qquad (2.11)$$

While the particle displacement is continuous across the disc, the traction vector $\underset{\sim}{t} = \underset{\sim}{n}\cdot\underset{\sim}{\tau}$ suffers a jump discontinuity which is defined as

$$\underset{\sim}{\Theta} = \lim_{\theta\downarrow\pi/2} \hat{\theta}\cdot\underset{\sim}{\tau} - \lim_{\theta\uparrow\pi/2} \hat{\theta}\cdot\underset{\sim}{\tau} = [\Theta_1,\Theta_2,\Theta_3], \text{ say} \quad . \qquad (2.12)$$

When (2.12) is used, (2.3) may be rewritten in the form

$$\underset{\sim}{u}^s(\underset{\sim}{x}') = -\int_S \underset{\sim}{G}(\underset{\sim}{x}'|\bar{\underset{\sim}{x}})\cdot\underset{\sim}{\Theta}(\bar{\underset{\sim}{x}})dS(\bar{\underset{\sim}{x}}) \quad , \qquad (2.13)$$

where the integration is carried out over the disc $0 \leqslant r < a$, $\theta = \pi/2$. In what follows we show how (2.13) may be used to calculate the T matrix of the disc.

3 SYMMETRY CONSIDERATIONS

In his calculation of the transition matrix for two dimensional elastic wave scattering by a rigid strip, Van den Berg (1982) explored the symmetry properties of the strip to considerable advantage. The rigid circular disc has a plane of symmetry $(\theta = \pi/2)$ and an axis of symmetry $(\phi = 0)$. Any vector $\underset{\sim}{f}$ may be decomposed into components which are even or odd with respect to the plane $\theta = \pi/2$ or the axis $\phi = 0$. Thus, we may write

$$\underset{\sim}{f} = \underset{\sim}{f}^{ee} + \underset{\sim}{f}^{eo} + \underset{\sim}{f}^{oe} + \underset{\sim}{f}^{oo} \quad , \tag{3.1}$$

where $\underset{\sim}{f}^{\sigma\nu}$ denotes the vector of parity σ (= e or 0) in θ and of parity ν (= e or 0) in ϕ. The components of $\underset{\sim}{f}^{\sigma\nu}$ transform as follows:

$$\begin{aligned}
f_r^{e\nu}(r,\bar\theta,\phi) &= f_r^{e\nu}(r,\theta,\phi) \quad , \\
f_\theta^{e\nu}(r,\bar\theta,\phi) &= - f_\theta^{e\nu}(r,\theta,\phi) \quad , \\
f_\phi^{e\nu}(r,\bar\theta,\phi) &= f_\phi^{e\nu}(r,\theta,\phi) \quad , \\
f_r^{o\nu}(r,\bar\theta,\phi) &= - f_r^{o\nu}(r,\theta,\phi) \quad , \\
f_\theta^{o\nu}(r,\bar\theta,\phi) &= f_\theta^{o\nu}(r,\theta,\phi) \quad , \\
f_\phi^{o\nu}(r,\bar\theta,\phi) &= - f_\phi^{o\nu}(r,\theta,\phi) \quad , \\
f_r^{\sigma e}(r,\theta,-\phi) &= f_r^{\sigma e}(r,\theta,\phi) \quad , \\
f_\theta^{\sigma e}(r,\theta,-\phi) &= f_\theta^{\sigma e}(r,\theta,\phi) \quad , \\
f_\phi^{\sigma e}(r,\theta,-\phi) &= - f_\phi^{\sigma e}(r,\theta,\phi) \quad , \\
f_r^{\sigma o}(r,\theta,-\phi) &= - f_r^{\sigma o}(r,\theta,\phi) \quad , \\
f^{\sigma o}(r,\theta,-\phi) &= - f_\theta^{\sigma o}(r,\theta,\phi) \quad , \\
f_\phi^{\sigma o}(r,\theta,-\phi) &= f_\phi^{\sigma o}(r,\theta,\phi) \quad ,
\end{aligned} \tag{3.2}$$

where $\bar\theta = \pi - \theta$.

In an isotropic material, the stress tensor is given by the expression

$$\underset{\sim}{\tau} = \underset{\sim}{\tau}(\underset{\sim}{u}) = \lambda I \nabla \cdot \underset{\sim}{u} + \mu(\nabla \underset{\sim}{u} + \underset{\sim}{u}\nabla) \tag{3.3}$$

and it follows that $\underset{\sim}{\tau}$ admits the decomposition

$$\underset{\sim}{\tau} = \underset{\sim}{\tau}^{ee} + \underset{\sim}{\tau}^{eo} + \underset{\sim}{\tau}^{oe} + \underset{\sim}{\tau}^{oo} \quad , \tag{3.4}$$

where $\underset{\sim}{\tau}^{\sigma\nu} = \underset{\sim}{\tau}(\underset{\sim}{u}^{\sigma\nu})$. It may be verified that, under the transformation $\theta \to \pi - \theta$, $\tau_{r\theta}^{e\nu}$ and $\tau_{\theta\phi}^{e\nu}$ change sign while the remaining components of $\underset{\sim}{\tau}^{e\nu}$ remain unaltered. On the other hand, $\tau_{r\theta}^{o\nu}$ and $\tau_{\theta\phi}^{o\nu}$ remain unaltered under this transformation while the remaining components of $\underset{\sim}{\tau}^{o\nu}$ change sign.

The symmetry properties of the basis vectors play a critical role in the remainder of our analysis. Thus, we may write

$$\underset{\sim}{\psi}_{j\sigma mn} = \underset{\sim}{\psi}_{je\sigma mn} + \underset{\sim}{\psi}_{jo\sigma mn} \quad , \tag{3.5}$$

where, e.g.,

$$\begin{aligned}
\underset{\sim}{\psi}_{10\sigma mn} &= \Gamma_{1mnr}\left\{\begin{matrix}\cos\\ \sin\end{matrix}\; m\phi\right\}\hat r \\
&\quad + \Gamma_{1mn\theta}\left\{\begin{matrix}\cos\\ \sin\end{matrix}\; m\phi\right\}\hat\theta \\
&\quad + \Gamma_{1mn\phi}\left\{\begin{matrix}\sin\\ -\cos\end{matrix}\; m\phi\right\}\hat\phi \quad , \; m+n \text{ odd} \\
\underset{\sim}{\psi}_{20\sigma mn} &= \Gamma_{2mn\theta}\left\{\begin{matrix}\cos\\ -\sin\end{matrix}\; m\phi\right\}\hat\theta \\
&\quad + \Gamma_{2mn\phi}\left\{\begin{matrix}\sin\\ \cos\end{matrix}\; m\phi\right\}\hat\phi \quad , \; m+n \text{ even} \\
\underset{\sim}{\psi}_{30\sigma mn} &= \Gamma_{3mnr}\left\{\begin{matrix}\cos\\ \sin\end{matrix}\; m\phi\right\}\hat r \\
&\quad + \Gamma_{3mn\theta}\left\{\begin{matrix}\cos\\ \sin\end{matrix}\; m\phi\right\}\hat\theta \\
&\quad + \Gamma_{3mn\phi}\left\{\begin{matrix}\sin\\ -\cos\end{matrix}\; m\phi\right\}\hat\phi \quad , \; m+n \text{ odd}
\end{aligned} \tag{3.6}$$

with, e.g.,

$$\begin{aligned}
\Gamma_{1mn\theta} &= k_p \Lambda^{1/2} \xi_{mn} [\sqrt{n(n+1)}/2n+1][h_{n-1}(k_p r) \\
&\quad + h_{n+1}(k_p r)]\bar B_{mn} \quad , \\
\Gamma_{2mn\theta} &= - k_s \eta_{mn} \sqrt{n(n+1)}\; h_n(k_s r)\bar{\bar B}_{mn} \quad , \\
\Gamma_{3mn\theta} &= k_s \eta_{mn} [n(n+1)/2n+1][(n+1)h_{n-1}(k_s r) \\
&\quad - n h_{n+1}(k_s r)]\bar B_{mn} \quad ,
\end{aligned} \tag{3.7}$$

$$\begin{aligned}
\bar B_{mn} &= [\sqrt{n(n+1)}/(2n+1)\sin\theta] \\
&\quad \{(\frac{n-m+1}{n+1}) P_{n+1}^m(\cos\theta) \\
&\quad - (\frac{n+m}{n}) P_{n-1}^m(\cos\theta)\} \quad , \\
\bar{\bar B}_{mn} &= - [m/\sqrt{n(n+1)}\; \sin\theta] P_n^m(\cos\theta) \quad ,
\end{aligned}$$

and in (3.6) and (3.7) we have only written down expressions for those terms which arise later on in our analysis.

4 CALCULATION OF TRANSITION MATRIX

Inside the region $|\underset{\sim}{x}| \leqslant a$, the source free incident field may be written

$$\begin{aligned}
\underset{\sim}{u}^i(\underset{\sim}{x}) &= \underset{\sim}{u}^{ie}(\underset{\sim}{x}) + \underset{\sim}{u}^{io}(\underset{\sim}{x}) \quad , \\
\underset{\sim}{u}^{ie}(\underset{\sim}{x}) &= \sum_{j,m,n} \{A_{mn}^{jee}\hat{\underset{\sim}{\psi}}_{jeemn} + A_{mn}^{jeo}\hat{\underset{\sim}{\psi}}_{jeomn}\} \quad , \\
\underset{\sim}{u}^{io}(\underset{\sim}{x}) &= \sum_{j,m,n} \{A_{mn}^{joe}\hat{\underset{\sim}{\psi}}_{joemn} + A_{mn}^{joo}\hat{\underset{\sim}{\psi}}_{joomn}\} \quad ,
\end{aligned} \tag{4.1}$$

while, outside the region $|\underset{\sim}{x}| \leqslant a$, the scattered field may be expressed in the form

$$\underset{\sim}{u}^{s}(\underset{\sim}{x}) = \underset{\sim}{u}^{se}(\underset{\sim}{x}) + \underset{\sim}{u}^{so}(\underset{\sim}{x}) \quad ,$$
$$\underset{\sim}{u}^{se}(\underset{\sim}{x}) = \sum_{j,m,n} \{\alpha^{jee}_{mn}\underset{\sim}{\psi}_{je} + \alpha^{jeo}_{mn}\underset{\sim}{\psi}_{jeomn}\} \quad , \tag{4.2}$$
$$\underset{\sim}{u}^{so}(\underset{\sim}{x}) = \sum_{j,m,n} \{\alpha^{joe}_{mn}\underset{\sim}{\psi}_{joemn} + \alpha^{joo}_{mn}\underset{\sim}{\psi}_{joomn}\} \ .$$

Our objective is to express the coefficients α of the scattered wave in terms of the coefficients A of the incident wave.

In the interests of brevity, we confine our remarks to the situation which arises when the incident field is odd, i.e., $\underset{\sim}{u}^{ie} \equiv \underset{\sim}{0}$. In this case, $\Theta_1 = \Theta_3 = 0$ and it follows that $\underset{\sim}{u}^{se} \equiv \underset{\sim}{0}$ while

$$\alpha^{jo\nu}_{mn} = -i\kappa k_s \int_S \hat{\psi}_{jo\nu mn\theta}(\underset{\sim}{\bar{x}})\Theta_2(\underset{\sim}{\bar{x}})\,dS(\underset{\sim}{\bar{x}}) \ . \tag{4.3}$$

There is no azimuthal coupling so it suffices to consider the cases for which $\Theta_2 = \Phi^e_{2m}(r)\cos m\phi$ or $\Theta_2 = \Phi^o_{2m}(r)\sin m\phi$. For the former case we find

$$\alpha^{joe}_{mn} = \begin{cases} -i\kappa k_s\pi \displaystyle\int_o^a \hat{\Gamma}_{jmn\theta}\Phi^e_{2m}(r)\,r\,dr \ , & j = 1,3,\ m+n \text{ even} \\ & j = 2,\ m+n \text{ odd} \\ 0, \ \text{otherwise} \ , \end{cases} \tag{4.4}$$

and here $\hat{\Gamma}_{jmn\theta}$ is obtained by replacing $h_n(\sim)$ in (3.7) by $j_n(\sim)$.

In order to calculate the α^{joe}_{mn} we need to find Φ^e_{2m}. When (2.13) is expressed in cylindrical polar coordinates we find, after some elementary analysis, that

$$u^{io}(r',\pi/2,\phi') = u^{ioe}_{\theta m}(r')\cos m\phi' \quad ,$$
$$u^{ioe}_{\theta}(r') = \frac{i\kappa\varepsilon_m}{4}\int_o^a \mathcal{G}^{22}_m(r,r')\,r\,\Phi^e_{2m}(r)\,dr \ , \quad 0 < r' < a \ , \tag{4.5}$$
$$\mathcal{G}^{22}_m(r,r') = \int_o^\infty \Big(\frac{\lambda^2}{\gamma_s} + \gamma_p\Big)\lambda J_m(\lambda r)J_m(\lambda r')\,d\lambda \ .$$

$(4.5)_2$ is the desired integral equation for Φ^e_{2m} and it is similar to the equation which arises in the problem of acoustic scattering from a rigid circular disc (Van den Berg, 1980). Let us write

$$\frac{i\varepsilon_m\kappa}{4}\Phi^e_{2m}(r) = \sum_{\ell=m}^{\infty} c^e_{2mp}(a^2-r^2)^{-1/2}u^m_p(r) \tag{4.6}$$

where

$$u^m_\ell(r) = \begin{cases} i^{m-\ell}\beta^m_\ell P^m_n([1-r^2/a^2]^{1/2}) \ , & m+n \text{ even} \\ 0, \ \text{otherwise} \end{cases} \tag{4.7}$$

and here $\beta^m_n = (2\pi/\varepsilon_n)^{1/2}\xi_{mn}$. It follows easily that

$$\frac{1}{a}\int_o^a (a^2-r'^2)^{-1/2}u^m_s(r')\,r'\,u^{ioe}_{\theta m}(r')\,dr' = \sum_{p=m}^{\infty} \overset{o}{Q}{}^m_{sp}c^e_{2mp} \tag{4.8}$$

with

$$\overset{o}{Q}{}^p_{mn} = \begin{cases} a\beta^p_m\beta^p_n P^p_m(0)P^p_n(0)\Big[\displaystyle\int_o^{k_s}\lambda^3(k^2-\lambda^2)^{-1/2} j_>(\lambda a)h_<(\lambda a)\,d\lambda \\ \quad + \displaystyle\int_o^{k_p}\lambda(k^2-\lambda^2)^{1/2} j_>(\lambda a)h_<(\lambda a)\,d\lambda \ , & m+n \text{ even} \ , \\ 0, \ m+n \text{ odd} \ , \end{cases} \tag{4.9}$$

where $> = \max(m,n)$, $< = \min(m,n)$ and in (4.8) we have used the fact that $u^{ioe}_\theta = u^{ioe}_{\theta m}(r)\cos m\phi$.

Substitution from (4.6) into (4.4) yields the formula

$$\alpha^{joe}_{mn} = \begin{cases} -\dfrac{4\pi k_s}{\varepsilon_m}\displaystyle\sum_{p=m}^{\infty} c^e_{2mp}\int_o^a (a^2-r^2)^{-1/2} r\hat{\Gamma}_{jmn\theta}(r)u^m_p(r)\,dr & j = 1,3,\ m+n \text{ odd} \\ & j = 2,\ m+n \text{ even} \\ 0, \ \text{otherwise} \ . \end{cases} \tag{4.10}$$

When (4.1) is used in (4.8) we find that

$$c^e_{2pm} = \sum_{s,n,j} (\overset{o}{Q}{}^p)^{-1}_{ms}\{\tfrac{1}{a}A^{joe}_{pn}L_{jpns}\} \ , \tag{4.11}$$
$$L_{jpns} = \int_o^a (a^2-r^2)^{-1/2}u^p_s(r)\,r\hat{\Gamma}_{jpn\theta}(r)\,dr \ .$$

When (4.11) is used in (4.10), we obtain the formula

$$\alpha^{joe}_{mn} = \sum_{k,q} T^{jkoe}_{mnq}A^{koe}_{mq} \ , \tag{4.12}$$

with

$$T^{jkoe}_{mnq} = -\frac{4\pi k_s}{a\varepsilon_m}\sum_{r,s,q} L_{jmnr}(\overset{o}{Q}{}^m)^{-1}_{rs}L_{kmqs} \ . \tag{4.13}$$

In a similar fashion when the incident field is odd in both θ and ϕ we find that the only nonvanishing α is α^{joo}_{mn}, and that

$$\alpha^{joo}_{mn} = \sum_{k,q} T^{jkoo}_{mnq}A^{koo}_{mq} \ , \qquad T^{jkoo}_{mnq} = (-1)^{j+k}T^{jkoe}_{mnq} \ . \tag{4.14}$$

5 CONCLUSIONS

We have shown how to calculate the T matrix for elastic wave scattering by a rigid disc when the incident field is odd. The derived expressions are in a form suitable for numerical computation. A matrix whose elements are integrals over a finite range has to be inverted. In the low frequency regime this may be done using techniques of the type used by Kristensson and Waterman (1982). In the high frequency regime, asymptotic forms of the Bessel functions may be employed to evaluate the integrals.

Similar results are valid when the incident field is even. The results may be extended in a straightforward way in order to obtain the transition matrix for scattering by a circular crack

ACKNOWLEDGMENT

This work was supported in part by a contribution from the Shell Companies Foundation in support of a Shell Distinguished Chair at the University of California, Berkeley, and in part by a Grant No. MEA-820534 from the National Science Foundation (Solid Mechanics Program) to the University of California, Berkeley.

REFERENCES

Kristensson, G. & P. C. Waterman 1982 The T matrix for acoustic and electromagnetic scattering by circular discs. J. Acoust. Soc. Amer. 72, 1612.

Pao, Y. H. & V. Varatharajulu 1976 Huygen's principle, radiation conditions, and integral transforms for the scattering of elastic waves. J. Acoust. Soc. Amer. 59, 1361.

Morse, P. M. & H. Feshbach 1953 Methods of Theoretical Physics. New York: McGraw-Hill.

Van den Berg, P. M. 1980 The transition matrix in acoustic scattering by a disc. Lecture presented at Institute of Theoretical Physics, Göteborg, Sweden, December 1980.

Van den Berg, P. M. 1982 Scattering of two dimensional elastodynamic waves by a rigid plane strip or a plane crack of finite width: The transition matrix. J. Acoust. Soc. Amer. 72, 1038.

Varatharajulu, V. & Y. H. Pao 1976 Scattering matrix for elastic waves. 1. Theory. J. Acoust. Soc. Amer. 60, 556.

5th International Symposium on Continuum Models of Discrete Systems / Nottingham / 14-20 July 1985

Incompatibility tensors and compatibility equations in orthogonal curvilinear coordinates

Dai Tian-Min
Department of Mathematics, Liaoning University, Shenyang, People's Republic of China

ABSTRACT: In the present paper we derive the specific expressions of incompatibility tensors as well as infinitesimal strain compatibility equations and stress compatibility equations in orthogonal curvilinear coordinates by use of physical frame for nonholonomic system.

1 MATHEMATICAL PRELIMINARIES

In physical frame for nonholonomic system (e.g. see [1]) an arbitrary tensor $\mathbf{A}$ may be written undistinguishly as follows:

$$\mathbf{A} = A_{i\cdots j}\,\bar{\mathbf{g}}_{i\cdots j} \quad , \tag{1}$$

where $A_{i\cdots j}$ are the physical components of $\mathbf{A}$, $\bar{\mathbf{g}}_i = \mathbf{g}_i/H_{\underline{i}}$ and $\bar{\mathbf{g}}_{i\cdots j} = \bar{\mathbf{g}}_i\cdots\bar{\mathbf{g}}_j$ are defined as the unit base vector and the unit base tensor [2] in the nonholonomic system. Throughout this paper a bar over and under a letter denotes that quantity belonging to the nonholonomic system and that index making no sum, respectively.

The partial derivative operator and the Christoffel symbols in the nonholonomic system are

$$\bar{\partial}_i = \frac{1}{H_{\underline{i}}}\partial_i = \frac{1}{H_{\underline{i}}}\frac{\partial}{\partial y_i} \tag{2}$$

and

$$\bar{\Gamma}_{ijk} = \bar{\partial}_i\bar{\mathbf{g}}_j\cdot\bar{\mathbf{g}}_k = -\bar{\partial}_i\bar{\mathbf{g}}_k\cdot\bar{\mathbf{g}}_j = -\bar{\Gamma}_{ikj} \quad , \tag{3}$$

$$\bar{\Gamma}_{ijk} = \frac{1}{H_{\underline{i}}H_{\underline{j}}H_k}\Gamma_{ijk} - \frac{1}{H_{\underline{i}}H_{\underline{j}}}\partial_i H_{\underline{j}}\,\delta_{jk} \quad . \tag{4}$$

The only nonvanishing Christoffel symbols are as follows:

$$\bar{\Gamma}_{\underline{j}i\underline{j}} = -\bar{\Gamma}_{\underline{j}\underline{j}i} = \bar{\partial}_i \ln H_{\underline{j}} \quad . \tag{5}$$

In cylindrical coordinates they are

$$\bar{\Gamma}_{212} = -\bar{\Gamma}_{221} = \frac{1}{r} \quad . \tag{6}$$

In spherical coordinates they are

$$\bar{\Gamma}_{212} = -\bar{\Gamma}_{221} = \bar{\Gamma}_{313} = -\bar{\Gamma}_{331} = \frac{1}{r} \quad ,$$
$$\bar{\Gamma}_{323} = -\bar{\Gamma}_{332} = \frac{\cot\vartheta}{r} \tag{7}$$

The Hamiltonian operator $\boldsymbol{\nabla}$ is defined as

$$\boldsymbol{\nabla} = \bar{\boldsymbol{g}}_i \, \bar{\partial}_i \quad . \tag{8}$$

Let $\boldsymbol{A}$ be a tensor of second order, then the invariant differential operators are as follows:

$$(1) \qquad \mathrm{grad}\,\boldsymbol{A} = \boldsymbol{\nabla A} = (\boldsymbol{\nabla A})_{ijk}\, \bar{\boldsymbol{g}}_{ijk} \ , \tag{9}$$

where

$$(\boldsymbol{\nabla A})_{ijk} = \bar{\partial}_i A_{jk} + \bar{\Gamma}_{imj} A_{mk} + \bar{\Gamma}_{imk} A_{jm} \ ; \tag{10}$$

$$(2) \qquad \mathrm{div}\,\boldsymbol{A} = \boldsymbol{\nabla}\cdot\boldsymbol{A} = (\boldsymbol{\nabla}\cdot\boldsymbol{A})_i \, \bar{\boldsymbol{g}}_i \ , \tag{11}$$

where

$$(\boldsymbol{\nabla}\cdot\boldsymbol{A})_i = \bar{\partial}_j A_{ji} + \bar{\Gamma}_{mjm} A_{ji} + \bar{\Gamma}_{jmi} A_{jm} \ ; \tag{12}$$

$$(3) \qquad \mathrm{curl}\,\boldsymbol{A} = \boldsymbol{\nabla}\times\boldsymbol{A} = (\boldsymbol{\nabla}\times\boldsymbol{A})_{ij}\, \bar{\boldsymbol{g}}_{ij} \ , \tag{13}$$

where

$$(\boldsymbol{\nabla}\times\boldsymbol{A})_{ij} = \epsilon_{mni} (\bar{\partial}_m A_{nj} + \bar{\Gamma}_{mkj} A_{nk} + \bar{\Gamma}_{mkn} A_{kj}) \ ; \tag{14}$$

$$(4) \qquad \nabla^2 \boldsymbol{A} = \boldsymbol{\nabla}\cdot\boldsymbol{\nabla A} = (\nabla^2 \boldsymbol{A})_{ij}\, \bar{\boldsymbol{g}}_{ij} \ , \tag{15}$$

where

$$\begin{aligned}(\nabla^2 \boldsymbol{A})_{ij} = {} & \nabla^2 A_{ij} + \bar{\partial}_m (\bar{\Gamma}_{mkj} A_{ij} + \bar{\Gamma}_{mki} A_{kj}) \\ & + \bar{\Gamma}_{nmn} (\bar{\Gamma}_{mkj} A_{ik} + \bar{\Gamma}_{mki} A_{kj}) \\ & + \bar{\Gamma}_{mli} (\bar{\partial}_m A_{lj} + \bar{\Gamma}_{mkj} A_{lk} + \bar{\Gamma}_{mkl} A_{kj}) \\ & + \bar{\Gamma}_{mlj} (\bar{\partial}_m A_{il} + \bar{\Gamma}_{mkl} A_{ik} + \bar{\Gamma}_{mki} A_{kl}) \ ; \end{aligned} \tag{16}$$

$$(5) \qquad \boldsymbol{\nabla\nabla} f = (\boldsymbol{\nabla\nabla} f)_{ij}\, \bar{\boldsymbol{g}}_{ij} \ , \tag{17}$$

where

$$(\boldsymbol{\nabla\nabla} f)_{ij} = (\bar{\partial}_i \bar{\partial}_j + \bar{\Gamma}_{ikj} \bar{\partial}_k) f \quad . \tag{18}$$

The material derivative of $\boldsymbol{A}$ is

$$\dot{\boldsymbol{A}} = \frac{D\boldsymbol{A}}{Dt} = \left(\frac{D\boldsymbol{A}}{Dt} \right)_{ij} \bar{\boldsymbol{g}}_{ij} \ , \tag{19}$$

where

$$\left(\frac{D\boldsymbol{A}}{Dt} \right)_{ij} = \frac{\partial A_{ij}}{\partial t} + v_k (\partial_k A_{ij} + \bar{\Gamma}_{kmi} A_{mj} + \bar{\Gamma}_{kmj} A_{im}) \ . \tag{20}$$

2 INCOMPATIBILITY TENSOR

Let $\mathbf{E}$ be the strain tensor, then the incompatibility tensor $\boldsymbol{\eta}$ which was introduced by Prof. Kröner (e.g. see [3] - [5]) is

$$\begin{aligned} \boldsymbol{\eta} &= inc\,\mathbf{E} = \boldsymbol{\nabla}\times\mathbf{E}\times\boldsymbol{\nabla} \\ &= [\boldsymbol{\nabla}\times(\boldsymbol{\nabla}\times\mathbf{E})^T]_{ij}\,\bar{\mathbf{g}}_{ij} = \eta_{ij}\,\bar{\mathbf{g}}_{ij} \quad . \end{aligned} \tag{21}$$

In what follows we shall derive its specific expressions in the orthogonal curvilinear coordinates. From (14) we have

$$\begin{aligned} \eta_{ij} &= (\boldsymbol{\nabla}\times\mathbf{M}^T)_{ij} \\ &= \epsilon_{mni}\,(\bar{\partial}_m M^T_{nj} + \bar{\Gamma}_{mKj} M^T_{nK} + \bar{\Gamma}_{mKn} M^T_{Kj}) \quad , \end{aligned} \tag{22}$$

where

$$\begin{aligned} M^T_{ij} &= M_{ji} = (\boldsymbol{\nabla}\times\mathbf{E})^T_{ij} \\ &= \epsilon_{mnj}\,(\bar{\partial}_m E_{ni} + \bar{\Gamma}_{mKi} E_{nK} + \bar{\Gamma}_{mKn} M^T_{Ki}) \quad . \end{aligned} \tag{23}$$

Taking into consideration the only non-vanishing Christoffel symbols (5) and using the following relation

$$\bar{\partial}_i E_{jK} - \bar{\Gamma}_{\underline{jj}i} E_{jK} = \frac{1}{H_{\underline{j}}}\,\bar{\partial}_i(E_{jK} H_{\underline{j}}) \quad , \tag{24}$$

we may obtain from (23) the following expressions:

$$M^T_{11} = \frac{1}{H_3}\bar{\partial}_2(E_{31}H_3) - \frac{1}{H_2}\bar{\partial}_3(E_{21}H_2) + \bar{\Gamma}_{221}E_{32} - \bar{\Gamma}_{331}E_{23} \quad , \tag{25a}$$

$$M^T_{22} = \frac{1}{H_1}\bar{\partial}_3(E_{12}H_1) - \frac{1}{H_3}\bar{\partial}_1(E_{32}H_3) + \bar{\Gamma}_{332}E_{13} - \bar{\Gamma}_{112}E_{31} \quad , \tag{25b}$$

$$M^T_{33} = \frac{1}{H_2}\bar{\partial}_1(E_{23}H_2) - \frac{1}{H_1}\bar{\partial}_2(E_{13}H_1) + \bar{\Gamma}_{113}E_{21} - \bar{\Gamma}_{223}E_{12} \quad , \tag{25c}$$

$$\begin{aligned} M^T_{21} &= \frac{1}{H_3}\bar{\partial}_2(E_{32}H_3) - \frac{1}{H_2}\bar{\partial}_3(E_{22}H_2) \\ &\quad + \bar{\Gamma}_{212}E_{31} + \bar{\Gamma}_{232}E_{33} - \bar{\Gamma}_{332}E_{23} \quad , \end{aligned} \tag{25d}$$

$$\begin{aligned} M^T_{13} &= \frac{1}{H_2}\bar{\partial}_1(E_{21}H_2) - \frac{1}{H_1}\bar{\partial}_2(E_{11}H_1) \\ &\quad + \bar{\Gamma}_{131}E_{23} + \bar{\Gamma}_{121}E_{22} - \bar{\Gamma}_{221}E_{12} \quad , \end{aligned} \tag{25e}$$

$$\begin{aligned} M^T_{32} &= \frac{1}{H_1}\bar{\partial}_3(E_{13}H_1) - \frac{1}{H_3}\bar{\partial}_1(E_{33}H_3) \\ &\quad + \bar{\Gamma}_{323}E_{12} + \bar{\Gamma}_{313}E_{11} - \bar{\Gamma}_{113}E_{31} \quad , \end{aligned} \tag{25f}$$

$$ {}^{\prime}M_{12}^{T} = \frac{1}{H_1}\bar{\partial}_3(E_{11}H_1) - \frac{1}{H_3}\bar{\partial}_1(E_{31}H_3) $$

$$ - \bar{\Gamma}_{121}E_{32} - \bar{\Gamma}_{111}E_{33} + \bar{\Gamma}_{331}E_{13} \quad , \tag{25g} $$

$$ M_{31}^{T} = \frac{1}{H_3}\bar{\partial}_2(E_{13}H_3) - \frac{1}{H_2}\bar{\partial}_3(E_{23}H_2) $$

$$ - \bar{\Gamma}_{313}E_{21} - \bar{\Gamma}_{323}E_{32} + \bar{\Gamma}_{223}E_{32} \quad , \tag{25h} $$

$$ M_{23}^{T} = \frac{1}{H_2}\bar{\partial}_1(E_{32}H_2) - \frac{1}{H_1}\bar{\partial}_2(E_{12}H_1) $$

$$ - \bar{\Gamma}_{232}E_{13} - \bar{\Gamma}_{212}E_{11} + \bar{\Gamma}_{112}E_{21} \quad . \tag{25i} $$

From (22) and (25) the specific expressions of η in cylindrical coordinates may be derived as follows:

$$ \eta_{rr} = \frac{1}{r^2}\frac{\partial^2 E_{zz}}{\partial\theta^2} - \frac{2}{r}\frac{\partial^2 E_{\theta z}}{\partial\theta\partial z} + \frac{\partial^2 E_{\theta\theta}}{\partial z^2} - \frac{2}{r}\frac{\partial E_{zr}}{\partial z} + \frac{1}{r}\frac{\partial E_{zz}}{\partial r} \quad , \tag{26a} $$

$$ \eta_{\theta\theta} = \frac{\partial^2 E_{zz}}{\partial r^2} + \frac{\partial^2 E_{rr}}{\partial z^2} - 2\frac{\partial^2 E_{zr}}{\partial z\partial r} \quad , \tag{26b} $$

$$ \eta_{zz} = \frac{1}{r}\left[\frac{1}{r}\frac{\partial}{\partial r}\left(r^2\frac{\partial E_{\theta\theta}}{\partial r}\right) + \left(\frac{1}{r}\frac{\partial^2}{\partial\theta^2} - \frac{\partial}{\partial r}\right)E_{rr} + \frac{2}{r}\frac{\partial}{\partial r}\left(r\frac{\partial E_{r\theta}}{\partial\theta}\right)\right] , \tag{26c} $$

$$ \eta_{r\theta} = \frac{1}{r}\frac{\partial^2 E_{zr}}{\partial\theta\partial z} - \frac{\partial}{\partial r}\left(\frac{1}{r}\frac{\partial E_{zz}}{\partial\theta}\right) + r\frac{\partial}{\partial r}\left(\frac{1}{r}\frac{\partial E_{z\theta}}{\partial z}\right) - \frac{\partial^2 E_{r\theta}}{\partial z^2} \quad , \tag{26d} $$

$$ \eta_{zr} = \frac{1}{r}\left(\frac{1}{r}\frac{\partial}{\partial r}\left(r\frac{\partial E_{z\theta}}{\partial\theta}\right) - \frac{\partial}{\partial r}\left(r\frac{\partial E_{\theta\theta}}{\partial z}\right) - \frac{1}{r}\frac{\partial^2 E_{zr}}{\partial\theta^2} + \frac{\partial^2 E_{\theta r}}{\partial\theta\partial z} + \frac{\partial E_{rr}}{\partial z}\right] , \tag{26e} $$

$$ \eta_{\theta z} = \frac{1}{r^2}\frac{\partial}{\partial r}\left(r^2\frac{\partial E_{r\theta}}{\partial z}\right) - \frac{1}{r}\frac{\partial^2 E_{rr}}{\partial\theta\partial z} - \frac{\partial}{\partial r}\left[\frac{1}{r}\frac{\partial(rE_{\theta z})}{\partial r}\right] + \frac{\partial}{\partial r}\left(\frac{1}{r}\frac{\partial E_{zr}}{\partial\theta}\right) . \tag{26f} $$

From (22) and (25) the specific expressions of η in spherical coordinates may be derived as follows:

$$ \eta_{rr} = \frac{1}{r^2\sin^2\vartheta}\left[\frac{\partial^2 E_{\theta\theta}}{\partial\varphi^2} + \frac{\sin^2\vartheta}{r}\frac{\partial(r^2E_{\theta\theta})}{\partial r} - \sin\vartheta\cos\vartheta\frac{\partial E_{\theta\theta}}{\partial\theta}\right. $$

$$ + \frac{\partial}{\partial\vartheta}\left(\sin^2\vartheta\frac{\partial E_{\varphi\varphi}}{\partial\theta}\right) + r\sin^2\vartheta\frac{\partial E_{\varphi\varphi}}{\partial r} - 2\frac{\partial^2(\sin\vartheta E_{\theta\varphi})}{\partial\vartheta\partial\varphi} $$

$$ \left. - 2\sin\vartheta\frac{\partial(\sin\vartheta E_{r\theta})}{\partial\vartheta} - 2\sin\vartheta\frac{\partial E_{r\varphi}}{\partial\varphi} - 2\sin^2\vartheta E_{rr}\right] , \tag{27a} $$

$$ \eta_{\theta\theta} = \frac{1}{r^2}\left[\frac{\partial}{\partial r}\left(r^2\frac{\partial E_{\varphi\varphi}}{\partial r}\right) - \frac{1}{\sin^2\vartheta}\frac{\partial^2 E_{rr}}{\partial\varphi^2} - r\frac{\partial E_{rr}}{\partial r} + \cot\vartheta\frac{\partial E_{rr}}{\partial\theta}\right. $$

$$ \left. - \frac{2}{\sin\vartheta}\frac{\partial^2(rE_{r\varphi})}{\partial r\partial\varphi} - 2\cot\vartheta\frac{\partial(rE_{r\theta})}{\partial r}\right] , \tag{27b} $$

$$ \eta_{\varphi\varphi} = \frac{1}{r^2}\left[\frac{\partial}{\partial r}\left(r^2\frac{\partial E_{\theta\theta}}{\partial r}\right) - 2\frac{\partial^2(rE_{r\theta})}{\partial r\partial\theta} + \frac{\partial^2 E_{rr}}{\partial\theta^2} - r\frac{\partial E_{rr}}{\partial r}\right] , \tag{27c} $$

$$\eta_{r\theta} = -\frac{1}{r^2\sin^2\vartheta}\left[r\sin\vartheta \frac{\partial^2(\sin\vartheta E_{\varphi\varphi})}{\partial r\partial\theta} - r\sin\vartheta\frac{\partial^2 E_{\theta\varphi}}{\partial\varphi\partial r} - \frac{\partial^2(\sin\vartheta E_{r\varphi})}{\partial\varphi\partial\vartheta}\right.$$

$$\left. + \frac{\partial^2 E_{r\theta}}{\partial\varphi^2} + 2\sin^2\theta E_{r\theta} - \sin^2\vartheta\frac{\partial E_{rr}}{\partial\theta} - r\sin\theta\cos\vartheta\frac{\partial E_{\theta\theta}}{\partial r}\right], \quad (27d)$$

$$\eta_{\varphi r} = -\frac{1}{r^2\sin^2\vartheta}\left[r\frac{\partial^2 E_{\theta\theta}}{\partial r\partial\varphi} - \frac{r}{\sin\vartheta}\frac{\partial^2(\sin^2\theta E_{\theta\varphi})}{\partial\theta\partial r} + \frac{\partial^2(\sin\vartheta E_{r\varphi})}{\partial\theta^2}\right.$$

$$\left. - \cot\vartheta\frac{\partial(\sin\theta E_{r\varphi})}{\partial\theta} + 2\sin\vartheta E_{r\varphi} - \sin\vartheta\frac{\partial^2}{\partial\vartheta\partial\varphi}\left(\frac{E_{r\vartheta}}{\sin\theta}\right) - \frac{\partial E_{rr}}{\partial\varphi}\right], \quad (27e)$$

$$\eta_{\theta\varphi} = -\frac{1}{r}\left[\frac{\partial^2}{\partial\varphi\partial\vartheta}\left(\frac{E_{rr}}{\sin\vartheta}\right) + \frac{\partial}{\partial r}\left(r^2\frac{\partial E_{\vartheta\varphi}}{\partial r}\right) - \frac{1}{\sin\theta}\frac{\partial^2(rE_{\theta r})}{\partial\varphi\partial r}\right.$$

$$\left. - \sin\vartheta\frac{\partial^2}{\partial\theta\partial r}\left(\frac{rE_{r\varphi}}{\sin\vartheta}\right)\right]. \quad (27f)$$

3 INFINITESIMAL STRAIN COMPATIBILITY EQUATIONS

$$\boldsymbol{\eta} = inc\,\boldsymbol{E} = \boldsymbol{0}, \quad (28)$$

or

$$\eta_{ij} = 0. \quad (29)$$

4 STRESS COMPATIBILITY EQUATIONS

In the absence of body forces the stress compatibility (Beltrami-Michell) equations have the following symbolic form:

$$\boldsymbol{\nabla}\cdot\boldsymbol{\nabla}\boldsymbol{T} + \frac{1}{1+\nu}\boldsymbol{\nabla}\boldsymbol{\nabla}\Theta = \boldsymbol{0} \quad (30)$$

where ν is Poisson ratio and $\Theta = T_{ii}$. With regard to (16) and (18), the Beltrami-Michell equations can be written in the following component form:

$$\nabla^2 T_{ij} + \bar{\partial}_m(\bar{\Gamma}_{mkj}T_{ik} + \bar{\Gamma}_{mki}T_{kj}) + \bar{\Gamma}_{nmn}(\bar{\Gamma}_{mkj}T_{ik} + \bar{\Gamma}_{mki}T_{kj})$$

$$+ \bar{\Gamma}_{mli}(\bar{\partial}_m T_{lj} + \bar{\Gamma}_{mkj}T_{lk} + \bar{\Gamma}_{mkl}T_{kj}) + \bar{\Gamma}_{mlj}(\bar{\partial}_m T_{il} + \bar{\Gamma}_{mkl}T_{ik} + \bar{\Gamma}_{mki}T_{kl})$$

$$+ \frac{1}{1+\nu}(\bar{\partial}_i\bar{\partial}_j + \bar{\Gamma}_{ikj}\bar{\partial}_k)\Theta = 0. \quad (31)$$

From (31) the specific expressions of Beltrami-Michell equations in the cylindrical and spherical coordinates can be derived.

In the absence of body forces the stress equation of equilibrium is satisfied by letting

$$\boldsymbol{T} = inc\,\boldsymbol{\chi}, \quad (32)$$

where $\boldsymbol{\chi}$ is called the stress function tensor.

5 INCOMPATIBILITY TENSORS IN COSSERAT CONTINUUM

From (11) - (16) of [6] for the rectangular coordinates the incompatibility tensors for Cosserat continuum in general symbolic representations may be written as follows:

$$\nabla \times \boldsymbol{\beta}^E - (\boldsymbol{\epsilon} \cdot \boldsymbol{\varkappa}^E) : \boldsymbol{\epsilon} = \mathbf{D} \,, \tag{33a}$$

$$\nabla \times \boldsymbol{\varkappa}^E = \mathbf{B} \,, \tag{33b}$$

$$\nabla \mathbf{v} - \dot{\boldsymbol{\beta}}^E - \boldsymbol{\epsilon} \cdot \mathbf{w} = \mathbf{J} \,, \tag{33c}$$

$$\nabla \mathbf{w} - \dot{\boldsymbol{\varkappa}}^E = \mathbf{S} \,, \tag{33d}$$

where $\mathbf{D}$ is the dislocation density, $\mathbf{B}$ the disclination density, $\mathbf{J}$ the dislocation current density and $\mathbf{S}$ the disclination current density; the other notations are the same as those in [6], too. By use of (10), (14) and (20) and then with regard to non-vanishing Christoffel symbols the specific expressions of the incompatibility tensors for Cosserat continuum in the cylindrical and spherical coordinates may be derived.

ACKNOWLEDGEMENT

The author wishes to express his thanks to Mr. Chen Mian of our University for valuable discussions.

REFERENCES

[1] Guo, Z.H., Theory of nonlinear elasticity (in Chinese), Scientific Publishing House, Beijing (1980)

[2] Zheng, Q.S. and Dai, T.M., Appl. Math. Mech. (English edition) 5 (1984) 1.

[3] Kröner, E., Kontinuumstheorie der Versetzungen und Eigenspannungen, Springer-Verlag (1958).

[4] Teodosiu, C., Elastic Models of Crystal Defects, Springer-Verlag (1982).

[5] Gairola, B.K.D., Nonlinear elastic problems, in: Dislocations in Solids, Vol.1, F. R. N. Nabarro (ed.), North-Holland (1979)

[6] Kluge, G., in: Continuum Models of Discrete Systems 4, O. Brulin & R. K. T. Hsieh (eds.), North-Holland (1981)

5th International Symposium on Continuum Models of Discrete Systems / Nottingham / 14-20 July 1985

Environmentally induced damage in composites

Y. Weitsman
Texas A&M University, College Station, USA

ABSTRACT: A mathematical formulation is proposed for the modelling of damage in fiber-reinforced composite materials due to moisture and temperature. Damage was observed to occur as profuse micro-cracking at the fiber/matrix interfaces and as matrix cracking transversing entire plies. These forms of damage are modelled by a continuum theory, which incorporates "damage" as an internal state variable.

1. INTRODUCTION

It is well known that polymeric-matrix composites absorb moisture when exposed to ambient humid environments. The moisture transport process was modelled, with some success, by the classical (Fickean) diffusion theory, [1], [2] although it has been noticed that both diffusivity and saturation levels were stress-dependent [3], [4]. Several investigators [5]-[6] noticed that moisture caused damage in the form of interfacial cracking between fibers and matrix. This damage was attributed to chemical impurities at the fiber/matrix interfaces which attract water, giving rise to concentrated pockets of high osmotic pressure [7]-[9].

Typical micrographs of moisture-induced damage in AS4/3502 graphite/epoxy composites are shown in Figs. 1 and 2 below. Note that the microcracks meander within the composite so as to embrace the fiber/matrix interfaces. These micro cracks start as isolated, crescent-shape, cracks which coalesce into continuous fracture-lines upon repeated cycling of ambient relative humidity.

Additional experiments on cross-ply laminates have shown that the spacing and intensity of transverse cracks caused by thermal excursions depend not only on the temperature difference ΔT but also on the rate of temperature fluctuation $\dot{T}$ or, alternately, on the temperature gradient ∇T.

In the following presentation the moisture-sorbing composite material is treated as a thermodynamically open system and the distributed, micro-mechanical, damage is represented as an internal state variable. In addition, general forms of damage growth laws are derived for isotropic and transversely isotropic composites.

Fig. 1: Newly forming microcracks in a laminate that was fully saturated at 75% RH and 130°F then dried at 0% RH and 130°F. 3000X

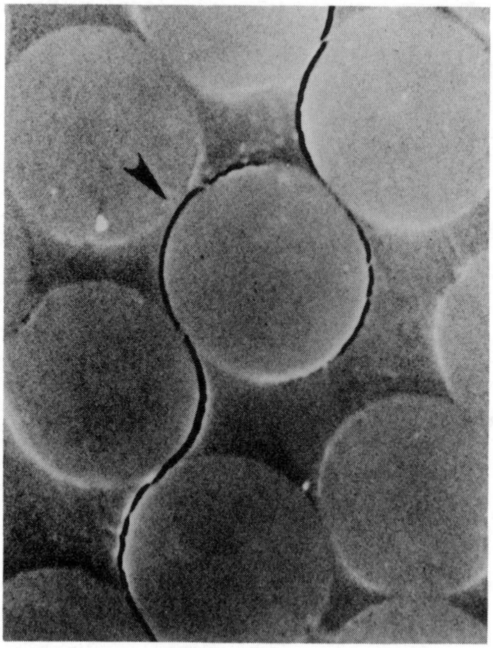

Fig. 2: Coalescence of microcracks in a laminate that was initially saturated at 95% RH and 130°F and then cycled at 9 day intervals between 95% RH and 0% RH at 130°F. 4500X

2. BASIC RELATIONS

Consider a solid contained within a material volume V_s bounded by the surface A_s with solid-mass density ρ_s. The solid contains vapor with a distributed mass m per unit volume of the solid. The vapor within the solid is considered to be in thermodynamic equilibrium with the vapor in a hypothetical, external reservior [10] [11] with pressure $\tilde{p}$, density $\tilde{\rho}$ and internal energy and entropy densities $\tilde{u}$ and $\tilde{s}$, respectively.

Let $\underline{f}$ and $\underline{q}$ denote fluxes of vapor and heat across A_s and let $\underline{v}$ designate the velocity of the solid particles. In addition, let u and s denote the internal energy and entropy densities of the solid-vapor mixture within V_s per unit solid volume.

We obtain the following relations:

2.1 Conservation of mass:

$$\dot{\rho}_s = 0 \qquad (1)$$

$$\dot{m} = -\underline{\nabla} \cdot \underline{f} \qquad (2)$$

2.2 Conservation of energy:

$$\frac{d}{dt}\int_{V_s} \rho_s u \, dV_s = \int_{A_s} \sigma_{ij} n_i v_j dA_s - \int q_i n_i dA_s - \int_{A_s} \tilde{p}\left(\frac{f_i}{\tilde{\rho}}\right) n_i dA_s - \int_{A_s} \tilde{u} f_i n_i dA_s \qquad (3)$$

In (3) σ_{ij} is the Cauchy stress caused by mechanical loads, $\underline{f}/\tilde{\rho}$ is equivalent to the velocity of the vapor across A_s and $\tilde{u}\,\underline{f}$ expresses the vapor-borne energy.

2.3 Entropy inequality:

$$\frac{d}{dt}\int_{V_s} \rho_s s \, dV_s \geq -\int_{A_s} \frac{q_i n_i}{T} dA_s - \int_{A_s} \tilde{s} f_i n_i dA_s \qquad (4)$$

In (4) T denotes temperature and $\tilde{s}\underline{f}$ is the vapor-borne entropy.

Introduce now the Helmholtz free energy of the solid-vapor mixture $\psi = u - Ts$, the enthalpy $\tilde{h}$ and the chemical-potential $\tilde{\mu}$ of the vapor

$$\tilde{h} = (\tilde{p}/\tilde{\rho}) + \tilde{u} \ , \ \tilde{\mu} = \tilde{h} - T\tilde{s}$$

then, upon employment of the divergence theorem and straightforward algebraic substitutions (including eqn. (2)) we obtain the "reduced" entropy inequality

$$-\rho_s \dot{\psi} - \rho_s s \dot{T} + \sigma_{ij} v_{i,j} - (\underline{g} \cdot \underline{q})/T + \tilde{\mu}\dot{m} - \underline{f} \cdot \nabla \tilde{\mu} - \tilde{s}\,\underline{g} \cdot \underline{f} \geq 0 \qquad (5)$$

The last three terms in eqn. (5) are due to the diffusion of vapor into the solid, causing the system to be thermodynamically "open". Also, in (5), $\underline{g} = \nabla T$.

3. CONSTITUTIVE RELATIONS

Assume elastic behavior. Let x_i be the current position of a solid material point which occupied the position X_I in the reference configuration, and let $F_{iJ} = \partial x_i/\partial X_J$ denote deformation gradients. Recognizing that σ_{ij} depends on F_{kL}, q_i depends on g_i and f_i depends of $\tilde{\mu}_{,k} = z_k$, and that the entire response depends also on some damage parameter Δ, we assume the following functional form for the free energy

$$\psi = \tilde{\psi}\ (F_{iJ},\ g_i,\ z_k,\ T,\ m,\ \Delta) \qquad (6)$$

At this point we must make a choice regarding the physical meaning of the damage Δ. A reasonable selection is to assume that Δ is related to the area of cracks or other imperfections within the solid [12],[13]. This implies that Δ is a skew-symmetric second rank tensor (or, equivalently, an axial vector).

Inspection of eqn. (6) indicates that ψ depends on F_{iJ} which is dimensionless, on T and m which can be made dimensionless (thru division by some reference values T_o and m_o, respectively), on g_i and z_k of dimension L^{-1} and on $\Delta_{[ij]}$ of dimension L^2. Considerations of mathematical convenience, such as a desire to linearize eqn. (6) in its various arguments, suggest that it is advantageous to non-dimensionalize $\underline{g}$, $\underline{z}$, and $\underline{\Delta}$. Alternately, such non-dimensionalization can be motivated on physical grounds by considering actual or hypothetical damage processes.

Assume that "damage" represents a statistically homogeneous profuse microcracking in the material, with a density that attains some saturation level beyond which localization takes place. When such localization occurs the distributed cracks coalesce to form some dominant crack, and the damage process is no longer modelled by internal state variable $\underline{\Delta}$ alone. To develop the argument further, assume in addition that the abovementioned saturation level is a material property which is independent of the cause for microcracking. Specifically, the saturation level is independent of F_{iJ}, g_i, z_k, T and m.

Under these assumptions it is possible to consider a characteristic cell, of dimensions $\ell_1 \times \ell_2 \times \ell_3$ which, for convenience, is associated with microcracks at their highest density. Such cells are sketched in Fig. 3 for randomly distributed cracks and in Fig. 4 for transverse cracking in a cross-ply, fiber-reinforced, composite laminate.

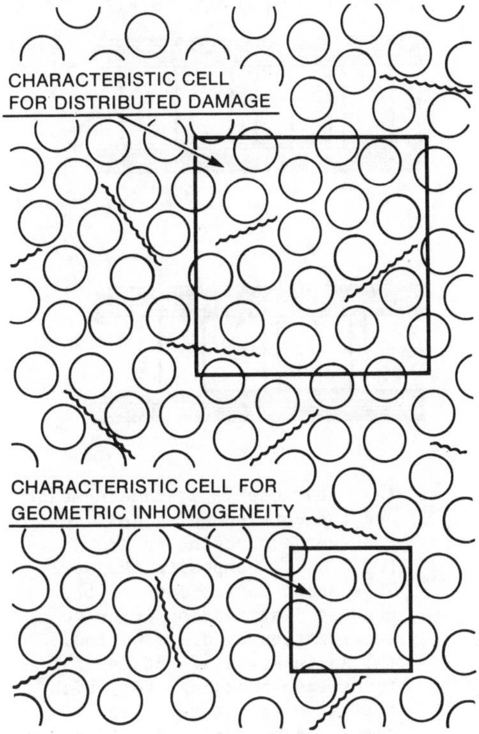

Fig. 3: A schematic drawing showing characteristic cells for damage by micro-cracking and for geometric spacing in fiber-reinforced composites

It is now possible to consider non-dimensional quantities in relation to the abovementioned characteristic cell, these are

$$\tilde{g}_i = \ell_i g_i = \frac{\partial T}{\partial (x_i/\ell_i)}$$

(no sum on i)

$$\tilde{z}_i = \ell_i z_i = \frac{\partial \mu}{\partial (x_i/\ell_i)}$$

$$\text{and } d_{[ij]} = \frac{\Delta_{[ij]}}{|\ell_i \times \ell_j|} \quad \text{(no sum on i or j).}$$

In the sequel we shall retain g_i and z_i as non-dimension gradients.

Substitution of (6) into (5) yields, in view of familiar arguments [14], that ψ cannot depend on $\underline{g}$ and on $\underline{z}$. Together with considerations of frame indifference [15] we obtain

$$\psi = \hat{\psi}\ (E_{KL},\ T,\ m,\ D_{[PQ]})$$

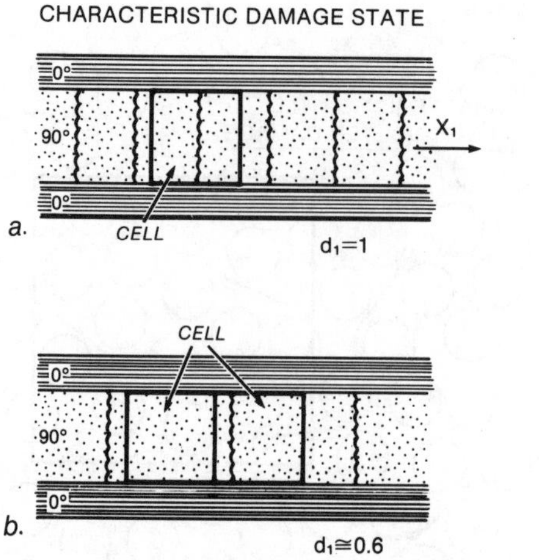

Fig. 4: (a) "characteristic damage state" and the corresponding cell for transverse cracking in a $[0_m°/90_n°]_s$ composite laminate. This state corresponds to the densest distribution prior to localization by delamination. (b) An intermediate state of transverse cracking, for which the value of the relative-damage parameter is $d_1 \simeq 0.6$.

$$s = -\frac{\partial \hat{\psi}}{\partial T}$$

$$\tilde{\mu} = \rho_s \frac{\partial \hat{\psi}}{\partial m} \qquad (7)$$

$$S_{KL} = \frac{\partial(\rho_s \hat{\psi})}{\partial E_{KL}}$$

where $E_{KL} = \frac{1}{2}(F_{iK}F_{iL} - \delta_{KL})$ is the Lagrangian strain, S_{KL} is the Kirchhoff stress (symmetric) and $D_{[PQ]}$ are the components of $d_{[ij]}$ in the reference configuration.

In view of (7) it is obvious that s, $\tilde{\mu}$ and S_{KL} do not depend on $\underline{g}$ and $\underline{z}$.

Furthermore, the entropy inequality (5), also yields

$$\tilde{\mu}_{,i} f_i + g_i\left(-\frac{q_i}{T} + \tilde{s} f_i\right) + r_{[ij]}\,\phi_{[ij]} \le 0 \qquad (8)$$

where $r_{[ij]} = \rho_s \frac{\partial \hat{\psi}}{\partial d_{[ij]}}$ is the affinity to $\phi_{[ij]} = \dot{d}_{[ij]}$

Considerations of frame indifference also lead to the following constitutive relations

$$Q_A = \hat{Q}_A(E_{KL}, G_C, Z_B, D_{[PQ]}, T, m)$$

$$F_A = \hat{F}_A(E_{KL}, G_C, Z_B, D_{[PQ]}, T, m) \qquad (9)$$

$$\Phi_{[IJ]} = \hat{\Phi}_{[IJ]}(E_{KL}, G_C, Z_B, D_{[PQ]}, T, m)$$

In (9) Q_A, F_A, G_C, Z_B and $\Phi_{[IJ]}$ are the counterparts of q_i, f_i, g_i, z_k and $\phi_{[ij]}$ referred to the undeformed configuration.

In view of the inequality (8) it is clear that the thermodynamic process considered herein is associated with internal dissipation and can be related through a dissipation function [16]-[18]. If we view G_C, Z_B and $R_{[KL]}$(the components of $r_{[ij]}$ in the reference configuration) as the components of the generalized thermodynamic force $\underline{\chi}$ and F_A, $Q_A/T + \tilde{s}\,F_A$, $\Phi_{[IJ]}$ as the components of a generalized thermodynamic flux $\underline{J}$, then according to an extended Onsager's principle [18] we have

$$\frac{\partial J_L}{\partial \chi_M} = \frac{\partial J_M}{\partial \chi_L} \qquad (10)$$

The forms of eqns. (9) are therefore restricted by the requirements of eqn. (10).

4. ISOTROPY

For symmetry under the full orthogonal group ψ depends [19] (pp. 288-293) on the three principal invariants of E_{KL}: I_{1E}, I_{2E}, I_{3E}, on the traces of D^2, D^2E, D^2E^2 and D^2EDE^2, on 22 invariants of the form $U_I^{(\alpha)} U_I^{(\alpha)}$, $U_I^{(\alpha)} E_{IJ} U_J^{(\alpha)}$, $U_I^{(\alpha)} (E^2)_{IJ} U_J^{(\alpha)}$, $U_I^{(\alpha)} (D^2)_{IJ} U_J^{(\alpha)}$, $U_I^{(\alpha)} (DE)_{IJ} U_J^{(\alpha)}$, ..., $U_I^{(\alpha)} (DE^2D^2)_{IJ} U_J^{(\alpha)}$ where $\alpha = 1,2$ with $\underline{U}^{(1)} = \underline{G}$ and $\underline{U}^{(2)} = \underline{Z}$, seven invariants of the form $G_I(DE)_{IJ} Z_J + Z_I(DE)_{IJ} G_J$ and five invariants of the form $G_I(ED)_{IJ} Z_J - Z_I(ED)_{IJ} G_J$. All in all 41 invariants.

Employment of procedures of refs. [19] and [20] yields the following expressions for the damage growth rate and the fluxes of heat and vapor mass:

$$\begin{aligned}\Phi_{[IJ]} = {} & \Phi^{(1)} D_{[IJ]} + \Phi^{(2)}(G_I Z_J - Z_I G_J) \\ & + \Phi^{(3)}(D_{[JK]}E_{KI} - D_{[IK]}E_{KJ}) \\ & + \Phi^{(4)}\left[G_K(Z_J D_{[KI]} - Z_I D_{[KJ]}) + Z_K(G_J D_{[KI]} - G_I D_{[KJ]})\right] \\ & + \ldots\ \Phi^{(51)}(\ldots\ldots)\end{aligned} \qquad (11)$$

$$Q_I = Q^{(1)} G_I + Q^{(2)} F_I + Q^{(3)} E_{IJ}\, G_J + \ldots$$

$$F_I = F^{(1)} G_I + F^{(2)} F_I + F^{(3)} E_{IJ}\, G_J + \ldots$$

In (11) $\Phi^{(i)}$, $Q^{(j)}$ and $F^{(j)}$ depend on the abovementioned 41 invariants.

5. TRANSVERSE ISOTROPY

Consider a fiber-reinforced composite with fibers parallel to the x_3 axis. Assume the material to possess rotational symmetry about x_3 and reflective symmetry about the $x_1 - x_2$ and $x_2 - x_3$ planes. In this case [21] ψ depends on the following 14 invariants: E_{33}, $E_{\alpha\alpha}$, $(E_{11} - E_{22})^2 + 4E_{12}^2$, $E_{3\alpha}E_{3\alpha}$, $E_{3\alpha}D_\alpha$, $D_\alpha D_\alpha$, D_3^2, $\det E_{IJ}$, $(E_{11} - E_{22})(E_{31}D_1 - E_{32}D_2) + 2E_{12}(E_{13}D_2 + E_{23}D_1)$, $(E_{11} - E_{22})(D_1^2 - D_2^2) + 4E_{12}D_1D_2$, $D_3(E_{31}D_2 - E_{32}D_1)$, $D_3[(E_{11} - E_{22})E_{31}E_{32} - E_{12}(E_{31}^2 - E_{32}^2)]$, $D_3[(E_{11} - E_{22})(E_{31}D_2 + E_{32}D_1) - 2E_{12}(E_{31}D_1 - E_{32}D_2)]$, and $D_3[(E_{11} - E_{22})D_1D_2 - E_{12}(D_1^2 - D_2^2)]$. $(\alpha = 1,2)$ Denote $\Phi_K = \frac{1}{2} e_{IJK}\Phi_{[IJ]}$, then the rates of damage growth are given by

$$\Phi_1 = f^{(1)}D_1 + f^{(2)}(D_3E_{31} + D_2E_{21}) + f^{(3)}[D_1(Z_1^2 - Z_2^2) + \ldots] + \ldots + f^{(40)}[(E_{11}-E_{22})(D_3Z_3G_2 + D_1Z_1G_2)]$$

$$\Phi_2 = f^{(1)}D_2 - f^{(2)}(D_3E_{32} + D_1E_{21}) + f^{(3)}[D_2(Z_1^2 - Z_2^2) + \ldots] + \ldots + f^{(40)}[\ldots\ldots]$$

$$\Phi_3 = h^{(1)}D_3 + h^{(2)}(D_1E_{31} - D_2E_{32}) + \ldots + h^{(7)}(\ldots\ldots) \qquad (12)$$

In (12) $f^{(i)}$ $(i=1,2,\ldots 40)$ and $h^{(i)}$ $(i=1,\ldots 7)$ depend on 84 invariants formed from E_{KL}, G_K and $D_M = \frac{1}{2} E_{MPQ} D_{[PQ]}$.

Some of the terms, like $\Phi^{(1)}$ in eqn. $(11)_1$ and $f^{(1)}$, $h^{(1)}$ in eqns. (12) lend themselves to physical interpretation. Specifically $\Phi^{(1)}$ is related to the critical energy release rate associated with self-similar damage growth, while $f^{(1)}$ and $h^{(1)}$ are related to similar release rates associated with damage growths in planes parallel and normal to the fiber direction.

6. CONCLUDING REMARKS

This presentation provided some tentative steps towards the modelling of the degradation of composite materials. In view of the highly complex internal structure of composite materials, with the multitude of interactions between fibers and resin and between plies of different orientation, this phenomenon appears to be more complicate than, say, the multiple fracturing which occurs in rocks. The major complexity appears to be associated with the appropriate forms of the damage growth relations, namely of functions like $f^{(i)}$ and $h^{(i)}$ in eqns. (12). An extensive experimental program, associated with the concepts of continuum damage modelling, may hopefully lead to a distilled and sufficiently simple method for predicting the durability of composites.

Acknowledgements

This work was performed, in part, under Contract N00014-82-K-0562 from the Mechanics Division, Engineering Sciences Directorate of ONR and, in part, under Grant 84-0069 from the Directorate of Aerospace Sciences of AFOSR.

REFERENCES

[1] P. Bonniau and A.R. Bunsell: "A Comparative Study of Water Absorption Theories Applied to Glass Epoxy Composites". In Environmental Effects on Composite Materials G.S. Springer, Editor. Volume 2. 1984, pp. 209-229.

[2] C.D. Shirrell: "Diffusion of Water Vapor in Graphite/Epoxy Composites" In Advanced Composite Materials - Environmental Effects. J.R. Vinson, Editor ASTM STP 658, 1977, pp. 21-42.

[3] O. Gillat and L.J. Broutman: "Effect of an External Stress on Moisture Diffusion and Degradation in a Graphite Reinforced Epoxy Laminate" ibid, pp. 61-83.

[4] M.C. Henson and Y. Weitsman: "Experimental Observations of Stress Assisted Diffusion in a Fiber Reinforced Composite and its Resin". Texas A&M University Report (Forthcoming).

[5] R.J. DeIasi, J.B. Whiteside and W. Wolter: "Effects of Varying Hygrothermal Environments on Moisture Absorption in Epoxy Composites" Proc. Army, AF, Navy, NASA 4th Conf. on Fibrous Composites in Structural Design. San Diego CA Nov. 1978.

[6] S.P. Jackson and Y. Weitsman: "Moisture Effects and Moisture Induced Damage in Composites" Proc. of 5th Intern. Conf. on Comp. Mat. (ICCM V) San Diego, CA July 30-Aug. 2, 1985 (Forthcoming)

[7] K.H.G. Ashbee and R.C. Wyatt: "Water Damage in Glass Fibre/Resin Composites" Proc. Roy. Soc. A. Vol. 312. 1969 pp. 553-564.

[8] E. Walter and K. Ashbee: "Osmosis in Composite Materials" Composites, Oct. 1982, pp. 365-368.

[9] L.T. Drzal, M.J. Rich and M.F. Koenig: "Adhesion of Graphite Fibers to Epoxy Matrices, III: The Effect of Hygrothermal Exposure". J. of Adhesion (to appear)

[10] M.A. Biot: "Theory of Finite Deformations of Porous Solids" Indiana U. Math. J. Vol. 21, No. 7, 1972, pp. 597-620.

[11] J.R. Rice and M.P. Cleary: "Some Basic Stress Diffusion Solutions for Fluid-Saturated Elastic Porous Media With Compressible Constituents" Rev. Geophysics and Space Res. Vol. 14, No. 2, 1976, pp. 227-241.

[12] L. Davison and A.L. Stevens: "Thermomechanical Constitution of Spalling Bodies" J. App. Phys. Vol. 44, 1973, pp. 667-674.

[13] D. Krajcinovic and G.V. Fonseka: "The Continuum Damage Theory of Brittle Materials" J. App. Mech. Vol. 48, 1981, pp. 809-812.

[14] B.D. Coleman and M.E. Gurtin: "Thermodynamics with Internal State Variables" J. Chem. Phys. Vol. 47, No. 2 (1967) pp. 597-613.

[15] W. Jaunzemis: "Continuum Mechanics" MacMillan Co. (1967), pp. 286-288.

[16] J. Kestin and J. Bataille: "Irreversible Thermodynamics of Continua with Internal Variables" in Continuum Models for Discrete Systems, Univ. of Waterloo Press, 1977, pp. 39-67.

[17] J. Bataille and J. Kestin: "Irreversible Processes and Physical Interpretation of Rational Thermodynamics" J. Non-Equil. Thermodyn. Vol. 4, 1979, pp. 229-258.

[18] D.G.B. Edelen: "On The Existence of Symmetry Relations and Dissipation Potentials", Arch. Rat. Mech. Anal. Vol. 51, 1973, pp. 218-227.

[19] A.J.M. Spencer: "Theory of Invariants" in Continuum Physics Vol. 1 A.C. Eringen Editor, Academic Press, 1971, pp. 239-353.

[20] A.C. Pipkin and R.S. Rivlin: "The Formulation of Constitutive Equations in Continuum Physics. I", Arch Rational Mech. Anal., Vol. 4, 1959, pp. 129-144.

[21] G.F. Smith: "On Transversely Isotropic Functions of Vectors, Symmetric Second-Order and Skew-Symmetric Second-Order Tensors". Quart. App. Math. Vol. 39 1982, pp. 509-516.

5th International Symposium on Continuum Models of Discrete Systems / Nottingham / 14-20 July 1985

Possibilities and limits of Nayfeh's equations for heat conduction in laminated composites

P.Mazilu
Technische Hochschule Darmstadt, FR Germany

ABSTRACT: By an accurate averaging of the heat equation over one dimension Nayfeh has derived a simple coupled system of parabolic equations. In recent years other scholars have reconsidered the same equations (without no reference to Nayfeh's work) for other purposes. In the present contribution the possibilities and the limits of applicatibility of Nayfeh's equations are investigated.

1 INTRODUCTION

The philosophical reluctance regarding the attribution to the same geometrical point of two or more different values of the same field is well known. Concerning the temperature, in his survey paper devoted to the inmiscible mixtures, Bedford(1983) states : "After stimulating debate and several collateral developments Dunwoody and Müller, Truesdell, Bowen and Garcia, Crain et al. presented theories of mixtures of continua which permited the constituent to have different temperatures and which yield results that agree with the classical thermodynamics."

Actually the hystory of the theories involving two temperatures is much longer and more stimulating than the founders of inmiscible mixtures doctrine seem to suppose.

The first mathematical model of a mixture involving two temperatures was suggested by Schumann(1929). His model considers a liquid which flows through a mass of crushed material. The equations are derived under the assumption that the heat imparted to a fluid volume by the solid is proportional to the temperature-difference.

If one denotes by T_g the temperature of the fluid and by T_s the temperature of the solid then the resulting equations are

$$\frac{\partial T_s}{\partial t} = k_1(T_g - T_s),$$
$$\frac{\partial T_g}{\partial t} + v\frac{\partial T_g}{\partial x} = -k_2(T_g - T_s). \qquad (1.1)$$

In these equations, compared to the transfer of heat from solid to fluid, the heat transfered by conduction is assumed to be so small that it can be neglected. According to the available informations the first who completes Schumann's equations by considering the conduction was Rubinstein(1948). In order to describe the heat transfer through heterogeneous materials Rubinstein suggested the following system of parabolic equations

$$\frac{\partial T_1}{\partial t} = k_1 \Delta T_1 - a(T_1 - T_2),$$
$$\frac{\partial T_2}{\partial t} = k_2 \Delta T_2 + a(T_1 - T_2). \qquad (1.2)$$

The subsequent evolution of the problem of two-temperatures model can be followed only in the framework of the analogous problems in infiltration theory. Various authors have postulated the equations (1.2) as proper to describe the flow through fissured porous media .Such theories, connected to the names of

Barenblatt, Zheltov, Kochina, Pilatovski and Stchelkachev have been critically examined in the Gheorghitza's monograph (1966). As it was pointed out, none of the russian authors had pointed out the procedure showing how and for which type of heterogeneous media the system (1.2) has been derived.

In the available literature the first exact derivation of the equations (1.2) seems to be given by Nayfeh (1975). He states rigurously that these equations correspond to the heat transfer through laminated composites. In spite of this publication, the subject of heat transfer into laminated composites seems to remain classified till the end of 1983. According to Schemmer (1983) some "widely distributed" military journals published details regarding the so called "active" and "reactive armor" which seems to represent the main application of the equations of heat transfer through laminated composites.

In the following we shall give a short presentation of Nayfeh's results.

Let us consider the heat flowing in a direction parallel to the layers of a periodic array of homogeneous and isotropic laminate, perfectly bounded at their interfaces.
The pair of "fiber" and "matrix" constituents in the generating unit cell is denoted by M_1 and M_2 respectively. The central role in the Nayfeh's approach is played by the averaging operator

$$\overline{(\ )}_\alpha = \frac{1}{h_\alpha}\int_0^{h_\alpha} (\ )_\alpha \, dy \ ,$$

where h_α denotes the thickness of component α (α=1,2). The procedure consists in applying the average to the energy balance and to the constitutive equation for the heat conduction. The resulting equation equations are

$$h_\alpha k_\alpha \frac{\partial^2 \overline{T}_\alpha}{\partial x} - h_\alpha \rho_\alpha c_\alpha \frac{\partial \overline{T}_\alpha}{\partial t} = -q_{y\alpha} \ , \alpha=1,2 \qquad (1.3)$$

where $q_{y\alpha}$ denotes the flux accros interface. According to the usual continuity conditions accros laminated interfaces, the equalities

$$q_{y2}(x,h_2,t) = -q_{y2}(x,-h_2,t) = q_{y1}(x,h,t) = q^*(x,t)$$

hold.
Assuming the heat flux to be linear with respect to y_α Nayfeh derived the following two equations

$$T_1(x,h_1,t) - \overline{T}_1(x,t) + (h_1/3k_1)q^* = 0$$
$$T_2(x,h_2,t) - \overline{T}_2(x,t) - (h_2/3k_2)q^* = 0. \qquad (1.4)$$

It follows

$$q^* = K(\overline{T}_1 - \overline{T}_2), \qquad (1.5)$$

where

$$\frac{1}{K} = \frac{h_1}{3k_1} + \frac{h_2}{3k_2} .$$

Setting (1.5) in (1.3) one finds

$$h_1 k_1 \frac{\partial^2 \overline{T}_1}{\partial x^2} - h_1 \rho_1 c_1 \frac{\partial \overline{T}_1}{\partial t} = K(\overline{T}_1 - \overline{T}_2)$$
$$h_2 k_2 \frac{\partial^2 \overline{T}_1}{\partial x^2} - h_2 \rho_2 c_2 \frac{\partial \overline{T}_2}{\partial t} = K(\overline{T}_1 - \overline{T}_2) \qquad (1.6)$$

In the case of the diffusion in both x- and z-direction (1.6) become

$$h_1 k_1 \Delta \overline{T}_1 - h_1 \rho_1 c_1 \frac{\partial \overline{T}_1}{\partial t} = K(\overline{T}_1 - \overline{T}_2)$$
$$h_2 k_2 \Delta \overline{T}_2 - h_2 \rho_2 c_2 \frac{\partial \overline{T}_2}{\partial t} = -K(\overline{T}_1 - \overline{T}_2) \qquad (1.7)$$

Remark 1. Since the heat flux $q_{y\alpha}$ are linear in y and $q(x,0,t) = q(x,0,t) = 0$ from (1.4) it follows

$$T(x,h_1,t) - \overline{T}_1(x,t) = O(h_1^2),$$
$$T(x,h_2,t) - \overline{T}_2(x,t) = O(h_2^2).$$

If h_1 and h_2 are suficiently small so that h_1^2 and h_2^2 can be neglected, then $\overline{T}_1 = \overline{T}_2 = \overline{T}$ hold. The system (1.7) reduces then to the classical one-temperature equation

$$(h_1 k_1 + h_2 k_2)\Delta \overline{T} - (h_1 \rho_1 c_1 + h_2 \rho_2 c_2) \frac{\partial \overline{T}}{\partial t} = 0$$

For such thin layers two-temperatures model remains valid only if the components are separeted by a bonding layer. This case has been considered

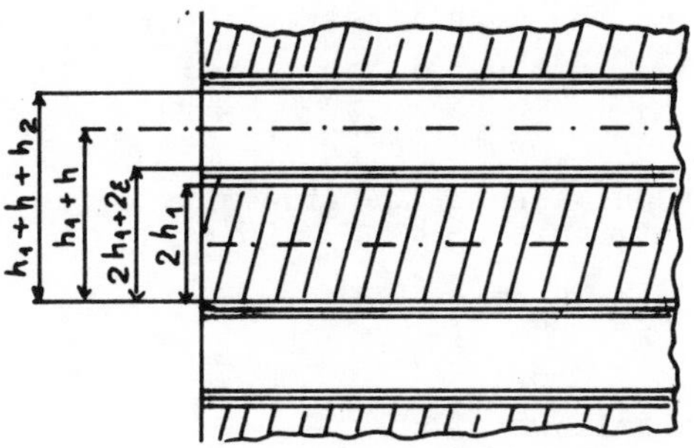

Fig.1 Composite Geometry

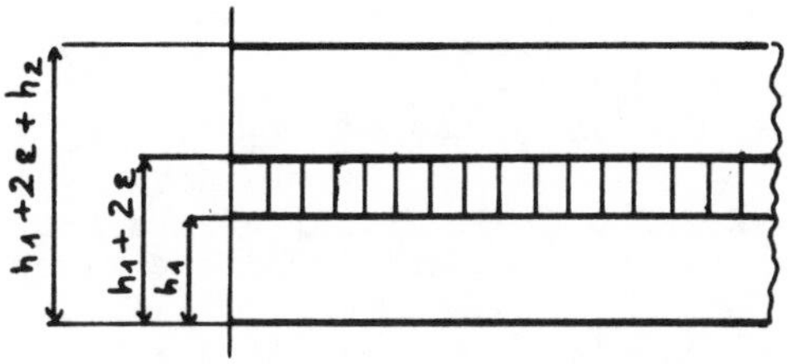

Fig.2 Sandwich Plates Geometry

by Nayfeh (1980),(fig.1).
Remark 2. Because of the periodicity of the solution, on the mean planes of the constituents, the heat flux is null. For this reason there is an analogy between the heat transfer in laminated composites and the heat transfer through a system of sandwich plates fully insulated on their upper and lower faces (fig.2).

2 GENERALIZATION OF NAYFEH'S EQUATIONS

The systems of equations (1.6) and (1.7) represent the equations for two-temperatures models corresponding to one- and two-dimensional flow respectively. The procedure given by Nayfeh can be easly applied to n-dimensional heat equation

$$\frac{\partial T}{\partial t} = a(x_n)\Delta_n T, \qquad (2.1)$$

where a(.) denotes a piece wise constant function

$$a(z) = \begin{cases} a_2 = \text{const. for } -h_2-\varepsilon<z<\varepsilon \\ a_o = \text{const. for } -\varepsilon<z<\varepsilon \\ a_1 = \text{const. for } \varepsilon<z<h_1+\varepsilon \end{cases}$$

This was done by Mazilu (1984). The following equations were derived

$$\begin{aligned} \frac{\partial \overline{T}_1}{\partial t} &= a_1\Delta_{n-1}\overline{T}_1 - K_1(\overline{T}_1 - \overline{T}_2), \\ \frac{\partial \overline{T}_2}{\partial t} &= a_2\Delta_{n-1}\overline{T}_2 + K_2(\overline{T}_1 - \overline{T}_2) \end{aligned} \qquad (2.2)$$

where K_1, K_2 are constants. This system represents a direct generalization to (n-1)-dimensions of the equations (1.7) derived by Nayfeh for one and two dimensions.
Recently Margolis (1978), Aifantis (1979), Aifantis and Hill(1980), Aifantis and Beskos(1980) and Aifantis (1980) have used Nayfeh's equations for such purposes like description of "the degradation of a thermocline in a packed bed thermal storage tank", "diffusion in the presence of a continuum distribution of high-diffusivity paths, such as grain boundaries and dislocations" or extraction of "geothermal energy". Unfortunately, without any exception, the authors of these ambitious projects make no reference to Nayfeh's work.
Henceforth we shall investigate the possibilities and the limits of applicability of Nayfeh's equations. First we shall give a physical interpretation of the equations (2.2) for n=4.

3 INTERPRETATION OF THREE-DIMENSIONAL NAYFEH'S EQUATIONS

Let us consider a three-dimensional orthogonal nettwork of circular cylindrical bars having the core and the mantel made from different materials (fig.3). Between the mantel and the core there is a bonding layer. This bonding layer is assumed to have a very small thermal conductivity k_o. We assume the heat flow in any particular bar to have axial symmetry. Let such a bar be paralel to x-direction. If by $T_x(x,r,t)$ one denotes the temperature then the equation

$$c_\alpha \rho_\alpha \frac{\partial T_x}{\partial t} = k_\alpha \frac{\partial^2 T_x}{\partial x^2} + \frac{k_\alpha}{r}\frac{\partial}{\partial r}\left(r\frac{\partial T_x}{\partial r}\right), \quad \alpha=0,1,2 \qquad (3.1)$$

holds. By $k_\alpha, \rho_\alpha, c_\alpha$ one denotes the

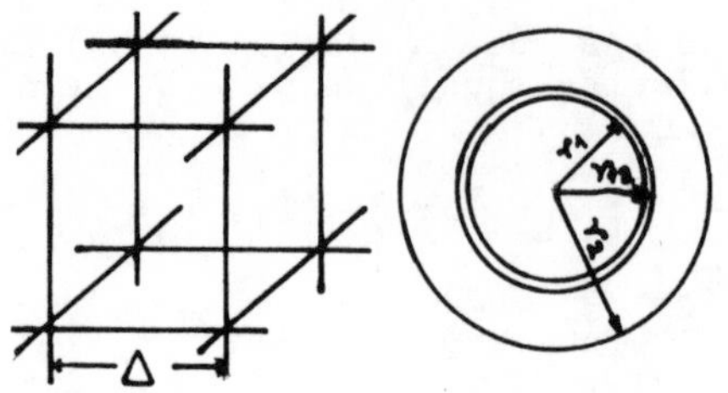

Fig.3 Three-dimensional Nettwork

thermal conductivity, density and thermal capacity in core (for $\alpha=1$), mantel (for $\alpha=2$) and bonding material (for $\alpha=0$). The interface continuity conditions are

$$T_x = \text{continuous for } r=r_1 \text{ and } r=r_{12}$$

$$k_1\frac{\partial T}{\partial r}\Big|_{r=r_1} = k_o\frac{\partial T}{\partial r}\Big|_{r=r_1}, \qquad (3.2)$$

$$k_o\frac{\partial T}{\partial r}\Big|_{r=r_{12}} = k_2\frac{\partial T}{\partial r}\Big|_{r=r_{12}}.$$

The boundary conditions are assumed to be

$$k_1\frac{\partial T}{\partial r}\Big|_{r=o} = 0 , \qquad (3.3)$$

$$k_2\frac{\partial T}{\partial r}\Big|_{r=r_2} = 0 .$$

By using the procedure suggested by Nayfeh one can describe such a process by means of a two-temperatures model.

Let us denote

$$\overline{T}_{x_1} = \frac{1}{\pi r_1^2}\int_o^{r_1}\int_o^{2\pi} rT_x(x,r,t)drd\theta ,$$

$$\overline{T}_{x_2} = \frac{1}{\pi(r_2^2-r_{12}^2)}\int_{r_{12}}^{r_1}\int_o^{2\pi} rT_x(x,r,t)drd\theta$$

the mean temperatures over the core and over the mantel respectively. By applying these average operators to the equations (3.1) it follows

$$c_1\rho_1\frac{\partial\overline{T}_{x1}}{\partial t} = k_1\frac{\partial\overline{T}_{x1}}{\partial x^2} + \frac{2k_1}{r_1}\frac{\partial T_x}{\partial r}\Big|_{r=r_1},$$

$$(3.4)$$

$$c_2\rho_2\frac{\partial\overline{T}_{x2}}{\partial t} = k_2\frac{\partial\overline{T}_{x2}}{\partial x^2} - \frac{2r_{12}}{r_2^2-r_{12}^2}k_2\frac{\partial T_x}{\partial r}\Big|_{r=r_{12}}.$$

From the constitutive law written in the bonding layer

$$q_r = -k_o\frac{\partial T}{\partial r},$$

if one assume the linearity of the heat flux

$$q_r = -k_o(T_x\big|_{r_{12}} - T_x\big|_{r_1}) + O(\varepsilon)$$

it follows

$$k_1\frac{\partial T}{\partial r}\Big|_{r_1} = k_2\frac{\partial T}{\partial r}\Big|_{r_{12}} + O(\varepsilon) =$$

$$= k_o(T_x\big|_{r_{12}} - T_x\big|_{r_1}) + O(\varepsilon). \qquad (3.5)$$

By developing in Taylor series and by taking into account (3.3) one finds

$$T_x(r) = T_x(o)+O(r_1^2), o<r<r_1 ,$$

$$(3.6)$$

$$T_x(r)=T_x(r_2)+O(|r_2-r_{12}|^2), r_{12}<r<r_2,$$

from where, by averaging, one obtains

$$\overline{T}_{x1} = T_x(o) + O(r_1^2) ,$$

$$\overline{T}_{x2} = T_x(r_2)+O(|r_2-r_{12}|^2). \qquad (3.7)$$

From (3.5),(3.6),(3.7) and under the assumption that $\varepsilon, r_1^2, (r_2-r_{12})^2$ are negligible one finds

$$k_1\frac{\partial T_x}{\partial r}\Big|_{r_1} = k_2\frac{\partial T_x}{\partial r}\Big|_{r_{12}} = k_o(T_{x2}-T_{x1}).$$

$$(3.8)$$

Setting (3.8) in (3.4) follows (one identifies $r_{12}=r_1$)

$$c_1\rho_1\frac{\partial\overline{T}_{x1}}{\partial t} = k_1\frac{\partial^2\overline{T}_{x1}}{\partial x^2} + \frac{2k_o}{r_1}(\overline{T}_{x2}-\overline{T}_{x1}) ,$$

$$(3.9)$$

$$c_2\rho_2\frac{\partial\overline{T}_{x2}}{\partial t} = k_2\frac{\partial\overline{T}_{x2}}{\partial x^2} - \frac{2r_1}{r_2^2-r_1^2}k_o(\overline{T}_{x2}-\overline{T}_{x1}).$$

Similar system of equations can be derived for y- and z-direction. The equations (3.9) and those which are corresponding to y-and z-direction do not regard any kind of reciprocal interaction. Actually in knots the temperatures T_x, T_y, T_z are equalized by the interchanged heat fluxes. This implies that the equations(3.9) and the similar ones remain valid only between knots. Since in knots $T_x = T_y = T_z$ holds it follows

$$\overline{T}_{x\alpha} = \overline{T}_{y\alpha} = \overline{T}_{z\alpha} = \overline{T}_{\alpha} \quad .$$

The heat supplied in knots has two origins: the heat $Q_{x\alpha}, Q_{y\alpha}, Q_{z\alpha}$ transfered through the same phase, or that interchanged between phases $R_{x12}, R_{x21}, \ldots\ldots, R_{z12}, R_{z21}$.
From the equations describing the heat transfer in x-,y-and z-direction (i.e.(3.9) and similar ones) we find

$$Q_{x\alpha} = k_\alpha \frac{\partial^2 T_{x\alpha}}{\partial x^2} \quad ,$$

$$R_{x12} = \frac{2k_o}{r_1} (\overline{T}_{x2} - \overline{T}_{x1}) = \frac{2k_o}{r_1} (\overline{T}_2 - \overline{T}_1) \quad ,$$

$$R_{x21} = -\frac{2r_1}{r_2^2 - r_1^2} k_o (\overline{T}_{2x} - \overline{T}_{1x}) =$$

$$= -\frac{2r_1}{r_2^2 - r_1^2} k_o (\overline{T}_2 - \overline{T}_1)$$

and so on.
From the energy balance one finds

$$c_1\rho_1 \frac{\partial \overline{T}_1}{\partial t} = k_1 \left(\frac{\partial^2 \overline{T}_{x1}}{\partial x^2} + \frac{\partial^2 \overline{T}_{y1}}{\partial y^2} + \frac{\partial^2 \overline{T}_{z1}}{\partial z^2} \right) + \frac{6k_o}{r_1} (\overline{T}_2 - \overline{T}_1) \quad , \qquad (3.10)$$

$$c_2\rho_2 \frac{\partial \overline{T}_2}{\partial t} = k_2 \left(\frac{\partial^2 \overline{T}_{x2}}{\partial x^2} + \frac{\partial^2 \overline{T}_{y2}}{\partial y^2} + \frac{\partial^2 \overline{T}_{z2}}{\partial z^2} \right) - \frac{6k_1}{r_2^2 - r_1^2} k_o (\overline{T}_2 - \overline{T}_1) \quad .$$

Let (x,y,z) be the coordinates of a knot and Δ the distance between two adjacent knots. Since

$$\frac{\partial^2 T_{x\alpha}}{\partial x^2} \simeq \frac{1}{\Delta} \left[\frac{\overline{T}_\alpha(x+\ ,y,z) - \overline{T}_\alpha(x,y,z)}{\Delta} - \frac{T_\alpha(x,y,z) - T_\alpha(x-\Delta,y,z)}{\Delta} \right] \simeq \frac{\partial^2 T_\alpha}{\partial x^2}$$

and the similar ones hold approximately, the equations (3.10) become

$$c_1\rho_1 \frac{\partial \overline{T}_1}{\partial t} = k_1 \Delta \overline{T}_1 + \frac{6k_o}{r_1} (\overline{T}_2 - \overline{T}_1) \qquad (3.11)$$

$$c_2\rho_2 \frac{\partial \overline{T}_2}{\partial t} = k_2 \Delta \overline{T}_2 - \frac{6r_1}{r_2^2 - r_1^2} k_o (\overline{T}_2 - \overline{T}_1) \quad .$$

These equations are just Nayfeh's equations corresponding to three-dimensional case.

4 LIMITS OF APPLICABILITY

We shall establish the most general case wich still allows us the application of Nayfeh's procedure.
Let us denote by $S_\alpha, \alpha = 1,2,\ldots n$ the geometrical support of the phase α. The first condition requiered by the above averaging procedure consists of possibility of representing all S_α in form of Cartesian product

$$S_\alpha = S_\alpha \times S_o \quad , \qquad (4.1)$$

where S_0 is 1-D interval or 2-D domain.
In the case of the above considered laminated composites or sandwich plates S_o is a domain in the layer plane, while S_α are intervals orthogonal to this plane.
In the case of the three-dimensional nettwork, S_o are intervals parallel to x-,y-,or z-direction and S_α, disks or circular annuli normal to S_o.
The second condition require the proportionality of the interchanged heat $Q_{\alpha\beta}$ (between the phase α and β) and the difference of the mean temperatures

$$Q_{\alpha\beta} = \kappa_{\alpha\beta} (\overline{T}_\beta - \overline{T}_\alpha) \qquad (4.2)$$

with $\kappa_{\alpha\beta}$ constants.
We can summarize the above results in form of the following
Theorem. Nayfeh's procedure of averaging is applicable provided the conditions (4.1) and (4.2) hold.
We can now give counterexamples which show that Nayfeh's procedure is not allways applicable.

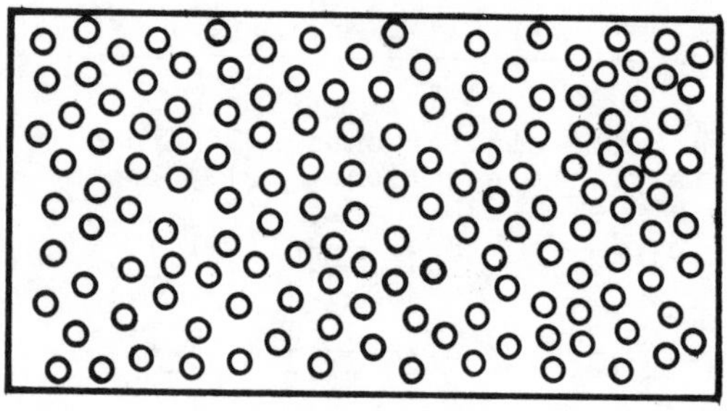

Fig.4 Inclusions with infinite Conductivity

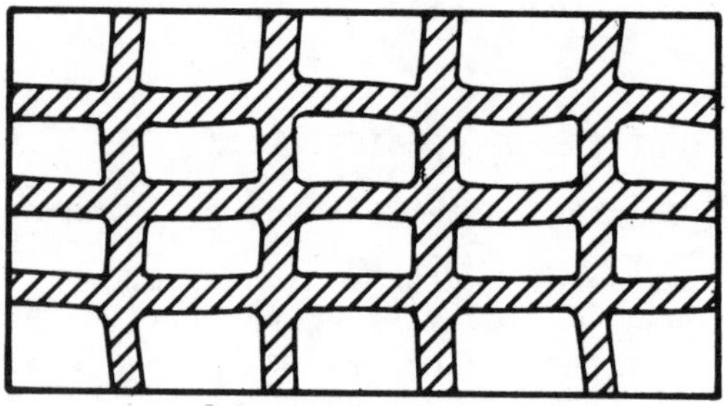

Fig.5 Bands with Zero Conductivity

First let us remark that between two-dimensional Nayfeh's equations and the problem of heat transfer through laminated sandwich plates there is a one-to- one correspondence. Having this in mind let us consider the following two limiting cases:

a) A plate containing inclusions with infinite conductivity (fig.4).

Let us assume that $k_m, 0<k_m<+\infty$ is the conductivity of the matrix and that $k_i=+\infty$ is the conductivity of the inclusions.The macroscopic conductivity of the heterogeneous medium will be $k^*, k_m<k^*<+\infty$. A sandwich plate system from the same constituents will be obviously of infinite conductivity $k^*=+\infty$.

b) A plate with zero conductivity bands (fig.5).

Let us denote by $k_m, 0<k_m<+\infty$ the conductivity of the matrix and by $k_b=0$ that of the insulating bands.The macroscopical conductivity will be,obviously $k^*=0$.However,a sandwich plate system of the same materials will have a finite conductivity $k^*, 0<k^*<+\infty$.

These counterexamples show that Nayfeh's equations are not adequate to describe the heat transfer through heterogenous materials of a) or b) type.

REFERENCES

Schumann,T.E.W. 1929. A liquid flowing through a porous prism,J.Franklin Institut,208:405-416.

Rubinstein,L.I. 1948. On the problem of the process of propagation of heat in heterogeneous materials (russ.),Izv.Akad.Nauk SSSR,Ser. Geogr,1.

Gheorghitza,St. 1966. Mathematical methods in the underground hydrodynamics (rum.), Publishing House of the Academy,Bucharest.

Nayfeh,A.H. 1975. A continuum mixture theory of heat conduction in laminated composites,J.Appl.Mech., June:399-404.

Margolis,S.B. 1978. Anharmonic analysis of a time-dependent packed bed thermocline,Quart.Appl.Math. 36:97-113.

Aifantis,E.C. 1979. Continuum basis for diffusion in regions with multiple diffusivity. J.Appl.Phys. 50:1334-1338.

Nayfeh,A.H. 1980. Simulation of the influence of bonding materials on the time-dependent processes in laminated composites,in Continuum Models of Discrete System 3, Eds. E.Kröner & H.-K.Anthony, University Waterloo Press.:503-519.

Aifantis,E.C. & J.M.Hill. 1980. On the theory of diffusion in media with double diffusivity, Quart. J. Mech. Appl. Math. 33:1-41.

Aifantis,E.C. & D.E. Beskos. 1980. Heat extraction from hot dry rocks, Mech. Research Com. 7 : 165-170.

Aifantis, E.C. 1980. Further comments on the problem of heat extraction from hot dry rocks. Mech. Research Com. 7 : 219-226.

Schemmer,B.I. 1983. Armed Forces Journal International,December :18-19.

Mazilu,P. 1984. Comment on "Continuum basis for diffusion in regions with multiple diffusivity",to be published.

5th International Symposium on Continuum Models of Discrete Systems / Nottingham / 14-20 July 1985

Bounds for the effective properties of a nonlinear lossy composite

D.R.S.Talbot
Coventry (Lanchester) Polytechnic, Coventry, UK

ABSTRACT: Diffusion of a population of defects in a random composite is considered. The composite is comprised of lossy spherical sinks embedded in a lossy matrix and the sink strengths of both spheres and matrix are assumed to be nonlinear functions of the defect concentration. The classical minimum energy principle for the system is given and, by introducing a suitable comparison material, new variational principles of Hashin-Shtrikman type are derived. These new principles provide both upper and lower bounds on the energy. Results are presented for a particular non-linear composite.

1. INTRODUCTION

The composite to be studied consists of a random distribution of spherical sinks embedded in a matrix. Point defects are introduced at a constant rate throughout the body and rearrange themselves by diffusion. Both spheres and matrix are assumed to be lossy, that is, they contain continuous distributions of sinks, and their sink strengths are taken to be nonlinear functions of the defect concentration. The case where the sink strengths are linear functions of concentration has been discussed extensively by Talbot and Willis (1984a,b). Specifically, the loss rate of the medium is taken to have the form $W'(c)$, where c is the defect concentration, prime denotes differentiation and W is strictly convex. Also, W has different forms in the spheres and in the matrix and so is a random function of position. Then, if the composite occupies a bounded region Ω, with boundary $\partial\Omega$ and having unit volume, the concentration c satisfies

$$\nabla^2 c - W'(c) + K = 0, \quad X\varepsilon\Omega, \qquad (1.1)$$

$$\frac{\partial c}{\partial n} = 0, \quad X\varepsilon\partial\Omega, \qquad (1.2)$$

where K is the generation rate of defects and prime denotes differentiation with respect to c.

For each W it is assumed that a homogeneous medium exists in which the defect concentration equals the average concentration, $\bar{c}$, in the actual material and which is characterized by a function $\tilde{W}(\bar{c})$. Thus $K = \tilde{W}'(\bar{c})$. The object of this work is to obtain bounds on the overall properties $\tilde{W}(\bar{c})$. To this end, the system (1.1), (1.2) is replaced by its classical weak formulation and then, through the introduction of uniform comparison materials, variational principles of Hashin-Shtrikman type are derived. These new principles have the same structure as those derived recently by Willis (1985) for a nonlinear dielectric problem and by Talbot and Willis (1985a) for a general class of nonlinear problems.

In the following sections, the overall properties are first defined and the new variational principles are then derived. The principles are applied explicitly allowing for the geometry of the composite and finally results are presented for a particular composite where W is a cubic function of c in the matrix and a quadratic function of c in the spheres. The development closely follows that of Talbot and Willis (1985b) where further details can be found.

2. DEFINITION OF OVERALL PROPERTIES

First, the minimum energy principle for the system (1.1), (1.2) is

$$\inf_{c\epsilon H^1(\Omega)} \int_\Omega dx\{\frac{1}{2}(\nabla c)^2 + W(c)-Kc\} \qquad (2.1)$$

We note that, with W strictly convex and coercive, the solution to (2.1) exists, is unique and, provided it is smooth enough, satisfies (1.1) and (1.2). The restriction to $H^1(\Omega)$ is for convenience and is discussed in Talbot and Willis (1985b). Now denote (2.1) by $-\tilde{V}(K)$ and define

$$\tilde{W}(\bar{u}) = \sup_K\{K\bar{u}-\tilde{V}(K)\} \qquad (2.2)$$

It is easy to see that $\tilde{V}(K)$ is convex, so

$$\tilde{V}(K) = \sup_{\bar{u}}\{K\bar{u}-\tilde{W}(\bar{u})\} \qquad (2.3)$$

and it follows that $\bar{u} = \tilde{V}'(K)$ if and only if $K = \tilde{W}'(\bar{u})$. Suppose now that c is the solution to (2.1) and satisfies (1.1),(1.2). Then c is a function of K and it is easy to show from its definition that $V'(K) = \bar{c}$. Thus at the solution

$$K = \tilde{W}'(\bar{c}) \quad , \qquad (2.4)$$

so the definition (2.2) is consistent with defining a homogeneous effective medium having total loss rate equal to that in the actual body and with concentration of defects equal to the average concentration in the actual material.

This section is concluded by remarking that, if $\hat{W}$ and $\hat{V}$ are related in the same way as $\tilde{W}$ and $\tilde{V}$ and $\hat{V}$ is a lower (upper) bound for $\tilde{V}$, then $\hat{W}$ is an upper (lower) bound for $\tilde{W}$.

3. NEW VARIATIONAL PRINCIPLES

We proceed by introducing a uniform comparison material characterized by $W_o(c)$. The function W_o is assumed to be strictly convex, differentiable and such that W_o-W is also strictly convex.

Now define

$$U_*(\pi) = \inf_{c\epsilon L^2(\Omega)} \{c\pi+W_o(c)-W(c)\} \qquad (3.1)$$

and it follows that, for any $c,\pi\epsilon L^2(\Omega)$

$$W(c) \leq \pi c + W_o(c) - U_*(\pi) \quad . \qquad (3.2)$$

Thus the integral in (2.1) can be bounded above for any $c\epsilon H'(\Omega)$ by replacing W by the right side of (3.2) and it follows that

$$-\tilde{V}(K) \leq \inf_{c^*\epsilon H'(\Omega)} \int_\Omega \{\frac{1}{2}(\nabla c^*)^2 + \pi^*c^* + W_o(c^*) - U_*(\pi^*) - Kc^*\}dx \qquad (3.3)$$

for any $\pi^*\epsilon L^2(\Omega)$. The bound in (3.3) is optimized by taking the infimum with respect to $\pi^*\epsilon L^2(\Omega)$ and it was shown by Talbot and Willis (1985a) that the inequality then becomes an equality. Thus, if $H_*(c^*,\pi^*)$ denotes the integral in (3.3), the minimum energy principle can be replaced by

$$\inf_{\pi^*\epsilon L^2(\Omega)} \inf_{c^*\epsilon H'(\Omega)} H_*(c^*,\pi^*). \qquad (3.4)$$

A lower bound on $-\tilde{V}(K)$ is obtained by taking W_o such that $W-W_o$ is strictly convex and defining

$$U^*(\pi) = \sup_{\pi^*\epsilon L^2(\Omega)} \{c\pi+W_o(c)-W(c)\} \qquad (3.5)$$

Similar reasoning to that above then yields, in an obvious notation,

$$\sup_{\pi^*\epsilon L^2(\Omega)} \inf_{c^*\epsilon H'(\Omega)} H^*(c^*,\pi^*), \qquad (3.6)$$

as an alternative to (2.1). Finally, it is remarked that at the minimum of H_* and at the saddle point of H^*, $\pi^*=W'-W_o'$ so that π^* has the status of a "polarization loss rate". In the case of a linear composite it coincides with the polarization introduced by Talbot and Willis (1984).

4. APPLICATION TO A MATRIX-INCLUSION COMPOSITE.

Although the comparison material introduced in the previous section need not be linear it is necessary to make this further restriction in order to make any significant progress. Let

$$W_o(c) = \frac{1}{2} k_o^2c^2. \qquad (4.1)$$

Then taking the infimum over c* in (3.4) or (3.6) shows that c* satisfies the linear equation

$$\nabla^2 c^* - k_o^{\,2} c^* - \pi^* + K = 0, \; x \in \Omega \qquad (4.2)$$

together with (1.2). The solution to (4.2) can be written

$$c^* = \Gamma(K - \pi^*) \qquad (4.3)$$

where Γ is the integral operator with kernel $G(x,x')$, the Green function for the operator $\nabla^2 - k_o^{\,2}$. Substituting into (3.4) or (3.6) now yields the problem

$$\underset{\pi^* \in L^2(\Omega)}{\mathrm{Inf\,(or\ Sup)}} \int_\Omega \{-\tfrac{1}{2}(\pi^* - K)\Gamma(\pi^* - K) - U(\pi^*)\}\, dx, \qquad (4.4)$$

where $U(\pi^*)$ is U^* or U_* according to whether $W - W_o$ is convex or concave. When the optimization in (4.4) is carried out over the whole of $L^2(\Omega)$, the value of (4.4) is exactly $-\hat{V}(K)$. However if π^* is restricted to lie in some subspace of $L^2(\Omega)$, (4.4) only provides bounds on $-\hat{V}(K)$. It is this case that we now consider when only statistical information about the distribution of spheres is available. With π^* suitably restricted the bounds obtained from (4.4) survive ensemble averaging so it is natural to seek bounds on the ensemble average of the energy.

Suppose now that the spheres have radius a and are centred at x_A, $A=1,\ldots,N$. Define

$$f_A(x) = \begin{cases} 1, & |x - x_A| < a, \\ 0, & |x - x_A| > a, \end{cases} \qquad (4.5)$$

and let

$$W(c,x) = W_1(c)\sum_A{}' f_A(x) + W_2(c)\left(1 - \sum_A f_A(x)\right) \qquad (4.6)$$

We consider trial fields which have the form

$$\pi^*(X) = \sum_A \pi_A f_A(x) + \pi_m(x)\left(1 - \sum_A f_A(x)\right) \qquad (4.7)$$

The function $\pi_m(x)$ depends on x only while two cases will be considered for the π_A: first, the functions π_A are taken to be constants with respect to x and to depend only on x_A and secondly, they are allowed to depend on x and x_A. In either case U has the form

$$U(\pi) = \sum_A{}' U_1(\pi_A) f_A(x) + U_2(\pi_m)\left(1 - \sum_A{}' f_A(x)\right) \qquad (4.8)$$

where U_1, U_2, with stars as appropriate are obtained from W_1, W_2 via (3.1) or (3.5).

The procedure now is to make the substitutions in (4.4), take expectation values and optimize with respect to π_A and π_m. The result of optimizing with respect to π_A when π_A is a function of x_A only is

$$\int dx\, f_A(x) P_A \Big\{ \frac{\partial U_1(\pi_A)}{\partial \pi_A} + \int dx' G(x,x')(\pi_A - \pi_m) f_A(x') + \int dx' G(x,x') \Big[\int dx_B P_{B|A} \pi_B f_B(x') + \pi_m \Big(1 - \int dx_B P_{B|A} f_B(x')\Big) \Big] \Big\}$$
$$= \int \frac{K}{k_o^{\,2}} f_A(x) P_A\, dx \,. \qquad (4.9)$$

In (4.9), P_A denotes the probability density for finding a sphere centred at x_A and $P_{B|A}$ denotes the probability density for finding a sphere centred at x_B given that one is centred at x_A. The mean value of G, namely $K/k_o^{\,2}$, has also been used in obtaining (4.9). When π_A is a function of x and x_A, the result has the same form, but with the integrals over x omitted.

The composite is now assumed to have no long range order and to be statistically isotropic. In this case $P_{B|A} - P_B$ is a function of $|x_B - x_A|$ only and can be assumed negligible if $|x_B - x_A| > \ell$ where ℓ is some "correlation length" such that $\ell \ll 1$. It is plausible that $\langle c^* \rangle$

(obtained by averaging (4.3)) is constant and coincides with $\bar{c}^*$. The constants π_A are independent of x_A, π_m is independent of x and G can be replaced by its infinite body form. On subtracting the mean value of (4.3) from (4.9), it now reduces to the algebraic equation

$$\frac{\partial U_1(\pi_1)}{\partial \pi_1} + H(\pi_1 - \pi_2) = \langle c \rangle , \qquad (4.10)$$

where π_1, π_2 are the values of π_A, π_m and H is given by

$$H = \frac{3}{4\pi a^3} \int dx\, f_A(x) \left\{ \int dx' G(x,x') f_A(x') + \int dx' G(x,x') \int dx_B f_B(x') (P_{B|A} - P_B) \right\} . \qquad (4.11)$$

In the second case π_A is a function of $|x - x_A|$ only and the result corresponding to (4.10) is the integral equation

$$\frac{\partial U_1(\pi_A)}{\partial \pi_A} + \int dx' G) x,x') f_A(x') (\pi_A - \pi_m) + \int dx' G(x,x') \int dx_B f_B(x') (P_{B|A} - P_B) (\pi_B - \pi_m) = \langle c^* \rangle, \quad |x - x_A| < a \qquad (4.12)$$

Finally, the result of optimizing with respect to π_m can be combined with (4.9) to yield the relation

$$\frac{3}{4\pi a^3} \int dx \left\{ \int dx_A \frac{\partial U_1}{\partial \pi_A}(\pi_A) P_A f_A(x) + \frac{\partial U_2}{\partial \pi_2} \left[1 - \int dx_A P_A f_A \right] \right\} = \langle c^* \rangle . \qquad (4.13)$$

The procedure now is to solve (4,10) (or 4.12) and (4.13) for π_1 (or π_A) and π_2 regarding $\langle c^* \rangle$ as known and substitute the result into the expectation value of (4.3):

$$\langle c^* \rangle = \frac{1}{k_o^2} (K - \langle \pi^* \rangle) , \qquad (4.14)$$

This yields a relation between K and $\langle c^* \rangle$. It is straightforward to show that the result of substituting into the expected value of the integral in (4.4) yields the bound $\hat{V}(K)$ on $\tilde{V}(K)$ where

$$\hat{V}(K) = K \langle c^* \rangle - \hat{W}(\langle c^* \rangle) , \qquad (4.15)$$

with

$$\hat{W}(\langle c^* \rangle) = \frac{1}{2} K \langle c^* \rangle + \int dx_A P_A f_A \left[\frac{1}{2} \pi_A \frac{\partial U_1}{\partial \pi_A} - U_1 \right] + \left(1 - \int dx_A P_A f_A\right) \left[\frac{1}{2} \pi_2 \frac{\partial U_2}{\partial \pi_2} - U_2 \right] . \qquad (4.16)$$

Thus, following the remarks at the end of section 2, $\hat{W}$ is a bound on $\tilde{W}$.

5. RESULTS

The example we consider has

$$W_1 = \frac{1}{2} k_1^2 c^2, \quad W_2 = \frac{1}{2} k_2^2 c^2 + \frac{1}{3} \alpha c^2 |c| , \qquad (5.1)$$

where k_1^2, k_2^2 and α are constants. These forms for W_1 and W_2 are motivated by studies of irradiation damage: with equal numbers of vacancies and interstitials, c is their concentration and the cubic term models recombination. For large values of k_1^2, the spheres may be thought of as voids and the lossy matrix as modelling a distribution of sinks of another type: eg. dislocations. A comprehensive review of this topic was given by Brailsford and Bullough (1981).

In order to obtain bounds, k_o^2 has to be chosen so that $W - W_o$ is convex (for a lower bound on $\tilde{W}$) or concave (for an upper bound). Now

$$(W - W_o)''(c) = k_1^2 - k_o^2, \quad x \varepsilon \text{ spheres},$$
$$= k_2^2 - k_o^2 + 2\alpha |c|, \quad x \varepsilon \text{ matrix} , \qquad (5.2)$$

so the first is achieved by choosing $k_2^2 = k_o^2$. However a suitable choice for k_o^2 to obtain an upper bound is not obvious, the problem being to keep $(W - W_o)''$ negative with increasing values of c. To overcome this difficulty first note that it can be shown that with c_1 and c_2 solutions to

$$\nabla^2 c - k_o^2 c + f(x) = 0,\ x\varepsilon\Omega,\ \frac{\partial c}{\partial n} = 0,\ X\varepsilon\partial\Omega, \quad (5.3)$$

corresponding to $f = f_1$ and $f = f_2$, if $f_1 > f_2$ then $c_1 > c_2$. Hence when $\pi_A = \pi_1$ the solution c* to (4.2) is bounded above by the solution c_u to (5.3) with $f = K - \min(\pi_1, \pi_2)$, that is by

$$c_u = \frac{1}{k_o^2}(K - \min(\pi_1, \pi_2)). \quad (5.4)$$

Hence choosing k_o^2 such that $k_o^2 \geq k_1^2$ and $k_2^2 - k_o^2 + 2\alpha c_u \leq 0$ ensures that $W - W_o$ is concave.

To summarize, when $\pi_A = \pi_1$ both upper and lower bounds can be obtained for $\hat{W}$ by choosing k_o^2 according to the prescription above and by taking $k_o^2 = k_2^2$. When π_A is allowed to vary over the sphere at x_A, a lower bound can be obtained by taking $k_o^2 = k_2^2$. However the extra work that would be involved in obtaining an upper bound does not appear to be justified by the improvement obtained.

The solution to (4.10) and (4.13) is straightforward if somewhat tedious to obtain; details are not recorded. In Figures 1 and 2 some specimen results are displayed when the spheres are distributed according to the Percus-Yevick (1958) form for the pair distribution function. Also shown are the simplest bounds obtainable from the classical energy principles. The upper (Voigt) bound is found by substituting c = <c> into (2.1). This gives the law of mixtures. The lower (Reuss) bound is found from the complementary energy principle obtained as the Legendre transformation of (2.1).

For both figures, $k_2 a = 1$, $k_1^2/k_2^2 = 10$ and $\hat{W}/W_2$ is plotted as a function of $\hat{c} = \alpha\langle c^*\rangle/k_2^2$. The volume concentrations of spheres used are 1/10 for Figure 1 and 1/2 for Figure 2. As can be seen, the new bounds show a considerable improvement on the Voigt and Reuss results. When π_A is allowed to vary, a lower bound can be obtained by choosing $k_o^2 = k_2^2$ and solving the integral equation (4.12). The result shows no significant improvement on the bounds in Figures 1 and 2 and very little improvement at higher values of k_1^2/k_2^2.

Finally, when $k_1^2/k_2^2 \to \infty$ the spheres become perfect sinks and serve as a model for voids. Results are shown in Figure 3 for $k_2 a = 1$ and volume concentration of 1/10 and in Figure 4 for $k_2 a = 1$ and volume concentrations 1/2. Allowing π_A to vary gives some improvement for low values of $\hat{c}$, but at high values the new bounds asymptote to the Reuss bound. It is perhaps worth noting that in this case $\hat{W}$ has the same functional form as W_2. No useful upper bound is obtainable.

Details of the solution of the integral equation (4.12) can be found in Talbot and Willis (1984a).

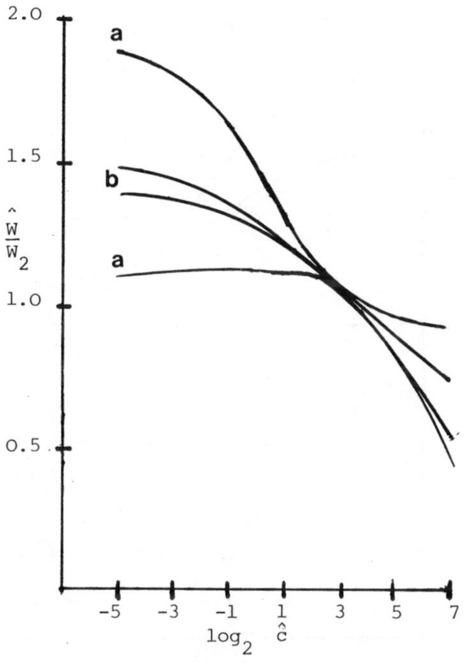

Fig.1 a) Simple classical bounds
b) New bounds.

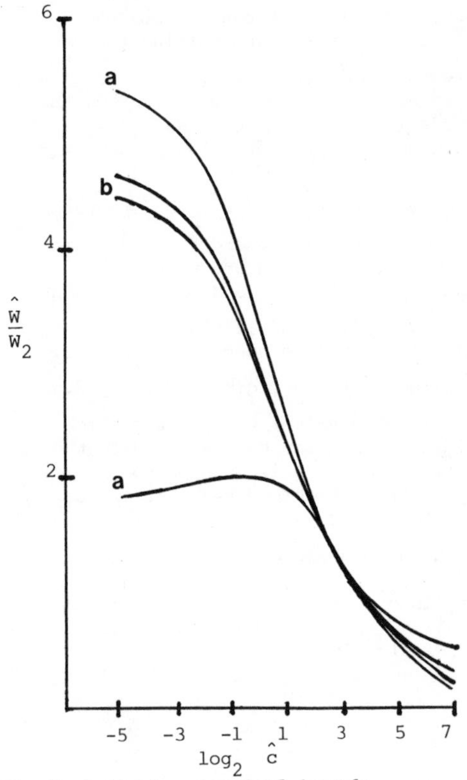

Fig.2 a) Simple classical bounds
b) New bounds

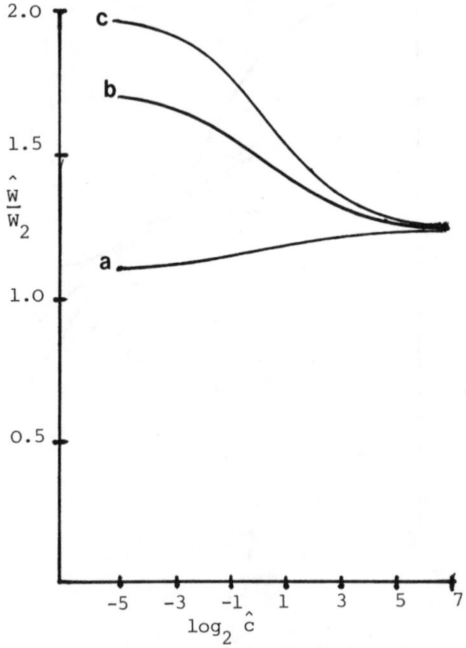

Fig.3 a) Simple classical bound
b) π_A constant c) π_A varying.

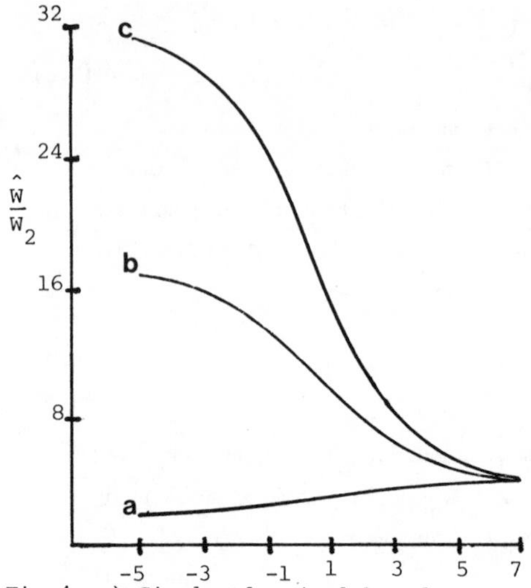

Fig.4 a) Simple classical bound
b) π_A constant c) π_A varying

REFERENCES

Brailsford, A.D. and Bullough, R. (1981) "The theory of sink strengths", Philos. Trans. Roy. Soc. A302, 87.

Talbot, D.R.S. and Willis, J.R. (1984a) "The overall sink strength of an inhomogeneous lossy medium. Part I: Self-consistent estimates", Mech. Mat. 3, 171.

Talbot, D.R.S. and Willis, J.R. (1984b) "ditto. Part II: Variational estimates", Mech. Mat. 3, 183.

Talbot, D.R.S. and Willis, J.R. (1985a) "Variational principles for inhomogeneous nonlinear media", IMA J. Appl. Math., 34, (to appear).

Talbot, D.R.S. and Willis, J.R. (1985b) "A variational approach to the stochastic homogenization of a class of nonlinear diffusion problems" submitted for publication.

Willis, J.R. "Variational estimates for the overall response of an inhomogeneous nonlinear dielectric", Proceedings, IMA Workshop on Homogenization and Effective Properties of Composite Materials, edited by D. Kinderlehrer, (to appear).

Percus, J.K. and Yevick, G.J. (1958) "Analysis of classical statistical mechanics by means of collective co-ordinates", Phys. Rev.110, 1.

5th International Symposium on Continuum Models of Discrete Systems / Nottingham / 14-20 July 1985

Dual estimates of overall instantaneous properties of elastic-plastic composites

J.L.Teply
Product Engineering Division, ALCOA Laboratories, Alcoa Center, Pa., USA
G.J.Dvorak
Department of Civil Engineering, Rensselaer Polytechnic Institute, Troy, N.Y., USA

A class of micromechanical models for inelastic analysis of fibrous and particulate composites is described. Periodic arrays of fibers or particles, separated by matrix interlayers, form the basis of the models. The case of a hexagonal array arrangement is discussed in detail. A representative volume element is selected from the periodic array and subdivided into a small number of finite elements. Piecewise uniform strain fields are introduced through linear shape functions. Variational principles of the finite element method are used to obtain upper and lower estimates on diagonalized overall instantaneous stiffnesses and compliances of elastic-plastic composites. Estimates of local fields are obtained as well.

1. INTRODUCTION

Elastic behavior of binary composite materials, both particulate and fibrous, is reasonably well understood. Evaluation of their overall thermoelastic properties, and of averaged local stress and strain fields can be routinely made, either in terms of rigorous bounds, or through self-consistent estimates [1-3]. The situation is quite different in the case of elastic-plastic aggregates. A bounding technique has not been developed, and the self-consistent method may give misleading results [4].

Reasons for this difference lie in analysis of local fields from which the overall properties are derived. In the elastic case the constituent phases are homogeneous, their properties are known constants, and actual local fields or their volume averages can be found for many microstructural geometries from solutions of certain inclusion problems in a representative volume element of the aggregate. The overall properties are, of course, also constant, and can be determined from superposition of results obtained for simple loading states applied to the representative volume. When at least one of the phases starts to deform plastically, its homogeneity is lost. Local properties are no longer constant or known within the phase. This loss of local homogeneity creates a major obstacle in evaluation of local fields. Solutions of inclusion problems become untractable, even in terms of stress and strain averages, because the local constitutive equation for averages of nonuniform fields is not initally know. This deficiency is sometimes circumvented by assuming that the averages are related in the same way as uniform fields, but this may cause serious errors in the presence of large stress and strain gradients.

In view of these difficulties, the elastic-plastic properties of composite media are often derived from micromechanical models which simplify the geometry of the microstructure in such a way that the local fields become piecewise uniform [5-8]. This makes it possible to find instantaneous overall properties of the composite model under incremental load, in terms of instantaneous properties of the phases. However, there is no assurance that model predictions obtained in this way are valid for the actual composite medium.

Under such circumstances, the elastic-plastic properties of composite media may be best derived from modeling

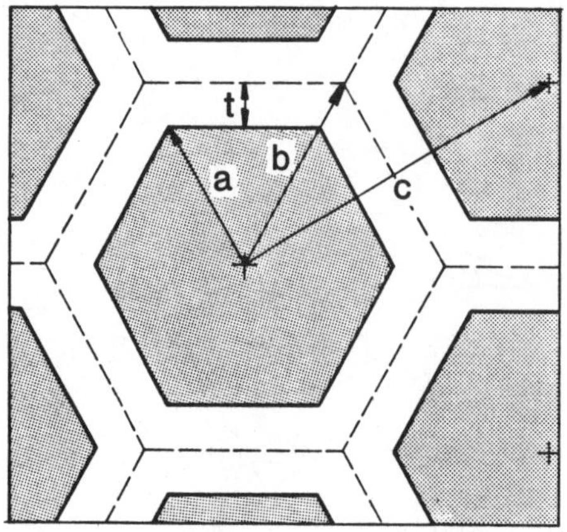

Figure 1. Characteristic Dimensions of Periodic Hexagonal Array

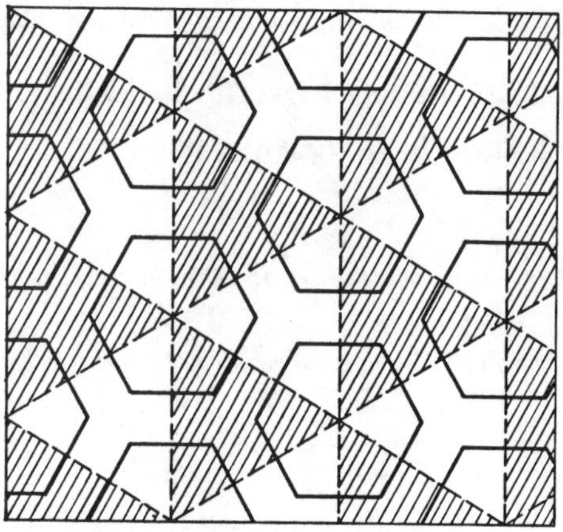

Figure 2. Mesh of Triangular Representative Volume Elements

techniques which utilize piecewise uniform fields, and at the same time give rigorous bounds on the results. An initial step toward development of such a class of micromechanical models was taken in our recent paper [9]. The models are based on variants of a periodic array geometry. Appropriate representative volume elements are chosen and subdivided into a small number of finite elements. Then, dual variational principles of the finite element method are applied to the elastic-plastic analysis of the periodic arrays. These principles lead to the two well-known finite element solutions, displacement and equilibrium. The results provide dual estimates on instantaneous overall moduli and other overall properties.

2. GEOMETRY OF PERIODIC MICROSTRUCTURE

The composite medium under consideration consists of a matrix which is reinforced by aligned cylindrical fibers. We assume that all fibers have identical cross sections which can be approximated by a regular n x 6-sided polygon. The centers of the fibers are arranged in a hexagonal periodic array. An example of such a geometry is shown in Figure 1 for the simplest hexagonal fiber cross section. If desired, the fiber can be represented by a 12, 24, etc. sided cross section. This does not affect the generality of the results that follow.

Overall properties of the composites are found when a uniform, overall stress or strain field is applied to the aggregate. This can be accomplished by prescribing uniform boundary transactions or displacements.

$$\bar{\underset{\sim}{t}} = \bar{\underset{\sim}{\sigma}}\,\underset{\sim}{n} \quad , \quad \bar{\underset{\sim}{u}} = \bar{\underset{\sim}{\varepsilon}}\,\underset{\sim}{x} \quad , \qquad (1)$$

where $\bar{u}$, and $\bar{t}$, are the boundary displacements, and traction vectors, respectively, x, is the position vector of boundary points, n is the unit outward normal; $\bar{\sigma}$, and $\bar{\varepsilon}$ are the tensors of uniform stress and strain.

Derivation of the overall properties is conveniently performed with the aid of a representative volume element (RVE). The RVE of periodic microstructures can be defined as a subregion which has the following properties:

a. When repeated in a certain pattern, it continuously covers the entire macroscopic volume of the composite.
b. When the composite is loaded with uniform stresses of strains (1), the local stresses and strains are identical within each RVE in the macroscopic volume.

For the periodic hexagonal array (PHA) model shown in Figure 1, a suitable selection of RVEs is indicated in Figure 2. The periodic array is divided into shaded and unshaded triangles, each of which will be shown to represent an RVE. To this end it is necessary to prove that the local stress and strain fields in the two sets of triangles are identical, as required by b above.

Consider that the area of each triangle, Figure 2, is equal to unity. Then, the dimensions shown in Figure 1 become; in terms of fiber volume fraction c_f:

$$a = \frac{2\sqrt{c_f}}{\sqrt[4]{3^3}} \qquad t = \frac{1-\sqrt{c_f}}{\sqrt[4]{3}}$$
$$b = \frac{2}{\sqrt[4]{3}^{\,3}} = \frac{c}{\sqrt{3}} \qquad c = \frac{2}{\sqrt[4]{3}} \tag{2}$$

Let the coordinates of an arbitrary point Y in a selected RVE be given as $y = [y_1, y_2, s]^T$. The periodicity of the array suggests that all points X with coordinates $x = [x_1, x_2, s]^T$ in the same set of RVE triangles, such that

$$\underset{\sim}{x} = \underset{\sim}{y} + \underset{\sim}{c} \tag{3}$$

where $c = c/2(i\sqrt{3},\ j,\ 2s/c)^T$, are equivalent to Y,or, more specifically, have identical local stress and strain states when the composite is loaded according to (1). The symbols i and j in (3) represent integers, both must be odd or even at the same time. The distance c is given by (2).

For eqivalence of the two sets of triangles, it is sufficient to show that the local stress and strain fields in one set of triangles are invariant with respect to the transformation

$$\underset{\sim}{x}' = -\underset{\sim}{I}\ (\underset{\sim}{y} + \underset{\sim}{c}) = \underset{\sim}{y}' + \underset{\sim}{c}' \tag{4}$$

where I is the identity matrix. This transformation converts the shaded set into unshaded, and vice versa. When (4) is applied to incremental forms of (1), it follows that

$$\Delta\bar{\underset{\sim}{t}}' = (-\underset{\sim}{I})\ \Delta\bar{\underset{\sim}{\sigma}}\ (-I^T)\ \bar{\underset{\sim}{n}}' = \Delta\bar{\underset{\sim}{\sigma}}\ \bar{\underset{\sim}{n}}'$$
$$\Delta\bar{\underset{\sim}{u}}' = (-I)\ \Delta\bar{\underset{\sim}{\varepsilon}}\ (-I^T)\ \bar{\underset{\sim}{x}}' = \Delta\bar{\underset{\sim}{\varepsilon}}\ \bar{\underset{\sim}{x}} \tag{5}$$

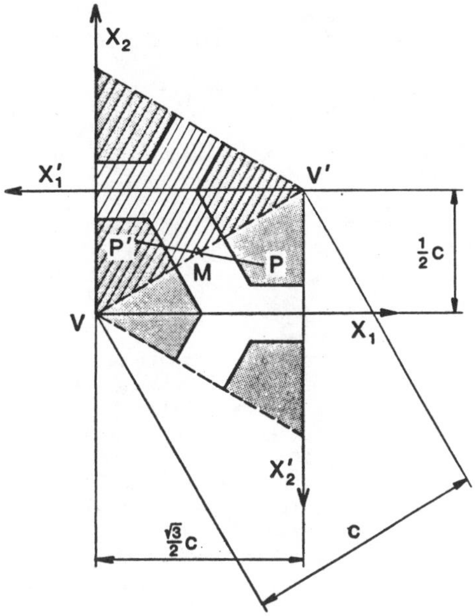

Figure 3. Two Adjacent Representative Volume Elements and Their Local Coordinate Systems

Therefore, the local fields are derived from the same boundary conditions as the local fields of the shaded set. The increments of local stress and strain fields are uniquely defined in strain hardening materials. Hence, the local fields are invariant with respect to transformation (3), as well as (4). That satisfies requirement b. above; the composite triangular prism, represented by the shaded or unshaded area of unit thickness ($0 < x_3 < 1$) can be admitted as a representative volume element.
For this choice of RVE one can derive certain connections between displacements of two adjacent RVEs, Figure 3. For a detail derivation see [10]; only results are summarized next.

Let $\underset{\sim}{c}$ equal to:

$$\underset{\sim}{c} = c/2(\sqrt{3},\ 1,\ 2s/c) \tag{6}$$

The displacements of the vertices V and V' in Figure 3 are then equal to:

$$\underset{\sim}{u}_{V'} - \underset{\sim}{u}_V = \bar{\underset{\sim}{\varepsilon}}\ \bar{\underset{\sim}{c}} \quad , \quad \underset{\sim}{u}'_V - u'_{V'} = \bar{\underset{\sim}{\varepsilon}}\ \underset{\sim}{c}' \tag{7}$$

Also, it can be shown that the displacement of the midpoint M M' is

$$\underset{\sim}{u}_M = 1/2(\underset{\sim}{u}_P + \underset{\sim}{u}_{P'}) = 1/2(\underset{\sim}{u}_V + \underset{\sim}{u}_{V'}) \quad (8)$$

where the points P and P' are equidistant from M.

Equations (7) and (8) combined yield:

$$\underset{\sim}{u}_M - \underset{\sim}{u}_V = 1/2 \bar{\underset{\sim}{\varepsilon}}\, \bar{\underset{\sim}{c}} \quad , \quad \underset{\sim}{u}'_M - \underset{\sim}{u}'_{V'} = 1/2\, \bar{\underset{\sim}{\varepsilon}}\, \underset{\sim}{c}'$$

The above results indicate a particularly simple way in which the overall strain $\bar{\underset{\sim}{\varepsilon}}$ can be related to the relative displacements of the vertices V, V', and of the midpoints M of the sides of the triangular RVE. In each side of the triangular RVE, these three points are translated during deformation in the same way as they would be in a uniformly strained solid.

3. DUAL ESTIMATES OF INSTANTANEOUS OVERALL PROPERTIES

Estimates of instantaneous overall moduli of the periodic hexagonal array model will be now determined by the dual finite element analysis of the RVE. The first estimate is obtained from the displacement approach, and the minimum principle for plastic strain rates, or minimum potential energy in the elastic case. The second estimate on moduli follow from the equilibrium model, static minimum principle in plasticity, or the elastic minimum complementary energy theorem.

Each bound is obtained in two separate steps:

a. The hexagonal array is replaced by a homogeneous material with unknown overall instantaneous moduli. A volume identical to the RVE is separated from the homogeneous medium and loaded by a selected overall load increment. The energy change in this equivalent homogeneous element (EHV) is expressed in terms of the unknown overall moduli and specified boundary displacements and surface tractions.

b. The hexagonal array is subdivided into finite elements. This subdivision also provides a finite element mesh for the RVE. In the subelements of the RVE, nodal displacements are chosen so that continuity requirements on stress, strain and displacement fields are satisfied for the displacement or equilibrium approaches. The simplest interpolation polynominals, linear for displacements and constant for stresses, can be used. The RVE is subjected to the overall load increment selected in a. above. Corresponding energy changes are calculated as functions of boundary displacements in all subelements. These individual energy contributions are summed up and made equal to those obtained under similar boundary conditions for the EHV. This yields an expression for evaluation of the overall instantaneous stiffness L in terms of RVE geometry and local instantaneous moduli L^k.

3.1 Instantaneous Material Properties

The modeling procedure can be used for composite materials consisting of different inelastic phases. However, in what follows we shall restrict our attention to elastic and elastic-plastic materials. Let $\bar{\underset{\sim}{\sigma}}_n$ and $\bar{\underset{\sim}{\varepsilon}}_n$ denote uniform stress and strains applied to the composite, at the end of loading step n, as in (1). Let $\underset{\sim}{\sigma}_n^k$ and $\underset{\sim}{\varepsilon}_n^k$ be the corresponding local quantities in element k. Strain increments during the loading step (n + 1) are

$$\Delta\bar{\underset{\sim}{\varepsilon}} = \bar{\underset{\sim}{\varepsilon}} - \bar{\underset{\sim}{\varepsilon}}_n \quad , \quad \Delta\underset{\sim}{\varepsilon}^k = \underset{\sim}{\varepsilon}^k - \underset{\sim}{\varepsilon}_n^k \quad (10)$$

where $\Delta\bar{\varepsilon}$, $\Delta\varepsilon^k$ are the instantaneous magnitudes. Then, the current stresses are equal to

$$\begin{aligned} \bar{\underset{\sim}{\sigma}}(\varepsilon) &= \bar{\underset{\sim}{\sigma}}_n + \underset{\sim}{L}\, \Delta\bar{\underset{\sim}{\varepsilon}} \\ \underset{\sim}{\sigma}^k(\underset{\sim}{\varepsilon}^k) &= \underset{\sim}{\sigma}^k + \underset{\sim}{L}_k\, \Delta\underset{\sim}{\varepsilon}^k \end{aligned} \quad (11)$$

where L, and L_k, are the overall, and local, instantaneous stiffness tensors, respectively. These instantaneous properties are taken as independent of the magnitude of stress or strain increments. On occasion, we shall write

$$\bar{\varepsilon}(\bar{\sigma}) = \bar{\varepsilon} + M \Delta\bar{\sigma}$$
$$\varepsilon^k(\sigma^k) = \varepsilon^k + M \quad \Delta\sigma^k \qquad (12)$$

where $M = L^{-1}$, and $M_k = L_k^{-1}$, if the inverses exist.

3.2 Estimate of L From the Displacement Model

Consider a macroscopic volume of a fibrous composite subjected to boundary conditions (1). Assume that the composite has been replaced by an effective homogeneous medium with instantaneous stiffness L of the composite, and that the equivalent homogeneous element (EHV), geometrically similar to the RVE, Figure 4, has been separated from the effective medium. The EHV volume V = 1.

Suppose that displacement has been applied to the composite aggregate according to (1). The The corresponding energy change in EHV is equal to

$$\Delta\bar{\pi} = \int_V \int_{\varepsilon_n}^{\bar{\varepsilon}_n+\Delta\bar{\varepsilon}} \bar{\sigma}^T \, d\bar{\varepsilon} dV$$
$$- \int_{S_F} \int_{u_n}^{\bar{u}_n+\Delta\bar{u}} \bar{t} \, d\bar{u} \, dS$$

The stress $\bar{\sigma}$ can be taken from (11), and $\Delta\bar{\pi}$ can be integrated to yield

$$\Delta\bar{\pi} = \bar{\sigma}_n^T \Delta\bar{\varepsilon} + 1/2\Delta\bar{\varepsilon}^T L \Delta\bar{\varepsilon}$$
$$- \int_{S_F} (\bar{t}_n + \Delta\bar{t})^T \Delta\bar{u} \, dS \qquad (13)$$

Since the overall and local stresses and strains are equal in the homogeneous EHV, the expression then represents $\Delta\bar{\pi}$ in terms of local quantities.

To evaluate (13) in EHV, it is advantageous to regard the EHV as a single finite element of the volume equal to unity. The result is

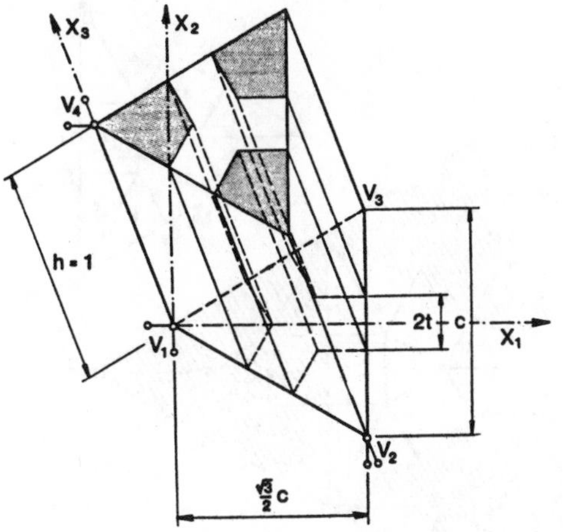

Figure 4. Dimensions of Representative Volume Element and Boundary Conditions for Displacement Model

$$\Delta\bar{\pi} = \bar{\sigma}_n^T \bar{B} \Delta\bar{a} + 1/2\Delta\bar{a}^T \bar{B}^T L \bar{B} \Delta\bar{a}$$
$$- (\bar{F}_n + \Delta\bar{F})^T \Delta\bar{a} \qquad (14)$$

where $\bar{B}$ is derived from a linear displacement field $\Delta\bar{\varepsilon} = \bar{B}\Delta\bar{a}$, and $\Delta\bar{a}$ is the vector of nodal displacements after rigid body motion contributions have been subtracted. The $\bar{F}_n$ and $\Delta\bar{F}$ are nodal force vectors equivalent to surface tractions $\bar{t}_n$ and $\Delta\bar{t}$, respectively.

Also, from the equation of virtual work,

$$\bar{F}^T = \bar{\sigma}_n^T \bar{B} \quad , \quad \Delta\bar{F}^T = \Delta\bar{\sigma}^T \bar{B} \qquad (15)$$

Therefore, the final expression for energy change $\Delta\bar{\pi}$ in the EHV is

$$\Delta\bar{\pi} = 1/2\Delta\bar{a}^T \bar{B}^T L \bar{B} \Delta\bar{a} - \Delta\bar{\sigma}^T \bar{B} \Delta\bar{a} \qquad (16)$$

b. Energy change in RVE

Evaluation of $\Delta\bar{\pi}$ in the RVE is quite similar to that outlined in Equations (13) to (16) above. The RVE is subdivided into subelements in such a way that the actual fiber and matrix stress and strain fields are approximated by piecewise uniform fields. An example of the subdivision

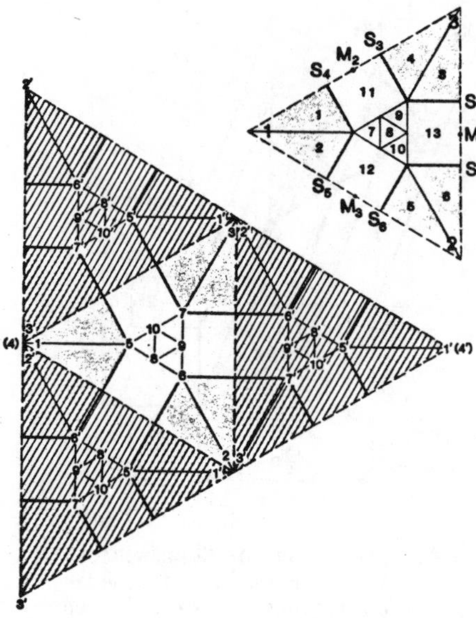

Figure 5. Transverse Plane Geometry and RVE for Displacement Model

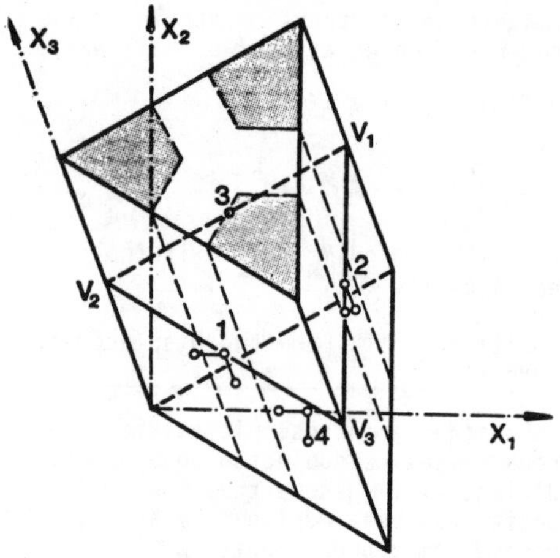

Figure 6. Boundary Conditions for Equivalent Homogeneous Volume and Representative Volume Element for Equilibrium Model

is shown in Figure 5. It represents a very simple choice, motivated, in part, by the requirement that the resulting upper bound mechanism should not be overconstrained in the plastic region. This condition, which leads to subdivision of the center triangle and of the matrix rectangles of the RVE, is discussed in Section 4.2. In each subelement, an interpolation of the displacement $\underset{\sim}{u}$ is designed in such a way that u is a continuous function in RVE. Appropriate choice of subelements and shape functions is important if the model is to be kept simple. Still, some heavy algebra is involved, hence discussion of the subject is presented elsewhere [10].

Energy contributions of individual subelements of the RVE are evaluated from equations analogous to (16). These energy terms are expressed in boundary displacements $\Delta\underset{\sim}{\bar{a}}$ of the RVE, and internal nodal displacements $\Delta\underset{\sim}{a}^r$. Then, contributions to are assembled. In order to find $\Delta\pi$ in terms $\Delta\underset{\sim}{\bar{a}}$ and $\Delta\underset{\sim}{\bar{F}}$, it is necessary to evaluate the variation of $\Delta\pi$ with respect to internal nodal displacements $\Delta\underset{\sim}{a}^r$. The minimum condition eliminates $\Delta\underset{\sim}{a}^r$, and the result then is:

$$\min(\Delta\pi)\big|_{\Delta\underset{\sim}{a}^r} = 1/2\,\Delta\underset{\sim}{\bar{a}}^T\,\underset{\sim}{K}\,\Delta\underset{\sim}{\bar{a}} - \Delta\underset{\sim}{\bar{F}}^T\,\Delta\underset{\sim}{\bar{a}} \qquad (17)$$

where $\underset{\sim}{K}$ is the assembled stiffness array of the RVE domain.

It is now possible to evaluate overall instantaneous moduli of the RVE through a comparison of (20) and (21). By definition,

$$\Delta\bar{\pi} = \min(\Delta\pi)\big|_{\Delta\underset{\sim}{a}^r} \qquad (18)$$

and also

$$\Delta\underset{\sim}{\bar{\sigma}}^T\,\underset{\sim}{\bar{B}}\,\Delta\underset{\sim}{\bar{a}} = \Delta\underset{\sim}{F}^T\,\Delta\underset{\sim}{\bar{a}} \qquad (19)$$

i.e., the composite aggregate is subjected to identical boundary conditions. Therefore, it is only necessary to make equal the first terms on the right hand sides of (16) and (17). This immediately yields the matrix of instantaneous overall moduli of the EHV, and of the composite aggregate

$$\underset{\sim}{L} = (\underset{\sim}{\bar{B}}^T)^{-1}\,\underset{\sim}{K}\,\underset{\sim}{\bar{B}}^{-1} \qquad (20)$$

Also, since $\Delta\underset{\sim}{\bar{\sigma}} = \underset{\sim}{L}\Delta\underset{\sim}{\bar{\varepsilon}}$, one obtains from (17) and (19)

$$\Delta\underset{\sim}{\bar{F}} = \underset{\sim}{K}\,\Delta\underset{\sim}{\bar{a}} \qquad (21)$$

which is, of course, the minimum condition for (17).

3.3 Estimate of $\underset{\sim}{L}$ From the Equilibrium Model

a. Energy Change in EHV

In analogy with the displacement approach, the composite is first replaced by an effective medium, and an EHV element of unit volume is considered, Figure 6.

Suppose that a stress of traction increment $\Delta\underset{\sim}{\bar{\sigma}}$ or $\Delta\underset{\sim}{\bar{t}}$ have been added to the current values of $\underset{\sim}{\bar{\sigma}}_n$, $\underset{\sim}{\bar{t}}_n$ applied to the composite EHV, in agreement with (1). The corresponding energy change is equal to

$$\Delta\bar{\pi}_C = \int_V \int_{\bar{\sigma}_n}^{\bar{\sigma}_n+\Delta\bar{\sigma}} \bar{\varepsilon}^T \, d\bar{\sigma} \, dV - \int_{S_u} \int_{\bar{t}}^{\bar{t}_n+\Delta\bar{t}} \bar{u}^T \, d\bar{t} \, dS \tag{22}$$

The evaluation of $\Delta\bar{\pi}_C$ starts with the substitution of $\Delta\underset{\sim}{\varepsilon}$ from (12) into (22). Then, the integration with respect to the prescribed loading paths $\Delta\underset{\sim}{\sigma}$ and $\Delta\underset{\sim}{\bar{t}}$ yields:

$$\Delta\bar{\pi}_C = \int_V (\bar{\varepsilon}^T \Delta\bar{\sigma} + 1/2\Delta\bar{\sigma}^T M \Delta\bar{\sigma}) dV - \int_{S_u} \Delta\bar{u}'^T \Delta\bar{t} \, dS - \int_S \bar{u}^T \Delta\bar{t} \, dS \tag{23}$$

The last term must be integrated over the entire boundary $S = S_F + S_U$ since $\underset{\sim}{\bar{u}}$ is known on S_F at the beginning of the increment.

In accordance with the equilibrium model, the components of $\Delta\underset{\sim}{\sigma}$ satisfy differential equations of stress equilibrium and also are in equilibrium with boundary tractions on S_F.

It is convenient to introduce independent stress and displacement trial functions into (23). The stress functions satisfy equilibrium equations inside EHV, and are balanced by boundary tractions with the help of Lagrange multipliers [13]. These multipliers can be readily identified with boundary displacements [14].

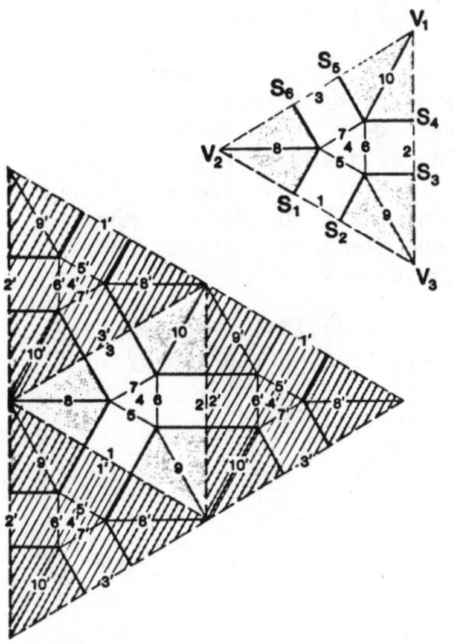

Figure 7. Transverse Plane Geometry and RVE for Equilibrium Model

The resulting energy change, which was obtained in more detail in [10], is

$$\Delta\bar{\pi}^*_C = 1/2\Delta\bar{\sigma}^T M \Delta\bar{\sigma} - \Delta\bar{\sigma}^T \int_{S_F} \bar{C} \, dS \, \delta\bar{a} + \Delta\bar{F}^T \Delta\bar{a} \tag{24}$$

where $\underset{\sim}{\bar{C}}$ is a matrix of directional cosines of outside boundary normals of the EHV. Since the stresses $\Delta\underset{\sim}{\bar{\sigma}}$ are defined only inside the element, the variation of $\Delta\bar{\pi}^*_C$ with respect to them is:

$$\frac{\partial(\Delta\bar{\pi}^*_C)}{\partial(\Delta\bar{\sigma})} = V M \Delta\bar{\sigma} - \int_{S_F} \bar{C}^T \, dS \, \Delta\bar{a} = 0 \tag{25}$$

from where

$$\Delta\bar{\sigma} = L \, 1/V \int_{S_F} \bar{C}^T \, dS \, \Delta\bar{a} \tag{26}$$

Substitution of (26) into (24) gives:

$$\Delta\bar{\pi}^*_C = - 1/2\Delta\bar{a}^T \bar{B} L \bar{B} \Delta\bar{a} + \Delta\bar{\sigma}^T \bar{B} \Delta\bar{a} \tag{27}$$

where $\underset{\sim}{\bar{B}}$ is defined as $\underset{\sim}{\bar{B}} = 1/V\int\underset{\sim}{\bar{C}}^T \, dS$.

b. Energy Change in RVE

The periodic composite geometry in the transverse plane is covered by a finite element mesh shown in Figure 7. The resulting division of the RVE into subelements is also shown there. Energy contributions of RVE subelements must be evaluated separately for boundary and interior subelements.

In analogy with (24), the boundary element contributions are given by:

$$\Delta\pi^*_{ck} = 1/2 \int_{V_k} \Delta\underset{\sim}{\sigma}^{k^T} \underset{\sim}{M} \Delta\underset{\sim}{\sigma}^k \, dV - \int_{V_k} \Delta\underset{\sim}{\sigma}^{k^T} \underset{\sim}{B}^k \Delta\underset{\sim}{a}^k \, dV + \Delta\underset{\sim}{\bar{F}}^{k^T} \Delta\underset{\sim}{a}^k \quad (28)$$

For the interior elements:

$$\Delta\pi^*_{ck} = 1/2 \int_{V_k} \Delta\underset{\sim}{\sigma}^{kT} \underset{\sim}{M}_k \, dV - \int_{V_k} \Delta\underset{\sim}{\sigma}^{kT} \underset{\sim}{B}^k \Delta\underset{\sim}{a}^k \, dV \quad (29)$$

In both equations, for each subelement k

$$\underset{\sim}{B}^k = \frac{1}{V_k} \int_{S_k} \underset{\sim}{C}^{k^T} \, dS \quad (30)$$

and $\Delta\underset{\sim}{a}^k$ are nodal displacements at the centroids of sides of k-th subelement. Overall stress increments $\Delta\bar{\sigma}^k$ in (28) do not enter into any other subelement, and therefore, the variation of $\Delta\pi^*_{ck}$ with respect to them can be found separately in each subelement and energy contributions (28) and (29) may be expressed only in nodal displacement.

The subsequent steps are identical to those outlined in the evaluation of the displacement model. It is necessary to evaluate a minimum of $\Delta\pi^*_c$ with respect to internal nodal displacements Δa^r. The result is:

$$\min(\Delta\pi^*_c)|_{\Delta a^r} = -1/2 \Delta\underset{\sim}{\bar{a}}^T \underset{\sim}{K} \Delta\underset{\sim}{\bar{a}} + \Delta\underset{\sim}{F}^T \Delta\underset{\sim}{\bar{a}} \quad (31)$$

From this point on, one can retrace the steps leading to (18)-(19), and obtain equations identical to (20) and (21).

4. PLASTICITY ANALYSIS

4.1 Constitutive Equations

Strain magnitudes in the fibrous composite laminates are limited by the failure strain of fibers, which seldom exceeds 0.01. Therefore, it is sufficient to consider only small strains in plasticity analysis. Also, most metal matrix systems have an aluminum matrix which hardens kinematically at small strains [16,17].

The Ziegler modification of Prager's hardening rule [18] can be used as an approximation of the actual behavior of aluminum at small strains [16,17].

On the basis of these considerations one can obtain the local elastic-plastic constitutive relations (11) and (12) in the form

$$\underset{\sim}{L}_k = \underset{\sim}{L}^e_k - \frac{\underset{\sim}{L}^e_k \, d\underset{\sim}{f}_k \, d\underset{\sim}{f}^T_k \, \underset{\sim}{L}^e_k}{d\underset{\sim}{f}^T_k \, \underset{\sim}{L}^e_k \, d\underset{\sim}{f}_k + 2/3H \, d\underset{\sim}{f}^T \underset{\sim}{f}_k}$$

and (32)

$$\underset{\sim}{M}_k = \underset{\sim}{M}^e_k + \frac{d\underset{\sim}{f}^T_k \, d\underset{\sim}{f}}{2/3H \, d\underset{\sim}{f}^T_k \, d\underset{\sim}{f}_k}$$

where $d\underset{\sim}{f}_k = \underset{\sim}{Q} \, (\underset{\sim}{\sigma}^k - \underset{\sim}{\sigma}^k)$

and

$$Q = \begin{pmatrix} \frac{2}{3} & -\frac{1}{3} & -\frac{1}{3} & 0 & 0 & 0 \\ -\frac{1}{3} & \frac{2}{3} & -\frac{1}{3} & 0 & 0 & 0 \\ -\frac{1}{3} & -\frac{1}{3} & \frac{2}{3} & 0 & 0 & 0 \\ 0 & 0 & 0 & 2 & 0 & 0 \\ 0 & 0 & 0 & 0 & 2 & 0 \\ 0 & 0 & 0 & 0 & 0 & 2 \end{pmatrix} \quad (33)$$

4.2 Constraints in Displacement Model

When a standard incremental procedure is applied to a particular composite system under increasing load, one obtains results of the type shown in Figure 8. This particular example was constructed for a tensile stress applied in the X_1 direction of the transverse plane, c.f., Figure 4, to a boron-aluminum composite system. The equilibrium estimate of Section 3 leads directly to the composite stress-strain curve shown in the figure. However, construction of the curve from displacement approach requires special precautions explained below. An unsuitable choice of the subelements and their displacement shape functions can lead to overconstraint in the displacement mechanism, and produce the unrealistic result represented by the steep curve in Figure 8. This curve was actually obtained with the division of the RVE shown in Figure 5, while the subelements 7 to 10 were considered as a single triangular subelement. During the loading process, the subelement stress increments were essentially isotropic and fully developed plastic flow in the RVE failed to materialize.

To correct this potential difficulty, the relations between constraints and degrees of freedom of nodal points need to be analyzed. For this purpose, the subelements 7 to 10 in Figure 5 are tentatively regarded as a single triangular element.

If the state of fully developed plastic flow exists in the RVE of Figure 5, then one can assume that the plastic and total strain increments are nearly equal. Strain increments in matrix subelements must satisfy the incompressibility condition

$$\Delta\varepsilon_{33}^{k} = -\Delta\varepsilon_{22}^{k} - \Delta\varepsilon_{11}^{k} \tag{34}$$

From the assumption of a perfect bond between the matrix and fibers it follows that:

$$\Delta\varepsilon_{33}^{k} = C_o \tag{35}$$

Without loss of generality, the constant C_o can be taken as equal to zero. Then (34) can be written in terms of the following nodal displacements in the transverse plane (x_1, x_2)

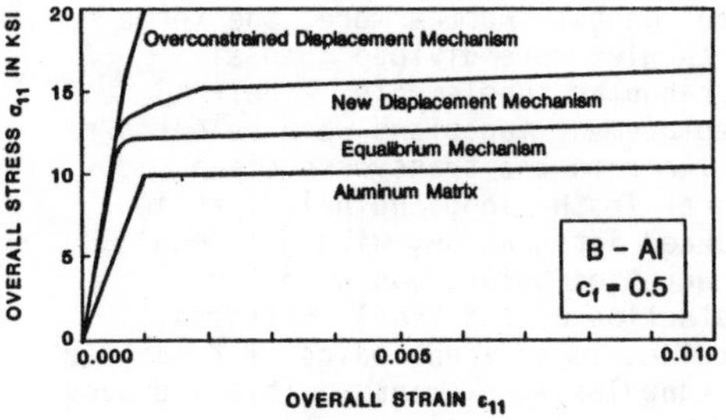

Figure 8. Comparison of Constrained and Unconstrained Upper Bound Mechanisms

$$\Delta\tilde{a}^{r} = (\Delta u_5, \Delta u_6, \Delta u_7, \Delta u_5, \Delta u_6, \Delta u_7)$$
$$\Delta\tilde{\bar{a}} = (\Delta\bar{u}_2, \Delta\bar{u}_3, \Delta\bar{v}_3) \tag{36}$$

The transverse displacements (36) must also satisfy six symmetry conditions derived from (8)

$$\Delta\bar{u}_{M1} = 1/2\Delta\bar{u}_2 \qquad \Delta\bar{u}_{M2} = 1/2(\Delta\bar{u}_2 + \Delta\bar{u}_3)$$
$$\Delta\bar{u}_{M3} = 1/2\Delta\bar{u}_3 \qquad \Delta\bar{v}_{M1} = 0 \tag{37}$$
$$\Delta\bar{v}_{M2} = 1/2\Delta\bar{v}_3 \qquad \Delta\bar{v}_{M3} = 1/2\Delta\bar{v}_3$$

where $\Delta\bar{u}_{M1}$, $\Delta\bar{u}_{M2}$, and $\Delta\bar{u}_{M3}$ can be taken from (9).

This means that the nine displacements (36) are subjected to four incompressibility constraints (34) and to six geometrical constraints (37).

These results suggest the following relationship between constraints and nodal displacements in x_1x_2-plane. Let p_{in} denote the number of incompressibility conditions (34) and p_{geo} the number of geometrical constraints (37) and n_{in} the number of nodal displacements (36). Then, nodal displacements and constraints of an displacement model which is not overconstrained in the plastic range must satisfy the inequality

$$p_{in} + p_{geo} \leq n_{in} \tag{38}$$

The displacement mechanism actually shown in Figure 5, which satisfies (38), was derived as follows: The middle triangle was divided into four triangles. Linear displacements were assumed in each of these new triangular subelements. Furthermore, the three rectangles were divided into six rectangular subelements. Their displacement functions were selected as bilinear in the transverse plane and linear in the longitudinal direction. Reduced integration, with only one integration point, was used for evaluation of the local stiffness matrices in adjacent pairs of these six rectangular subelements. This approach reduces possible linear stress distributions in subelements 11, 12, and 13 into constant ones.

5. SELECTIVE RESULTS

The first objective of this section is to compare the results obtained from PHA model with bounds on elastic moduli obtained from the composite cylinder assemblage (CCA) model [1]. It was shown in [9] and [10] that the dual analyses of the PHA models provide alternative bounds to those derived in [1].

A boron-aluminum system was selected for the comparison study. Material constraints of the constituents appear in the following table:

TABLE I: MATERIAL CONSTANTS OF FIBERS AND MATRIX

Material	E	ν	σ_y	c= 2/3H
		GPa		
Boron	414	0.21	assumed elastic	
Aluminum	72	0.33	0.77	0.07,0.7

In agreement with Figures 3 and 4, the fiber axis will always coincide with X_3, and X_1, X_2 will define the transverse plane.

Figure 9 shows bounds on selected elastic moduli as functions of fiber volume fraction c_f. Both PHA and CCA model results are presented. Comparison of these results shows that the PHA bounds are generally tighter than their CCA equivalents. In some instances, the PHA range is shifted with respect to the CCA estimates. The bounds on axial Young's modulus, E_{33}, coincide and are also identical in both models. The PHA bound on moduli E_{11}, G_{31}, and ν_{31} tend to bracket one of the bounds from on the CCA model. This behavior is probably

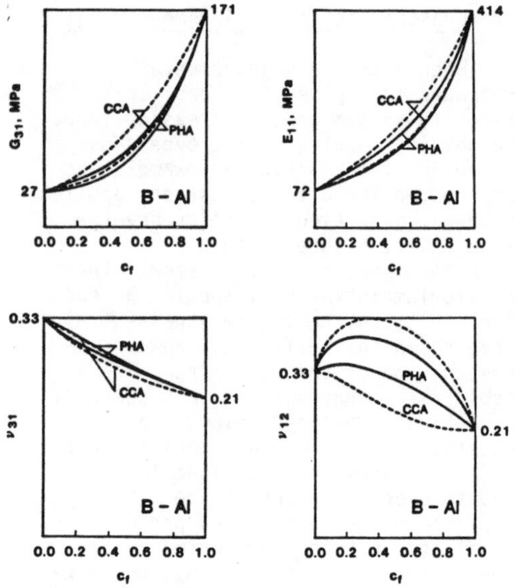

Figure 9

related to the choice of hexagonal fiber geometry in the PHA model. To bring the pHA bounds within the CCA bounds, it might be necessary to select a fiber cross section which is closer to a circular one, for example a twelve or twenty-four sided polygon.

Elastic-plastic behavior of the boron-aluminum aggregate is studied next with the aid of the dual PHA models. It is probably obvious to the reader that, under certain conditions, the bonding procedures of the elastic region be extended into the elastic-plastic region if the principles of the minimum plastic strains and stresses rates applied. However, the conditions under which the duel estimates will bracket instantaneous moduli are complex, and the authors intend to discuss them in a separate paper.

In This paper the dual responses of a single boron-aluminum lamina subjected to a combination of a transverse normal and longitudinal shear stress are presented. To illustrate the effect of these two stress components on ply response, a calculation of overall strains was made in a ply loaded by certain ratios of transverse tension and longitudinal shear.

The results are shown in Figure 10. The longitudinal shear stress has a very

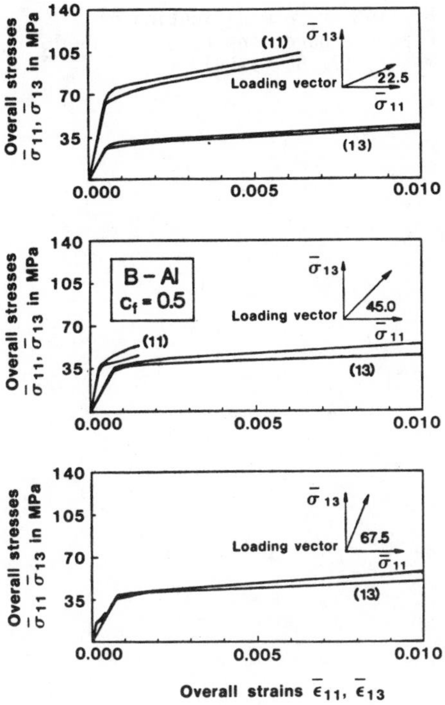

Figure 10

pronounced effect on ply response, and much larger magnitudes of $\bar\varepsilon_{13}$ strain are generated compared to $\bar\varepsilon_{11}$ strain, even if $\bar\sigma_{13} < \bar\sigma_{11}$. When $\bar\sigma_{11} < \bar\sigma_{13}$, ply deformation is essentially dominated by $\bar\varepsilon_{13}$, and only small $\bar\varepsilon_{11}$ magnitudes are recovered. These results are not entirely unexpected. A similar response would be obtained if fibers were absent. For the same reason, one also finds that both the initial yield and flow stress $\bar\sigma_{11}$ are reduced when $\bar\sigma_{11}$ and $\bar\sigma_{13}$ are applied simultaneously. In each case, the upper and lower bounds on flow stress are brought closer together under combined loading. For example, under overall strain $\bar\varepsilon_{11} = 0.005$, the ratio $\bar\sigma_{11}U/\bar\sigma_{11}L$ is equal to 0.15 to $\bar\sigma 13 = 0$, and to 0.06 at $\sigma_{13} = 0.4142\ \bar\sigma_{11}$.

Although limited in scope, the numerical results presented here show that the simplest variant of the PHA model provides relatively close bounds on the instantaneous stiffness of an elastic-plastic fibrous composite. The distance between bounds varies with applied stress state. The bounds are especially tight in the elastic case, and also in the plastic state when the fiber volume fraction is low, or when a combination of transverse tension and longitudinal shear is applied.

The bounds can be improved through refinement of the RVE mesh. A refinement, however, causes an increase in computing time necessary for calculation of instantaneous overall properties. An excessive increase in computing time can prevent incorporation of PHA subroutines into a finite element code, for stress analysis of composite structures might become prohibitively expensive.

For the present PHA model, the calculation of instantaneous mechanical properties requires inversion of two stiffness submatrices of the RVE. The orders of these submatrices are 12 by 12, and 6 by 6. This compares favorably with the computations required by other methods. For example, the self-consistent procedure leads to a solution of a system of 21 simultaneous, non-linear, algebraic equations. Thus, the proposed class of models provides both a rigorous and economical elastic-plastic analysis of fiber reinforced composites.

REFERENCES

1. Z. Hashin and B. W. Rosen, The elastic moduli of fiber reinforced materials. J. Appl. Mech., 31, 233 (1964).

2. R. Hill, Theory of mechanical properties of fiber-strengthened materials: III. Self-consistent model. J. Mech. Phys. Solids, 13, 189 (1965).

3. L. J. Walpole, On the overall elastic moduli of composite materials. J. Mech. Phys. Solids, 17, 235 (1969).

4. G. J. Dvorak and Y. A. Bahei-El-Din, Elastic-plastic behavior of fibrous composites. J. Mech. Phys. Solids, 27, 51 (1979).

5. W. Huang, Plastic behavior of some composite materials. J. Composite Materials, 5, 320 (1971).

6. G. J. Dvorak and Y. A. Bahei-El-Din, Plasticity analysis of fibrous composites. J. Appl. Mech., 49, 327 (1982).

7. G. J. Dvorak and C. J. Wung, Thermoplasticity of unidirectional metal matrix composites. In Mechanics of Material Behavior, G. J. Dvorak and R. T. Shield, editors, Elsevier, 87 (1984).

8. J. Aboudi, A continuum theory for fiber-reinforced elastic-viscoplastic composites. Int. J. Eng. Sci., 20, 605 (1982).

9. G. J. Dvorak and J. L. Teply, Periodic hexagonal array models for plasticity analysis of composite materials. Plasticity Today: Modelling, Methods and Applications, A. Sawczuk, editor, Elsevier, 625 (1984).

10. J. L. Teply, Ph.D. dissertation. University of Utah (1984).

11. J. B. Martin, Plasticity: fundamentals and general results. The MIT Press, Cambridge, MA, 931 (1975).

12. B. F. DeVeubeke, Displacement and equilibrium models in the finite element method. Chapter 9 of Stress Analysis, edited by O. C. Zienkiewicz and G. S. Holister (1978).

13. G. Strang and G. J. Fix, Analysis of the finite element method. Prentice-Hall, New Jersey (1973).

14. O. C. Zienkiewicz, The finite element method, McGraw-Hill Book Company, 3rd ed. (1977).

15. G. J. Dvorak, Metal matrix composites: Plasticity and fatigue, in Mechanics of Composite Materials, Recent Advances, edited by Z. Hashin and C. T. Herakovich, Pergamon Press, 73 (1983).

16. A. Phillips and M. Ricciutti, An experimental investigation concerning yield surfaces and loading surfaces, Acta. Mech., 27, 91 (1976).

17. A. Phillips and H. Moon, Fundamental experiments in plasticity and creep of aluminum-extension of previous results. Int. J. Solids Structures, 72, 159 (1977).

18. H. Ziegler, A modification of Prager's hardening rule. J. Appl. Mech., 17, 55 (1959).

5th International Symposium on Continuum Models of Discrete Systems / Nottingham / 14-20 July 1985

Random structure analysis and modelling by mathematical morphology

D.Jeulin
IRSID, Maizières-les-Metz, France

ABSTRACT : A quantitative approach of the description of the microstructure of materials is presented. Based on Mathematical Morphology it enables to determine correlation functions of physical porperties of materials from image analysis measurements or from predictions by random structure modelling.

1. INTRODUCTION

A prediction of overall physical properties of composite materials from their microstructure can be made on different ways : substitution of the complex actual structure by a simplified geometrical model (e.g. spheroïdal inclusions in (Ondraçek, Pejsa 1976)), use of correlation functions of the material properties, simulation of the microstructure by random models. In this contribution, we will present our quantitative approach to characterize heterogeneous structures from a random sets (Matheron 1975, Serra 1982) point of view. It enables to measure morphological data (including correlation functions) on materials by means of image analysis (using instruments as the texture analyzer (Serra 1965)), and to develop random structures modelling as will be illustrated by different examples.

2. RANDOM STRUCTURE ANALYSIS BY MATHEMATICAL MORPHOLOGY

Material microstructures, and their representation by random models, may be classified into the following categories : point processes (germs, defects, micro-inclusions...), random tessellations (granular structures), multiphase sets (porous structure, minerals), random functions (rough surfaces, distribution of crystallographic orientations, chemical concentration...).

The main purpose of mathematical morphology, developed in France by J. Serra, G. Matheron, is to produce a coherent and significant quantitative description of spatial structures. It was shown that any quantitative morphological parameter may be defined by two different steps : (a) a set transformation of the structure, and (b) a measurement on the result of the transformation.

A classical example is given by sieve analysis, where the transformation separates particles with regard to their size, and the size distribution can then be determined from weight measurements.

In addition to quantification of structures, mathematical morphology was extended to image processing such as binary or digital image segmentation, pattern recognition (Serra 1982). These procedures are of importance for automatic analysis of images, but will not be presented here.

2.1. Elementary operations and measurements of Mathematical Morphology

A common approach to shape description is to compare a structure with well defined shapes (spheres, segments, etc.). Such a procedure in mathematical morphology is associated with the concept of a structuring element : a geometrical set of points arranged in a well defined shape and of a well defined size is moved within the structure by translation. An answer to logical question about the relative position of the structure and the element is recorded for each position of its centre, to perform a morphological transformation of the structure. The following example is given :

Question : is the structuring element included in the studied object ?

The set of points corresponding to a positive answer is a new object.

This morphological transformation is called 'erosion'. It is obtained by giving value 1 to all the positions of the center of the structuring element B for which this last one is included inside analyzed set X (fig. 1a). The result, called $X \ominus B$ is presented in fig. 1b for an hexagonal structuring element.

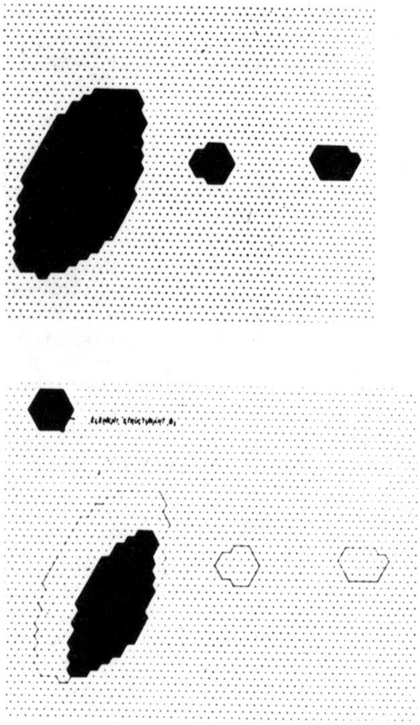

a)

b)

Fig. 1 Erosion of set X (1.a) by hexagonal structuring element (1.b)

- a second elementary morphological operation, called 'dilation' is obtained by giving value 1 to the center of B when it encounters object X.

By means of erosions by convex bodies with increasing size the whole picture is transformed step by step according to the size of the objects it contains. Conversely, dilations by convex bodies with increasing sizes will gradually fill up space according to the distribution of objects. The number of objects in the picture, or their area, as a function of the size of structuring element give quantitative information on sizes of different connected parts of X (erosion) or on their spatial distribution (dilation).

With a linear structuring element (segment with length ℓ) the analyzed structure is deformed in a given direction, which allows to study anisotropies.

- Closing and opening operations by structuring element B result from a combination of two operations : dilation and erosion (closing) or erosion and dilations (opening). It can be shown that measurements of the area of the picture after openings or closings by convex bodies with increasing sizes, gives estimation of sizes distributions for object X (opening) or X^c (closing). These granulometric distributions are a generalization of classical operations made in sieve analysis (which is perfectly simulated for a population of regular hexagons). They may be applied to objects with any shape.

2.2. Second an Third order morphological analysis of a stationary multiphase structure

2.2.1. Covariance analysis

An important particular case is the erosion of the image by a structuring element made of pairs of points at distance h from each other. The result of this operation is obtained by translation of image X by vector (- h) and intersection $X_{-h} \cap X$ (fig. 2). The result of this operation depends both on size and distribution of parts building X. The estimation of the volume fraction V_v of picture after erosions by couples of points with increasing distances h provides measurements of covariance function C (h) for set X (assumed to be a stationary random set)

$$C(h) = V_v \, (X \cap X - h) \qquad (1)$$

In the case of multiphase structures, it may be interesting to get morphological information on each phase separately. Another kind of operation is sensitive to mutual spatial arrangement of pairs of phases ; this is obtained by demanding to the pair of points (x, x + h) to answer to the following conditions : x belongs to phase Xi and x + h belongs to phase Xj.

This composite operation results from intersection between Xi and Xj (-h).

From this morphological operation are estimated crossed covariance Cij (h) between phases Xi and Xj ; they allow to estimate specific surface of contacts between phases, and vicinity indexes that summarize spatial agencement of phases in multi-component textures (Jeulin 1981 a).

$$Cij(h) = V_V(Xi \cap Xj - h) \qquad (2)$$

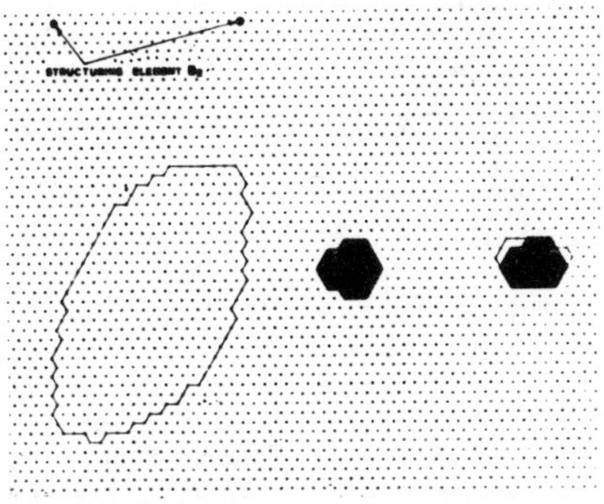

Fig. 2 Covariance function of set X : erosion of fig. 1a by a pair of points (x, x + h)

2.2.2. n-points morphological correlation functions

It is not difficult to extend covariance analysis to functions measured from erosions by a set of n points (x, x + h_1 ..., x + h_{n-1}). In the case of three points, we can define

$$T_{ijk}(h_1,h_2) = V_V(X_i \cap X_{j-h_1} \cap X_{k-h_2}) \qquad (3)$$

These functions allow to study mutual associations of triples of phases (Xi, Xj, Xk).

2.2.3. Experimental determination

Relations (2) and (3) show that functions Cij(h) and $T_{ijk}(h_1,h_2)$ can be estimated from image analysis performed on plane sections of investigated materials (in the case of anisotropic structures, orientations of vectors h, h_1, h_2 must be accounted for) after storage of the pictures of different phases in connectable binary memories of an image analyzer (the operations involved by relations (3) and (4) can be performed at T.V. rate on most instruments). Higher order morphological correlation functions require the knowledge of the structure in the three-dimensionnal space and therefore needs a special preparation of the materials.

Each kind of morphological measurement must be estimated from a representative number of fields of analysis in the microstructure, and must be controlled by statistical indications.

2.2.4. Morphological limitation of n-points correlation analysis

Morphological characterization by n-points correlation analysis is a powerful tool in the case of multiphase textures (e.g. mineralogical textures). However it is of no help when some subsets under study are degenerate (e.g. cracks in a solid, or grain boundaries in a polycristal). In that case, generalizations of covariance functions were proposed (Jeulin 1985 a).

2.3. Correlation functions of material properties

In the case of physical properties with linear laws (e.g. elastic properties, electrical and thermal conductivity, dielectric and magnetic properties...), statistical continuum theories (Beran 1968, Kröner 1971, Willis 1981) give estimates of composite materials overall properties from the correlations functions of these properties.

Order n correlation function of property a may be defined by

$$W(h_1, \dots h_{n-1}) = E[a(x)\, a(x+h_1) \dots a(x+h_{n-1})] \qquad (4)$$

(it is assumed that the material is stationary)

and

$$\overline{W}(h_1, \dots h_{n-1}) = E[(a(x) - \bar{a}) \dots (a(x+h_{n-1}) - \bar{a})] \qquad (5)$$

where E is the mathematical expectation (average value) and $\bar{a}$ is the average value of property a ($\bar{a} = E[a]$).

In the case of a multiphase material where the property a is supposed to be constant in each phase Xi (with value a_i), the property a (x) at point x is given by

$$a(x) = \sum_{i=1}^{i=n} ai\, ki(x)$$

where $ki(x) = 1$ if $x \in Xi$
$ki(x) = 0$ elsewhere

Correlation functions (4) and (5) of the property a may be deduced from the ai and from the previously introduced morphological correlation functions. For instance

$$W(h) = E[a(x)\, a(x+h)] = \sum_{i,j} aij\, Cij(h) \qquad (7)$$

$$\overline{W}(h) = \sum_{i,j} ai\, aj\, \overline{Cij}(h) = W(h) - \sum_{i,j} ai\, aj\, pi\, pj \qquad (8)$$

where $\overline{Cij}(h) = Cij(h) - pi\ pj$ (9)

and $pi = Cii(o) = P\{x \in Xi\}$ (10)

In the case of a two-phase material (e.g. porous material)

$$W(h) = (C_{11}(h) - p_1)(a_1 - a_2)^2 + p_1 a_1^2 + p_2 a_2^2 \quad (11)$$

with $p_1 + p_2 = 1$

and $\overline{W}(h) = (a_1 - a_2)^2 (C_{11}(h) - p_1^2)$ (12)

Similarly

$$W(h_1, h_2) = E[a(x)\ a(x+h_1)\ a(x+h_2)]$$
$$= \sum_{i,j,k} ai\ aj\ ak\ Tijk(h_1, h_2) \quad (13)$$

$$\overline{W}(h_1, h_2) = W(h_1, h_2) - \bar{a}^3 - \bar{a}[\overline{W}(h_1) + \overline{W}(h_2) + \overline{W}(h_2 - h_1)] \quad (14)$$

It is clear from relations (7) to (14) that when dealing with a set of different physical properties a on the same material, the n-point correlation functions associated with these properties can be experimentally estimated from the knowledge of morphological correlation functions up to order n. As seen in 2.2.3. these last properties may be obtained from image analysis measurements.

3. RANDOM STRUCTURE MODELLING BY MATHEMATICAL MORPHOLOGY

As explained earlier, the experimental determination of morphological correlation functions with order higher than 3 is difficult, mainly due to 3.D data acquisition on material microstructure. Furthermore, when increasing the order, the number of operations (translations and intersections), and the number of collected data become too high for convenient handling. An alternative to this approach is the use of random morphological modelling of heterogeneous materials. The main advantages of structure modelling are the following :

- it allows to condense available morphological data in a few parameters, that sum up the complexity of the structure. Properties that are difficult to be experimentally measured (e.g. 3D size distribution) may be predicted from these parameters
- it can provide means of simulation of the structure formation, for a better understanding of the material genesis.

3.1. Random structure properties

In part 2 were presented the main morphological criteria that can be used for a quantitative description of structures, in an image analysis approach (set transformation and measurement). It can be shown (Matheron 1975) that random structures are characterized by the following probability laws :

$$Q(B_x) = P\{B_x \subset X^c\} = P\{x \in X^c \ominus \check{B}\} \quad (15)$$

For multiphase structures (15) becomes

$$Q(B_x) = P\{B_{1x} \subset X_1^c, \ldots B_{nx} \subset X_n^c\}$$
$$= P\{x \subset X_1^c \ominus \check{B}_1 \cap \ldots \cap X_n^c \ominus \check{B}_n\} \quad (16)$$

Properties of a material with a continuous range of variations may be considered as realizations of random functions $Z_x(x)$: height of a surface, chemical concentration of a component... In that case, (15) becomes

$$Q(B_{x,z}) = F_{X \oplus \check{B}}(z, x) = P\{Z_{X \oplus \check{B}}(x) < z\} \quad (17)$$

where X is the graph of the function $Z_x(x)$.

Usually average properties of the structure do not depend on the position of x in space. They can be modelized by stationary random structures, for which probability laws (15) (16) and (17) do not depend on x. If furthermore the structure is ergodic, these laws can be estimated from volume fraction measurements, in a similar way as in part 2 :

- for sets : $Q(B) = V_V(X^c \ominus \check{B})$ (18)
- for multiphase structures :
  $Q(B) = V_V(X_1^c \ominus \check{B}_1 \cap \ldots \cap X_n^c \ominus \check{B}_n)$ (19)
- for random functions :
  $Q(B, z) = F_{X \oplus \check{B}}(z) = P\{Z_{X \oplus \check{B}} < z\}$ (20)

The n-points morphological correlation functions introduced in part 2 are particular cases of the probability laws (18) and (19). A complete characterization of a random structure requires the knowledge of probability laws of all orders (defined for B being a finite set of points or an infinite set of points such as a segment, a disk, ...).

3.2. Principle of random structure modelling

A model of structure is defined by assumptions involving a succession of elementary processes and geometrical modes of interaction, that set up the genesis

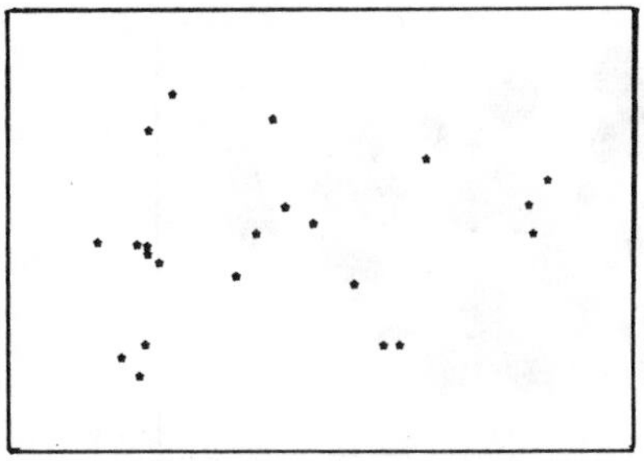

Fig. 3 Example of a plane Poisson stochastic point process

of the random structure.

In each case, the probability laws (18), (19), (20) are a function of the definition of the model, of the value of its parameters, and of the geometrical set of points B.

The parameters of a given model may be estimated, and the assumptions controlled, from measurements by image analysis on samples (performing combinations of erosion operations).

The probability laws of a model can be estimated as a function of the parameters either from theoretical calculations, or by image analysis measurements on simulations. Thus may be predicted high order correlation functions of the material properties from a restricted amount of information.

In the remaining of this part, we will now present briefly some examples of models of structure for different kinds of sets, and give some of their salient properties. More details are to be found in the references.

3.3. A model of random germs : the Poisson point process

Stochastic point processes provide models for random implantation of sites (for instance germs of a crystallization, or location of small inclusions...). The most popular stochastic point process is the Poisson point process, that simulates germination without spatial interaction (no repulsion or attraction) (see fig. 3).

The homogeneous Poisson point process with constant density θ has the following properties :

- the number of points N (V) and N (V') falling in two disconnected volumes V and V' are independant random variables
- N (V) is a Poisson random variable with density 0V :

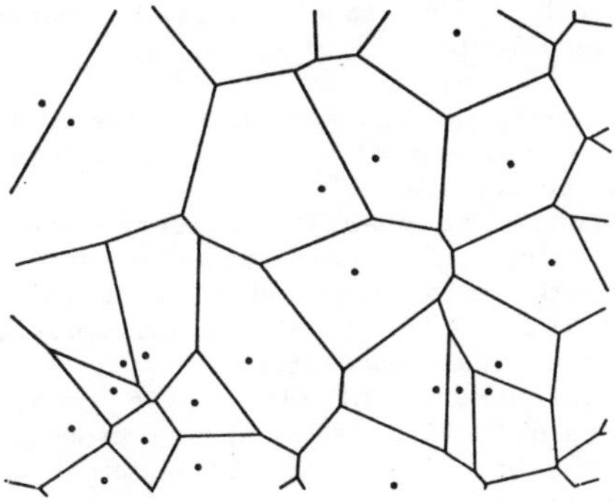

a) Voronoï polyhedra

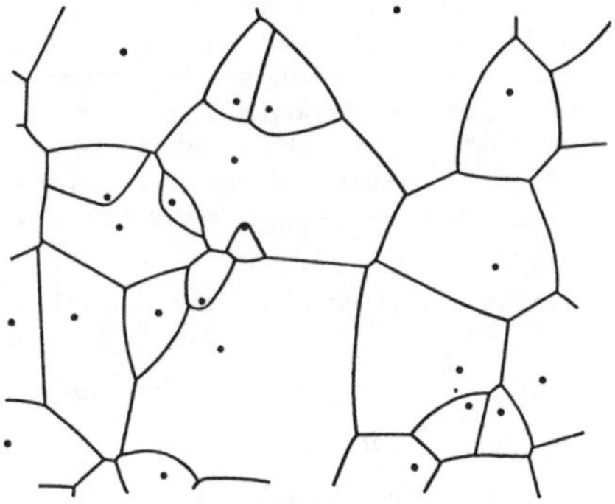

b) Johnson-Mehl grain model

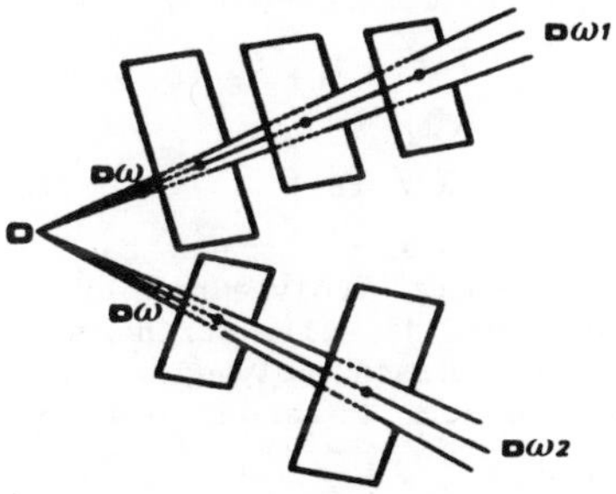

c) Poisson polyhedra

Fig. 4 Random tessellations of space

$$P\left\{N(V) = n\right\} = \frac{(\theta V)^n}{n} \exp(-\theta V) \qquad (21)$$

From (21) it comes :

$$Q(B) = P\left\{N(B) = 0\right\} = \exp(-\theta V(B)) \qquad (22)$$

3.4. Single phase random sets : random tessellations

Random tessellations of space give convenient modelling of granular structures such as polycrystals.

* Voronoī polyhedra (Gilbert 1962) are obtained from simultaneous growth of crystals in a liquid state from Poisson germs, with growth interruption, when two crystals encounter.

* Johnson-Mehl (1979) grain model is obtained with a similar process, but with continuous crystals germination during growth (fig. 4b).

It is easy to simulate these processes in the plane by construction of skeletons (Squiz) on random germs.

* Poisson polyhedra (Miles 1972, Matheron 1975) have planar boundaries obtained from random planes implantation (fig. 4c) : planes orthogonal to straight line $D\omega$ containing the origin O are issues from a Poisson point process on $D\omega$ with infinitesimal density $\lambda(\omega)\, d\omega$.

For Poisson polyhedra it can be shown that

$$Q(B) = \exp - \int_{2\pi ster} \lambda(\omega)\, b(B,\omega)\, d\omega \quad (23)$$

where $b(\omega)$ is the width of B in the direction ω.

When $\lambda(\omega) = \lambda$, polyhedra are isotropic, and (23) becomes for sphere with radius r, segment with length l, or planar object B (with perimeter $\mathcal{L}(B)$) :

$$Q(r) = \exp(-4\pi r \lambda) \quad (24)$$

$$Q(l) = \exp(-\pi \lambda l) \quad (25)$$

$$Q(B) = \exp(-\frac{\pi}{2} \lambda \mathcal{L}(B)) \quad (26)$$

These tessellations of space can be defined in 3D space (Polyhedra) or in 2D space (polygons) as well. Furthermore, 2D sections of Poisson Polyhedra (with density λ) induce Poisson polygons (with density $\lambda' = \frac{\pi}{2}\lambda$).

3.5. A two-phase random set : the Boolean model

This model, proposed by G. Matheron (1967), is convenient for the description of a second phase homogeneous dispersion in a matrix. It is built in two steps :

. Germs are randomly and homogeneously dispersed according to a Poisson point process with density θ (cf. fig. 3).

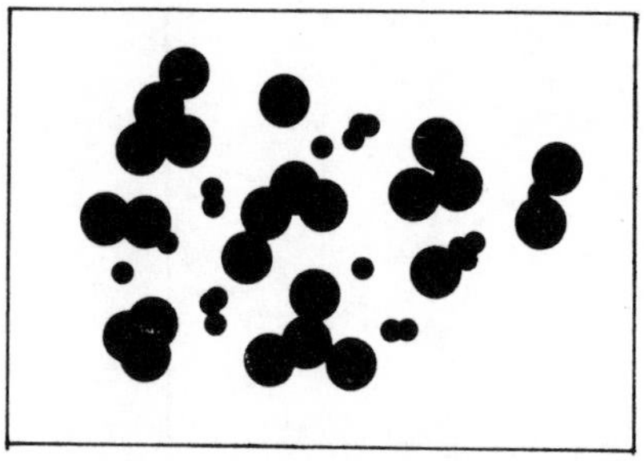

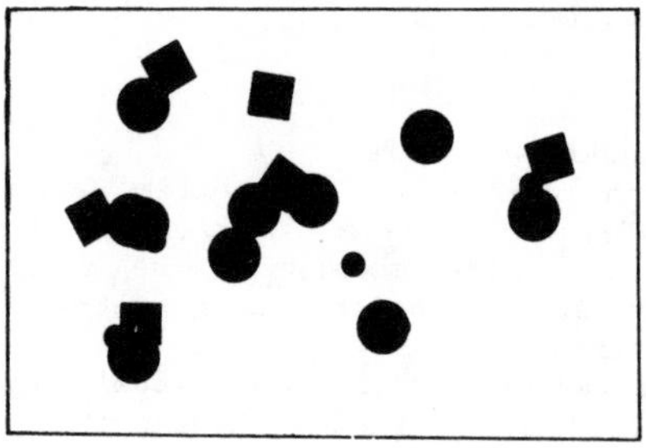

Fig. 5 Simulation of Boolean models

. Random primary grains A' are implanted in each germ (fig. 5). Oriented grains can simulate inclusions without clusters.

The union of all grains builts a phase X (e.g. a porous medium) standing out of a matrix. From the point of view of the genesis of the structure the construction of this model is similar to the Voronoī polyhedra but with some differences :

. the growth of a given germ is not homogeneous (a primary grain A' may have arbitrary shape and size)

. there is no interaction during the growth of germs, so that two grains can overlap.

In the 3D space

$$Q(B) = P(B \subset X^C) = \exp(-\theta \bar{V}(A' \oplus \check{B})) \quad (27)$$

where $V(A' \oplus \check{B})$ is the average volume of $A' \oplus \check{B}$ obtained over the realizations of random set A'.

Three particular cases of relation (27) are of interest :

- When B is a point, $q = V_V(X^C) = \exp(-\theta \bar{V}(A'))$ (28)
- when B is a pair of points x, x + h $Q(h) = q^2 \exp(\theta K(h))$ (29) where $K(h) = \bar{V}(A' \cap A'h)$

- when B is a triple of points $(x, x + h_1, x + h_2)$

$$T(h_1,h_2) = q^3 \exp\left(-\theta\left(\bar{V}(A' \cap A'_{h_1} \cap A'_{h_2}) - K(h_1) - K(h_2) - K(h_2-h_1)\right)\right) \quad (30)$$

If A' is a random convex set, we have for a segment of length l

$$Q(l) = q \exp(\theta K'(0)\, l) \quad (31)$$

Relation (31) is used to test the assumption "Boolean model with convex primary grains" from experimental measurements.

We used such a description for sinter porous structure, with random spheres or cylinders as primary grains (Jeulin 1979 - 1983).

Different one-parameter distribution of sphere diameters were used in order to synthetize the structure with two parameters : density θ, and a size parameter a defining the size distribution. These parameters are estimated from image analysis measurements (determination of experimental curve Q(l)), and allowed to compute Q(h). An example of experimental results is given in fig. 6, where it appears that the density of pores θ increases with Fe content in iron ore sinters, as a result of the pore coarsening effect when using flux additions (and therefore decreasing Fe content).

The order 2 and 3 correlation functions of material properties of a Boolean structure (assuming value a_1 for X and a_2 for X^c) may be deduced from application of relations (29) (30) to relations (7) to (14).

$$W(h) = (a_1-a_2)\, q^2 \exp(\theta K(h)) + a_1^2 + 2 a_1 q (a_2-a_1) \quad (32)$$

$$\bar{W}(h) = (a_1-a_2)^2 q^2 [\exp(\theta K(h)) - 1] \quad (33)$$

$$\bar{W}(h_1,h_2) = (a_2-a_1)^3 q^3 \left[\exp\left(-\theta\left(V(A' \cap A'_{h_1} \cap A'_{h_2}) - K(h_1) - K(h_2) - K(h_2-h_1)\right)\right) - \exp(\theta K(h_1)) - \exp(\theta K(h_2)) - \exp(\theta K(h_2-h_1))\right] \quad (34)$$

3.6. Multiphase structures built on random tessellations

A first class of multiphase modelling may be obtained by affectation of each class of a random tessellation (as those presented in § 3.4) to phase i with

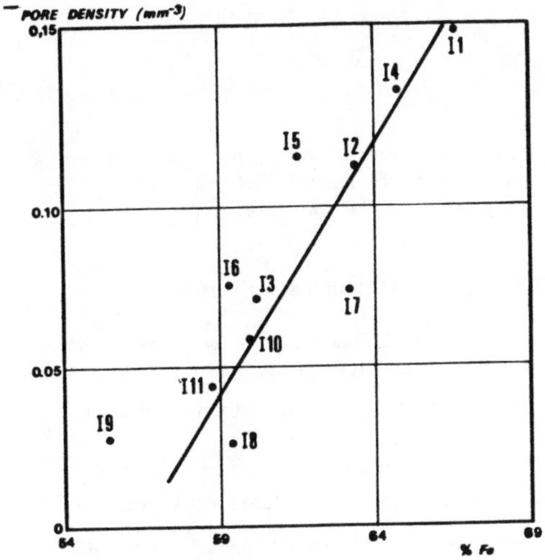

Fig. 6 Variation of pore density deduced from a Boolean model, with total iron content

probability pi (i = 1,2, ... n). The affectations of two separate classes are statistically independant. It can be shown (Jeulin 1979) that for each phase

$$Qi(B) = G_B(1 - pi) \quad (35)$$

$$Pi(B) = P(B \subset Xi) = G_B(pi) \quad (36)$$

where $G_B(s) = \sum_{n=0}^{n=\infty} P\{N(B) = n\}\, s^n$ (37),

N(B) being the (random) number of classes of the tessellation hit by B.

Relations (35) and (36) allow to compute morphological functions Cij(h) and Tijk(h) (Jeulin 1983).

For instance

$$Cii(h) = pi^2 + pi(1-pi)\, K(h)/K(0) \quad (38)$$

$$Cij(h) = pi\, pj\, [1 - K(h)/K(0)] \quad (39)$$

where $K(h) = \bar{V}(X' \cap X'_h)$ (X' being a class of the tessellation). In the case of anisotropic Poisson tessellation (cf. § 3.4) $K(h)/K(0) = \exp(-\pi \lambda h)$.

For such a definition of random structure, each phase of the material has similar geometrical properties (up to its volume fraction Vvi = pi). There is no distinction between a continuous background and second phases as it was the case for the Boolean model.

The order 2 and 3 correlation functions of material properties are expressed by

$$\overline{W}(h) = \frac{K(h)}{K(0)} E[(a-\bar{a})^2] \quad (40)$$

and $$\overline{W}(h_1, h_2) = \frac{\overline{V}(X' \cap X'_{h_1} \cap X'_{h_2})}{K(0)} E[(a-\bar{a})^3] \quad (41)$$

These expressions may be extended to an infinite number of phases (continuous variation of parameter a).

3.7. Sequential random multiphase textures

Genetic multiphase models were developed to simulate sequential crystallizations of the n different phases X_t^i from a melt $Z^c t$ ($Zt = \bigcup_{i=1}^{i=n} X_t^i$) (Jeulin 1979-1980) : in the time interval (t, t + dt) a set of new crystals $X_1^i(t,dt)$ appear in the melt as an infinitesimal Boolean model :

$$X_{t+dt}^i = X_t^i \cup (X_1^i(t,dt) \cap Z_t^c) \quad (42)$$

The choice of time dependent laws implied in the genesis of the structure must take into account physical data on germination and growth of the phases, as well as the existing cooling profile. For instance the effect of thermal conditions on the final microstructure may be simulated. From definition (42) we calculated theoretical expressions of morphological correlations functions $Cij(h,t)$ and $Tijk(h_1,h_2,t)$ (Jeulin 1979).

When restricting the model to an homogeneous process with time (densities θ^i and geometrical properties of implanted phases are supposed constant with time) there is no melt left as $t \to \infty$, and

$$Cii(h,\infty) = \frac{2(\theta^i K^i(0))^2 - \theta^i K^i(h)\,(2\theta^i K^i(0) - \theta K(0))}{\theta^2 K(0)\,[2K(0) - K(h)]} \quad (43)$$

$$Cij(h,\infty) = \frac{\theta^i \theta^j (2K^i(0)K^j(0) - K^j(h)K^i(0) - K^i(h)K^j(0)]}{\theta^2 K(0)\,[2K(0) - K(h)]} \quad (44)$$

where $K^i(h) = V(X_1^i(t,dt) \cap X_1^i(t,dt)_h)$ (45)

and $\theta K(h) = \sum_{i=1}^{i=n} \theta^i K^i(h)$

The second-order correlation function of material properties is given by

$$\overline{W}(h) = \Big\{ (K(h)/K(0) \Big(\sum_{i=1}^{i=n} ai\,\theta^i K^i(0)\Big)^2 - 2\Big(\sum ai\,\theta^i K^i(0)\Big)\Big(\sum ai\,\theta^i K^i(h)\Big) + \theta K(0)\Big(\sum ai^2 \theta^i K^i(h)\Big)\Big\} \Big/ \Big(\theta^2 K(0)(2K(0) - K(h)\Big) \quad (46)$$

A simplification of (46) is obtained for n phases with the same morphology but with different properties a_i :

$$W(h) = \frac{K^i(h)}{2K^i(0) - K^i(h)} E[(a-\bar{a})^2] \quad (47)$$

where $E[(a-\bar{a})^2] = \sum a_i^2/n - (\sum a_i/n)^2$

3.8. Boolean Random function

The models of structures that were presented up to now describe polycrystals or phases, i.e. structures with locally constant properties (e.g. constant crystallographic orientation inside a class of a tessellation that isolates a crystal) In this part we present a model of random function that was developped to simulate rough-planar surfaces (Jeulin 1981 b, Jeulin 1985 b). It may be applied to other properties such as chemical concentrations... It is defined in R^n, but for a simplified presentation we will give its definition in R^2 :

- as for the Boolean model (§ 3.5) a Poisson point process with density θ is considered in the plane
- independent realizations of a random function X' are implanted at each point of the process (fig. 7).

For a Boolean random function (20) may be expressed by

$$Q(B,z) = \exp\left(-\theta \overline{A}(X' \oplus \check{B}) \cap \Pi z\right) \quad (48)$$

where Πz is the horizontal plane at height z and $\overline{A}$ the average of the area.

For instance the distribution function of variable Z (x) is given by

$$F(z) = \exp\left(-\theta \overline{A}(X'z)\right) \quad (49),$$

X'z being the horizontal section of X' at level z.

Properties of such random functions are given in the references for different cases, including anisotropic random functions. We give here the second order correlation functions of material properties (noted z) :

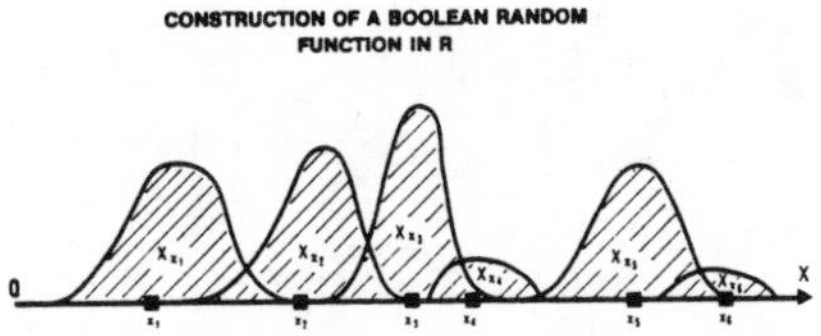

Fig. 7 Simulation of a random Boolean function (simplified representation on a line)

$$W(h) = \iint z_1 \, z_2 \, f(z_1, z_2, h) \, dz_1 \, dz_2 \qquad (50)$$

$$\overline{W}(h) = W(h) - (\overline{z})^2$$

$$\text{with } \overline{z} = \int_0^{\infty} \big(1 - F(z)\big) \, dz \qquad (51)$$

$f(z_1, z_2, h)$ being the probability density function of pair $Z(x)$ $Z(x+h)$. It is deduced from $F(z_1, z_2) = P\{Z(x) < z_1, Z(x+h) < z_2\} = \exp(-\theta \overline{A}(X'z_1 \cap X'z_2\, h))$ (52)

If we note $K(h, z_1, z_2) = \overline{A}(X'z_1 \cap X'z_2\, h)$, $K'z_1(h)$, $K'z_2(h)$ and $K''z_1z_2(h)$ its partial derivatives, and $\overline{A}(X'z) = A(z)$, $\frac{\partial}{\partial z}\overline{A}(X'z) = A'(z)$, it comes

$$f(z_1, z_2) = \theta^2 F(z_1, z_2) \, [A'(z_1) A'(z_2) + K''z_1z_2(h) + K'z_1(h) K'z_2(h) - K'z_1(h) A'(z_2) - K'z_2(h) A'(z_1)] \qquad (53)$$

4. CONCLUSION

The articulation of the different models of random structure that were introduced in this paper is summed up in fig. 8. The wide range of available models issued from the Mathematical Morphology approach, as well as the implementation of image analysis on sample of materials, should help to make some progress in the prediction of overall physical properties of heterogeneous materials from an accurate and appropriate (use of correlation functions) quantitative morphological description.

REFERENCES

Beran, M.J. 1968 - Statistical Continuum Theories - New-York : Interscience publishers.

Resulting Random Structure	First step of the random structure: Random germs (stochastic Poisson process)	Random Poisson flats	Random tessellation
Random tessellation	Zone of influence of germs - Voronoï - Johnson-Mehl	Poisson polyhedra	
Multi-phase random set	Union of random primary grains - Boolean model (2-phase) - Sequential multiphase model (n phase)		Random affectation of each class — multiphase textures
Random function	Random Boolean function		Random slot function

Fig. 8 Example of construction of random structures

Gilbert, E.N. 1962. Ann. Math. Stat. 33 : 958
Jeulin, D. 1979. Mathematical Morphology and physical properties of iron ore sinters and of metallurgical coke. Thesis. School of Mines of Paris.
Jeulin, D. 1980. Mikroskopie. 37,S : 130-137.
Jeulin, D. 1981 a. Mathematical Morphology and multiphase materials. Stereol. Iugosl. 3 Suppl. 1 : 265-286.
Jeulin, D., Jeulin, P. 1981 b. Synthesis of rough surfaces by random morphological models. Stereol.-Iugosl. 3 Suppl. 1 : 239-246.
Jeulin, D. 1983 a. Quantitative morphology and quality of iron ore sinter. 42nd Ironmaking Conference. Atlanta. IRSID Report RE 979.
Jeulin, D. 1983 b. Mém. Et. Sci. Rev. Mét. n° 5 : 251-266.
Jeulin, D. 1985 a. Study of spatial distributions in multicomponent structures by image analysis. To appear in Stereol. Iugosl.
Jeulin, D. 1985 b. Anisotropic rough surface modelling by random morphological functions. To appear in Stereol. Iugosl.
Johnson, W.A., Mehl, R.F. 1939. Trans. AIME 135 : 416-458.
Kröner, E. 1971. Statistical Continuum mechanics. Springer Verlag
Matheron, G. 1967. Eléments pour une théorie des milieux poreux. Paris : Masson.
Matheron, G. 1975 Random sets and integral geometry - New-York : Wiley.

Miles, R.E. 1972. Suppl. to Adv. in Appl. Prob. : 243-266.
Ondracek, G. Pejsa, R. 1967. J. Microsc. 107 : 335-441.
Serra, J. & IRSID. Brevet sur un dispositif d'analyse de textures, n° 1449.059, déposé en France le 2/07/65 et 1ère addition au brevet 1449.059, n° 92.537, déposé en France le 29/05/67.
Serra, J. 1982. Image analysis and mathematical morphology. London : Academic Press.
Willis, J.R. 1981. Variational and related methods for the overall properties of composites. Advances in applied mechanics. 21 : 1-78.

5th International Symposium on Continuum Models of Discrete Systems / Nottingham / 14-20 July 1985

Direct continuum model of a solid (Abstract)

H.Zorski
Polish Academy of Sciences, Warsaw

This paper is mainly a review of the work done by Z. Banach and H. Zorski on the so-called 'direct' continuum model of a solid, with mechanical and electromagnetic interactions. The most important references are as follows.

Banach, Z. 1981. On a certain wave motion in the spatially non-local and non-linear medium. Int. J. Eng. Sci. Vol. 19: 1047-1068.

Banach, Z. 1981. Direct continuum model of an elastically deformable polarizable and magnetizable body. I: Lagrangian function. II: Field equations. Arch. Mech. Vol. 33, No.1: 55-71; No.2: 195-214.

Zorski, H. 1977. Direct continuum model of interacting particles. Proc. 2nd CMDS Symposium. University of Waterloo Press.

Zorski, H. 1978. Continuum model of a particle system with three-point interactions. Lett. Appl. Eng. Sci. Vol. 16: 571-678.

Zorski, H. 1981. Non-local direct continuum with magnetic interactions. Arch. Mech. Vol. 33, No.6: 909-916.

We consider either a single continuum or two interpenetrating ionic continua. The following problems are considered.

1. The structure of the body force and the definition of the mechanical stress tensor on the basis of the static potential of the Lennard-Jones type. Both Cauchy and Piola-Kirchhoff stresses are examined in the general non-local case. Three-point forces are introduced. Generalized multi-point densities are used.

2. The most important electromagnetic interactions in a mixture of two continua and their relation to the mechanical forces. The Lagrangian formalism and the field and conservation equation are derived.

3. Simple static deformations and the generalization of the Carroll waves: the latter waves are investigated for the case of two interpenetrating continua in an electromagnetic field. The effects of non-locality and non-linearity and their influence on the dispersion equation are considered.

4. Definition of a new material tensor γ_{ij} describing polarization in a magnetized body, on the basis of the properties of singular integrals. The tensor γ_{ij} describes the magnitude and the shape of the cut-out small volume in the procedure used in the transition from a discrete system of charges to a continuum.

5th International Symposium on Continuum Models of Discrete Systems / Nottingham / 14-20 July 1985

A gauge fields approach to classical and continuum mechanics (Abstract)

B.Kunin
Case Western Reserve University, Cleveland, Ohio, USA

I.Kunin
University of Houston, Texas, USA

1. In modern physics, gauge theory is considered to be the most powerful method for establishing interactions between fundamental fields, see surveys [1,2].

2. Recently, gauge theory was used for describing some phenomena in macroscopic physics: irregularities in liquid crystals, magnets, spin glasses [3,4], defects in solids [5,6]. The question arises: is the application of gauge theory limited to these rather special phenomena or are there more universal connections with macroscopic physics and mechanics?

3. The purpose of the paper is to show that classical and continuum mechanics can be included in the framework of gauge theory.

4. Particles, rigid bodies, heavy tops are examples of Yang-Mills type gauge theories with zero curvature, where the Galilean group is the gauge group.

5. Elasticity, hydrodynamics are flat gauge theories of the general relativity type. Here the translation gauge group acting on a reference state plays the main role. Plasticity, defects in solids corresponds to non-flat gauge theory.

6. Some advantages of the gauge theoretical approach: For mechanics, it provides a unifying geometrical language closely related to modern physics. Notions of gauge invariance and spontaneous symmetry breaking may be of importance in such fields as dislocation theory, fracture mechanics, turbulence. For gauge theory, mechanics is an excellent testing ground for further development of gauge ideas.

REFERENCES

1. Drechsler, W., Mayer, M. E., Fibre bundle technique in gauge theories. Lecture Notes in Physics, No. 67, Springer, Verlag Berlin, 1977.

2. Aitchison, I. J. R. An informal introduction to gauge field theories. Cambridge University Press, Cambridge, 1982.

3. Dzyaloshinskii, I. E., Volovik, G. E., Annals of Physics, vol. 125, 67 (1980).

4. Rivier, N., Duffy, D. M. J. Physique, v. 43, 293, (1982).

5. Kadic, A., Edelen, D. Agauge theory of dislocations and disclinations. Lecture Notes in Physics, No. 174, Springer Verlag, Berlin, 1983.

6. Kunin, I., in The Mechanics of Dislocations, Eds. E. C. Aifantis, J. P. Hirth, American Society of Metals, 1983.

5th International Symposium on Continuum Models of Discrete Systems / Nottingham / 14-20 July 1985

Streaming motions in a bed of vibrationally-fluidized dry granular material (Abstract)

S.B.Savage
McGill University, Montreal, Canada

Experimental and theoretical studies of vibration-induced flow and mixing of dry granular materials are described. Tests were performed on rounded polystyrene beads contained in a rectangular box having transparent front and back walls and a flexible, nominally horizontal bottom which could be driven at various frequencies and amplitudes. The amplitude of the bottom vibrations was a maximum at the centre and decreased towards the vertical side walls. Slow recirculating flows were observed; they had the form of two vortices in which the velocity was upwards at the vertical centreline and downwards along the vertical sidewalls. The streaming velocities were measured as a function of bed vibration frequency and amplitude. An explanation proposed for the recirculating flows is that the vibrating base sends 'acoustic' waves upwards through the bed. These waves 'fluidize' the granular material but are in turn attenuated because of the dissipative nature of the collisions between the 'fluidized' particles. Thus the slow recirculating flows in the granular material are analogous to the more familiar 'acoustic streaming' in air. An approximate analysis of these streaming motions is developed by making use of a modification of the constitutive theory of Jenkins and Savage (1983). A number of simplifying assumptions are introduced to make the analysis tractable. The general flow pattern and the correct order of magnitude of the streaming motions are predicted, but the analysis is restricted to a rather narrow range of conditions.

List of participants

AIFANTIS, E. C. — Dept. of Mechanical Engineering and Engineering Mechanics,
Michigan Technological University,
Houghton,
Michigan 49931,
U.S.A.

ATKINSON, C. — Department of Mathematics,
Imperial College of Science and Technology,
Queen's Gate,
South Kensington,
London, SW7 2BZ,
ENGLAND.

ANDREADOU, A. — Department of Theoretical Mechanics,
University of Nottingham,
University Park,
Nottingham, NG7 2RD,
ENGLAND.

BANACH, Z. — Theoretische Physik,
Universität-Gesamthochschule-Paderborn,
Fachbereich 6,
Warburgerstrasse 100,
4790 Paderborn,
WEST GERMANY.

BAYLIS, E. R. — Department of Theoretical Mechanics,
University of Nottingham,
University Park,
Nottingham, NG7 2RD,
ENGLAND.

BOGY, D. — Department of Mechanical Engineering,
University of California,
Berkeley,
California 94720,
U.S.A.

BROOMHEAD, D. S.

Centre for Theoretical Studies,
Royal Signals and Radar Establishment,
St. Andrews Road,
Great Malvern,
Worcester, WR14 3PS,
ENGLAND.

BRULIN, O.

Division of Mechanics,
The Royal Institute of Technology,
100 44 Stockholm 70,
SWEDEN.

CAPRIZ, G.

Istituto di Matematica,
Università di Pisa,
c/o Tecsiel,
Via S.Maria 19,
56100 Pisa,
ITALY.

CHEN Yu

Department of Theoretical Mechanics,
University of Nottingham,
University Park,
Nottingham, NG7 2RD,
ENGLAND.

COHEN, H.

Department of Civil Engineering,
University of Manitoba,
Winnipeg,
Manitoba,
CANADA.

CROCHET, M. J.

Unité de Mécanique Appliquée,
Université de Louvain-la-Neuve,
2 Place du Levant,
B-1348 Louvain-la-Neuve,
BELGIUM.

DAI Tian-Min

Faculty of Mathematics,
Liaoning University,
Shenyang,
PEOPLE'S REPUBLIC OF CHINA.

DUKA, E.

Department of Theoretical Mechanics,
University of Nottingham,
University Park,
Nottingham, NG7 2RD,
ENGLAND.

FAULKNER, T. R.

Department of Theoretical Mechanics,
University of Nottingham,
University Park,
Nottingham, NG7 2RD,
ENGLAND.

GREEN, W. A.
Department of Theoretical Mechanics,
University of Nottingham,
University Park,
Nottingham, NG7 2RD,
ENGLAND.

HAVNER, K. S.
Department of Civil Engineering,
North Carolina State University,
Raleigh,
North Carolina 27607,
U.S.A.

HAYES, M.
Department of Mathematical Physics,
University College,
Belfield,
Dublin 4,
IRELAND.

HIBBERD, S.
Department of Theoretical Mechanics,
University of Nottingham,
University Park,
Nottingham, NG7 2RD,
ENGLAND.

HOLDEN, J. T.
Department of Theoretical Mechanics,
University of Nottingham,
University Park,
Nottingham, NG7 2RD,
ENGLAND.

HSIEH, R. K. T.
Department of Mechanics,
Royal Institute of Technology,
S-10044 Stockholm,
SWEDEN.

JANSONS, K. M.
Dept. of Applied Mathematics and Theoretical Physics,
University of Cambridge,
Silver Street,
Cambridge, CB3 9EW,
ENGLAND.

JEULIN, D.
Institut de Recherches de la Sidérurgie Francaise,
B.P. 64,
57210 Maizières-les-Metz,
FRANCE.

JONES, R.
Department of Theoretical Mechanics,
University of Nottingham,
University Park,
Nottingham, NG7 2RD,
ENGLAND.

KLEISER, T.	Institut für Material- und Festkörperforschung, Universität und Kernforschungzentrum Karlsruhe, Postfach 3640, D-7500 Karlsruhe 1, WEST GERMANY.
KRÖNER, E.	Institut für Theoretische und Angew. Physik, Universität Stuttgart, Pfaffenwaldring 57/VI, D-7000 Stuttgart 80, WEST GERMANY.
KUNIN, I.	Department of Mechanical Engineering, University of Houston, Houston, Texas 77004, U.S.A.
LANCHON, H. A. R.	Laboratoire d'Énergétique et de Mécanique Théorique et Appliquée, 2 Rue de la Citadelle, B.P. 850, 54011 Nancy, FRANCE.
LITEWKA, A.	Technical University of Poznan, ul. Piotrowo 5, 60-965 Poznan, POLAND.
McCARTHY, M. F.	Department of Mathematical Physics, University College, Galway, IRELAND.
McTAGGART, C. L.	Department of Theoretical Mechanics, University of Nottingham, University Park, Nottingham, NG7 2RD, ENGLAND.
MARKOV, K.	Centre for Mathematics and Mechanics, University of Sofia, P.O. Box 373, Sofia 1090, BULGARIA.
MAUGIN, G. A.	Laboratoire de Mécanique Théorique, Université Pierre et Marie Curie, Tour 66, 4 Place Jussieu, 75230 Paris 05, FRANCE.

MAZILU, P.

Institut für Mechanik,
Technische Hochschule Darmstadt,
Hochschulstrasse 1,
D-6100 Darmstadt,
WEST GERMANY.

MIDDLETON, D.

Department of Theoretical Mechanics,
University of Nottingham,
University Park,
Nottingham, NG7 2RD,
ENGLAND.

MISTURA, L.

Dipartimento di Energetica,
Università 'La Sapienza',
Via A.Scarpa 14,
00161 Rome,
ITALY.

MORRO, A.

Dipartimento di Ingegneria, Biofisica e Elettronica,
Facoltà di Ingegneria,
Università di Genova,
Viale Causa 13,
16145 Genova,
ITALY.

MURDOCH, A. I.

Department of Mathematics,
University of Strathclyde,
Livingstone Tower,
26 Richmond Street,
Glasgow, G1 1XH,
SCOTLAND.

MUSCHIK, W.

Institut für Theoretische Physik,
Technische Universität PN 7-1,
D-1 Berlin 12,
GERMANY.

O'NEILL, J. M.

Department of Theoretical Mechanics,
University of Nottingham,
University Park,
Nottingham, NG7 2RD,
ENGLAND.

PARKER, D. F.

Department of Theoretical Mechanics,
University of Nottingham,
University Park,
Nottingham, NG7 2RD,
ENGLAND.

PARRY, G. P.

School of Mathematics,
University of Bath,
Claverton Down,
Bath, BA2 7AY,
ENGLAND.

RAJAGOPAL, K. R.
Department of Mechanical Engineering,
University of Pittsburg,
Pittsburg,
Pennsylvania 15261,
U.S.A.

RIVLIN, R. S.
Center for the Application of Mathematics,
Lehigh University,
203 E.Packer Avenue,
Bethlehem,
Pennsylvania 18015,
U.S.A.

ROGERS, T. G.
Department of Theoretical Mechanics,
University of Nottingham,
University Park,
Nottingham, NG7 2RD,
ENGLAND.

SATAKE, M.
Department of Civil Engineering,
Tohoku University,
Aoba Aramaki,
Sendai 980,
JAPAN.

SAVAGE, S. B.
Department of Civil Engineering,
McGill University,
817 Sherbrooke Street West,
Montreal,
Quebec, H3A 2K6,
CANADA.

SEYMOUR, B. R.
Department of Mathematics,
University of British Columbia,
Vancouver 8,
British Columbia,
CANADA.

SHI Jingyu
Department of Theoretical Mechanics,
University of Nottingham,
University Park,
Nottingham, NG7 2RD,
ENGLAND.

SOLDATOS, K.
Department of Mathematics,
University of Ioannina,
Ioannina, GR-45332,
GREECE.

SPENCER, A. J. M.
Department of Theoretical Mechanics,
University of Nottingham,
University Park,
Nottingham, NG7 2RD,
ENGLAND.

STEFANIAK, J.
Institute of Applied Mechanics,
Technical University of Poznan,
ul. Piotrowo 3,
60-965 Poznan,
POLAND.

TALBOT, D. R. S.
Department of Mathematics,
Coventry (Lanchester) Polytechnic,
Priory Street,
Coventry, CV1 5FB,
ENGLAND.

TEODOSIU, C.
Institut für Physik,
Max-Planck-Institut für Metallforschung,
Heisenbergstrasse 1,
7000 Stuttgart 80,
WEST GERMANY.

TEPLY, J. L.
Product Engineering Division,
Alcoa Laboratories,
Aluminum Company of America,
Alcoa Center,
Pennsylvania 15069,
U.S.A.

THORNTON, C.
Department of Civil Engineering and Construction,
University of Aston,
Birmingham, B4 7ET,
ENGLAND.

TRZESOWSKI, A.
Institute of Fundamental Technological Research,
Polish Academy of Sciences,
ul. Swietokrzyska 21,
00-049 Warsaw,
POLAND.

WALPOLE, L. J.
School of Mathematics and Physics,
University of East Anglia,
Norwich, NR4 7TJ,
ENGLAND.

WEITSMAN, Y.
Department of Civil Engineering,
Texas A & M University,
College Station,
Texas 77843,
U.S.A.

ZARKA, J.
Laboratoire de Mécanique des Solides,
École Polytechnique,
91128 Palaiseau,
FRANCE.

ZORSKI, H. Institute of Fundamental Technological Research,
Polish Academy of Sciences,
ul. Swietokrzyska 21,
00-049 Warsaw,
POLAND.